BASICS OF BIOMEDICAL ULTRASOUND FOR ENGINEERS

BASICS OF BIOMEDICAL ULTRASOUND FOR ENGINEERS

HAIM AZHARI

Published by John Wiley & Sons, Inc., Hoboken, New Jersey
Published simultaneously in Canada

For general information on our other products and services or for technical support, please contact our Customer Care Department within the United States at (800) 762-2974, outside the United States at (317) 572-3993 or fax (317) 572-4002.

Wiley also publishes its books in a variety of electronic formats. Some content that appears in print may not be available in electronic formats. For more information about Wiley products, visit our web site at www.wiley.com.

Library of Congress Cataloging-in-Publication Data:

Azhari, Haim, 1955–
Basics of biomedical ultrasound for engineers / Haim Azhari.
p. cm.
Includes bibliographical references and index.
Summary: "Basics of Biomedical Ultrasound for Engineers is a structured textbook for university engineering courses in biomedical ultrasound and for researchers in the field. This book offers a tool for building a solid understanding of biomedical ultrasound, and leads the novice through the field in a step-by-step manner. The book begins with the most basic definitions of waves, proceeds to ultrasound in fluids, and then delves into solid ultrasound, the most complicated kind of ultrasound. It encompasses a wide range of topics within biomedical ultrasound, from conceptual definitions of waves to the intricacies of focusing devices, transducers, and acoustic fields"—Provided by publisher.
ISBN 978-0-470-46547-9
1. Ultrasonics in medicine. 2. Ultrasonics. I. Title.
R857.U48A94 2009
616.07′54–dc22

2009025404

Printed in the United States of America

10 9 8 7 6 5 4 3 2

To the memory of
Elad Grenadier,
who was killed by terrorists on July 17, 2002,
at the age of 21.
And to the memory of his father
and my friend, Dr. Ehud Grenadier.
Blessed be their souls.

CONTENTS

PREFACE

This book partially summarizes the knowledge that I have accumulated during my 25 years of research and teaching in this fascinating field. This is actually the third edition of the book, but the first one to appear in English (the first two editions were published in Hebrew). I have tried my best to correct and improve the text based on comments that were given to me by the readers. Nevertheless, I presume that there are still quite a number of things to correct and improve. Thus, I would highly appreciate any comments sent to my E-mail address: Haim@BM.Technion.Ac.IL Kindly indicate "Ultrasound book—Comments" in the subject. Thank you in advance.

HAIM AZHARI

Haifa, Israel
September 2009

ACKNOWLEDGMENTS

I thank G_D for helping me complete this book and for making me meet the many good and talented people who have helped me learn this fascinating field.

I thank Dr. Kenneth Jassbey, who was my first teacher of ultrasound and my physician friends, Dr. Ehud Grenadier and Dr. Diana Gaitini, from whom I have learned the clinical side of ultrasound.

Many thanks are due to my students, past and present, from whom I have learned much. As Rabbi Hanina states in the Talmud, Taanit 7:70a: "I have learned a lot from my teachers, more than that from my friends, but most of all from my students."

I also thank Yifat Levy, Simona Beilin, and Alexandra Alexandrovich, who helped me convert my lecture notes to digital format. I also thank the faculty of Biomedical Engineering at the Technion IIT, for their administrative support.

Finally, many many thanks to my beloved family.

H. A.

INTRODUCTION

PRELUDE AND BASIC DEFINITIONS

From the day we are born we are trained to treat each of our body systems as a separate unit capable of performing one task. We see light with our eyes (actually our visual system). We hear sounds with our ears (our auditory system). We sense pressure and temperature with our somatosensory system and usually move things with our limbs. And here comes ultrasound and allows us to perform all these tasks with one modality. With ultrasound we can "see" the internal structure of the body. We can "hear" the speed of moving objects. We can sense and apply pressure to the extent of "crushing" kidney stones, and we can warm and "cook" tumors to destroy them noninvasively.

There is no doubt that ultrasound has helped humankind in many areas, and perhaps the most prominent one is medicine. There are many medical applications of ultrasound (see some mentioned in the following), and the potential of developing new applications is still quite large. The aim of this book is to introduce this fascinating area to a novice in the field and provide the basic "toolkit" of knowledge needed to use it and conduct research.

Let us begin with the very basic definitions:

Ultrasound—*Acoustic waves* (sound or pressure waves) propagating within a *matter medium* at frequencies exceeding the *auditory band*.

Basics of Biomedical Ultrasound for Engineers, by Haim Azhari

As can be noted, this definition is based on three italic terms that need further clarification. Let us define the first term:

Acoustic Waves—A physical phenomenon during which *mechanical energy* is transferred through matter, without mass transfer, and which originates from a local *change* in the stress or pressure field within the medium.

The phenomenon that we are dealing with is in fact a process of "energy flow" from one place to another. This energy is *mechanical* in nature. It is embedded in the matter in the form of elastic strains (stemming from the spring-like behavior of the matter) and vibrations of molecules. Indeed, molecules of the medium do move when an acoustic wave passes through, but this motion is mostly localized. As stated above, this phenomenon is not associated with mass transfer. This corresponds mainly to solid or semisolid substances. It should be noted, however, that in fluids, a phenomenon called "acoustic streaming" may appear (see Chapter 12). But this is an effect stemming from the acoustic wave and not the wave itself (see Section 3.6 in Chapter 3).

The source of these waves is a *change* in the stress or pressure field within the medium—for example, a clap of hands or a knock with a hammer or an explosion. It is therefore understood that a *material* medium is needed to allow this energy to propagate and that acoustic waves, unlike electromagnetic waves, cannot propagate in vacuum. This implies that there is a strong relation between the properties of the medium (e.g., structure, elasticity, density, etc.) and the corresponding acoustic properties of the acoustic waves (e.g., speed of propagation, attenuation, possible wave types, etc.) passing through it.

Finally, we should explain why are these waves called ultrasonic. Well, in fact the physics describing ultrasonic waves is the same as the physics of sonic waves and subsonic waves. The spectral range of the human ear (the auditory band) is normally 20 Hz to 20 kHz (animals of course can hear other frequencies). It is merely more practical and a matter of convenience that we use in most of our medical applications frequencies exceeding 20 kHz. These frequencies are not detected by our auditory system and hence the range of higher frequencies is arbitrarily defined as "ultrasound." [It should be pointed out, however, that in Doppler-effect-based techniques (see Chapter 11), audible signals are the output of the process.]

THE ADVANTAGES OF USING ULTRASOUND IN MEDICINE

Ultrasound offers several advantages that make it attractive for medical applications. Let us note some of them:

A. *Hazardless Radiation.* As long as high intensities are not used (see Chapter 12), ultrasound is considered a safe and hazardless modality (see comments on safety in the following). This may be easily conceived by being aware of the fact that we are continuously "immersed" in an ocean of sounds. From the almost unheard sounds of our beating heart, through the music we hear from the radio to the loud noise of a passing jet airplane, we continuously absorb acoustic energy. Our body is "accustomed" to this energy. Even a baby in his mother's womb is exposed to sounds. It is therefore understood why ultrasonic medical examination is the most popular imaging modality used during pregnancy. The fact that ultrasound is hazardless allows repeated examinations of the same patient with no risk using standard equipment [see, for example, declarations and restrictions published by the American Institute for Ultrasound in Medicine (AIUM) on the internet].

B. *High Sampling Rate.* Because the speed of sound, C, in most soft tissues (see Appendix A at the end of the book) is around 1540 m/sec and the typical ranges used in medical examinations are relatively short (0.1–25 cm), the time it takes a wave to cover such ranges is very short (on the order of several tens of microseconds). Consequently, many waves can be transmitted within a short time and gain sufficient information to follow dynamic changes occurring in "real time" within the body.

C. *Compact Transducers.* Most of the ultrasonic transducers used in medicine are made of piezoelectric crystals that are very compact. A typical imaging probe may have the size of a match box. Furthermore, very small crystals can be manufactured. These miniature crystals are sufficiently small to be placed atop minimally invasive surgery devices, or even atop catheters (e.g., intravascular ultrasound (IVUS); see Chapter 9), or even implanted within the body.

D. *A Single Transducer Can Be Used to Transmit and Receive Waves.* As piezoelectric transducers can both transmit and detect waves (see Chapter 8), the same element can serve as a source of waves at one time and then switched to serve as a sensor for detecting echoes at another time. This fact enables us to probe almost any region within the body (in an invasive or partially invasive manner).

E. *A Wide Variety of Parameters Can Be Measured.* Since the acoustic waves properties are strongly related to the properties of the medium in which they travel, there is a wide variety of parameters that can be measured. For example: speed of sound, attenuation, acoustic impedance, dispersion, nonlinearity, elasticity, and more. These parameters can be used to map and characterize tissues.

F. *Cost Effective.* Ultrasound systems do not require any RF shielded rooms as MRI does. The transducers commonly do not wear out and there is no need for films (as in X-ray systems). There is also no need for expensive and dangerous materials such as in nuclear medicine. On the contrary, the components are relatively inexpensive. Thus, ultrasound is an excellent choice also from the economical point of view. Therefore, it is a very popular modality and is available in numerous clinical sites all over the world.

A GENERAL STATEMENT ON SAFETY

As stated above, ultrasound—if used properly—is safe. After so many years of use and gained experience, it can be categorically stated that diagnostic ultrasonic devices that comply with the current standards and regulations impose no danger to the examined patient. (This issue is further elaborated in Chapter 12.) However, like in many other things in life, abuse is not recommended, especially when babies and sensitive organs are examined. Generally speaking, it can be stated that the safe performance zone is determined mainly by two factors: (i) the intensity of the acoustic energy used (which is measured in W/cm^2) and (ii) exposure time. In Fig. I.1 a chart based on a graph published by the AIUM Bioeffects Committee in 1976 [1] is depicted. Although this chart is obsolete, it provides a good visual concept of safety limits. This chart maps the safe zone (shaded area) in the intensity–exposure time plane (the scale is logarithmic for both parameters). The exposure limits set at that time by the AIUM allowed a maximal transducer output intensities (see term: spatial peak temporal average in Chapter 12) of $100\,mW/cm^2$ for exposure durations exceeding 500 sec. Higher intensities up to $50\,W/cm^2$ could be used with shorter exposure times, provided that they are kept within the shaded area shown.

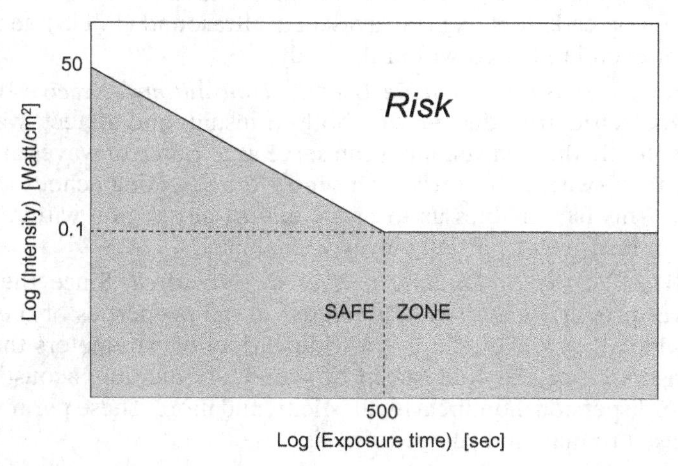

Figure I.1. Schematic chart based on a graph published by the AIUM in 1976 [1]. This chart maps the zone (shaded area) considered to be safe at that time in the intensity–exposure time plane. The scale is logarithmic for both parameters. For current limits see, for example, reference 2.

Current limits set by various regulatory institutes include more parameters and allow higher intensities for long exposure durations. The present limit set by the American FDA is 720 mW/cm^2 for a parameter called *spatial peak temporal average intensity*. (For additional indices see Chapter 12.) These limits naturally refer to imaging systems. For therapeutic applications, such as hyperthermia [3] or lithotripsy [4], much higher intensities must to be used. Consequently, some damage to normal tissues should be expected.

SOME COMMON APPLICATIONS OF ULTRASOUND

As noted above, there are currently many applications of ultrasound in medicine, and new applications are frequently suggested. The main application today is ultrasonic imaging. This modality offers rapid and convenient tools for acquiring images of soft tissues. Using ultrasonic images, sizes and distances within organs can be measured (e.g., the heart's left ventricular diameter and the diameter of the eye), pathological tissues (e.g., cysts and tumors) can be identified and characterized, kidney and gall stones can be detected, and volumes and areas can be calculated.

In pregnancy monitoring, ultrasound is a convenient key tool, which allows frequent follow-ups and assessment of the baby's development and allows the parents today to "see" their baby's face in 3D (see example in Chapter 9). Ultrasound also plays a major role in cardiovascular diagnosis. Using the Doppler effect, blood flow in various vessels and through the heart's valves can be imaged. Duplex imaging allows the physician to combine an anatomical image with a color-coded flow/motion map. Miniaturization allows intravascular (IVUS) imaging and invasive plaque analysis.

Hard tissues such as bones and teeth pose a technical challenge due to their high reflectivity and attenuation of acoustic waves. There are a few commercial systems available in the market today for bone assessment and for dental applications, but this market is still not dominant.

Ultrasound is also commonly used in physiotherapy where it is used for warming inner tissue layers in order to help accelerate the healing process. In the past few years a renaissance of an idea that had been presented in the 1940s has started. The idea is to use high-intensity focused ultrasound (HIFU) to ablate tumors noninvasively [3]. There are several commercially available systems today for treating the breast, the uterus, and the prostate. Other applications combining HIFU with MRI and other imaging modalities guidance are under investigation.

Another application of ultrasound is in lithotripsy, which has become a routine clinical procedure. By focusing high-intensity acoustic bursts on kidney stones, these stones are disintegrated and washed out of the body through the urinary tract.

Other applications utilize acoustic waves to disintegrate blood clots (acolysis) and to increase the permeability of the skin and internal membranes for drug delivery. In addition, it has been demonstrated that ultrasound can be used to insert large molecules into cells for gene therapy (see Chapter 12).

It has also been suggested to use ultrasound for cosmetic applications. There were several start-up companies who have developed ultrasonic hair-removing devices. And there are at least two companies today that have developed systems for removing fat (lypolyses) and for body contouring (i.e., reshaping the body).

These are just demonstrative applications of the potential of ultrasound in medicine. Although this field is almost a century old, it continues to flourish.

WHAT IS IT THAT WE NEED TO KNOW?

After we have been introduced to the field, let us define in general terms what it is that we need to know in order to be able to use ultrasound properly. For that purpose, consider the general schema shown in Fig. I.2. In this figure a schematic layout of ultrasonic waves "flow" through and out of the body is depicted. For simplicity, let us restrict our discussion to two dimensions. Usually, we have a source of acoustic waves or an array of such sources that transmit waves into the body (represented schematically by the "transmitter" box). In most cases we have control over the temporal profile and the timing of the transmitted waves (designated here as $F(t, x, y)$). It is important to note that in certain applications the source may be positioned within the body. As a result of the interaction between the waves and the tissue, part of the acoustic energy is absorbed as a function of time and location: $B(t, x, y)$. Part of the energy passes through the body to the other side: $T(t, x, y)$. And the rest is scattered to various directions: $S(t, x, y)$. The properties of the tissues combined

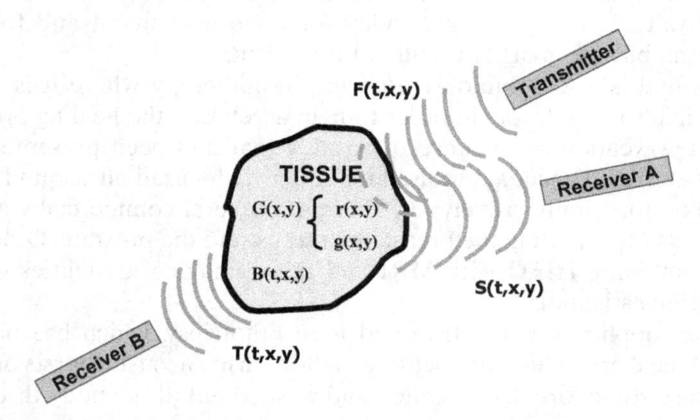

Figure I.2. Schematic layout of ultrasonic waves "flow" through and out of the body. A transmitted wave can pass though the organ and be detected on the other side. Scattered waves (echoes) can be detected at different positions around the body.

with its geometry $G(x, y)$ determine the energy division between these three types of mechanisms and their temporal and spatial profiles. By measuring $T(t, x, y)$ and/or $S(t, x, y)$, we try to characterize the tissue properties $G(x, y)$. For example, in a pulse–echo imaging process we try to map the geometry $g(x, y)$, whereas in a tissue characterization process we may focus our analysis on the reflectivity properties of the tissue: $r(x, y)$. And during a therapeutic process we may need to know the tissue's absorption properties.

In order to be able to perform these tasks, we must be very well familiarized with the connections and relations between the medium's properties and the properties of acoustic waves passing though it. Furthermore, we need to know what are the currently available techniques for transmitting and handling these waves in order to use them effectively. *And that is what this book is all about.*

REFERENCES

1. The American Institute of Ultrasound in Medicine (AIUM), The Bioeffects Committee, Statement on Mammalian In-Vivo Ultrasonic Biological Effects, August 1976.
2. Ter Haar G and Duck FA, editors, *The Safe Use of Ultrasound in Medical Diagnosis*, British Institute of Radiology, London, 2000.
3. Kennedy JE, Ter Haar GR, and Cranston D, High intensity focused ultrasound: Surgery of the future? *Br J Radiol* 76:590–599, 2003.
4. Srivastava RC, Leutloff D, Takayama K, and Gronig H, editors, *Shock Focusing Effect in Medical Science and Sonoluminescence*, Springer-Verlag, Berlin, 2003.

CHAPTER 1

WAVES—A GENERAL DESCRIPTION

Synopsis: This chapter deals with the basic definitions of waves in general and acoustic waves in particular. The objective of this chapter is to "touch base" with wave properties and their mathematical description. We shall start with a qualitative description and define the various types of waves. We shall examine what properties can be used to describe waves, introduce the wave equation in a homogeneous medium, learn about "group" and "standing" waves, and describe in detail spherical and cylindrical waves. Finally, we shall study the wave equation in a nonhomogeneous medium and the Born and Rytov approximations associated with its solution.

(*Note:* It is likely that many of the items presented in this chapter are trivial or familiar to most readers. Thus, a reader who is well aquatinted with the material may hop to the next chapter.)

1.1 GENERAL DEFINITIONS OF WAVES—A QUALITATIVE DESCRIPTION

When a mechanical wave propagates through matter, energy is transferred from one location to another [1, 2]. However, this energy is not associated with mass transfer. For example, one may consider the propagating waves formed when moving an end point of a jumping rope or when the first piece from a set of dominoes aligned in a row falls down. Clearly, energy is transferred by motion of matter, but mass is not really transferred from one location to another. The nature of the propagating wave is determined by the type of perturbation causing it to appear but also and mainly by the properties of the

Basics of Biomedical Ultrasound for Engineers, by Haim Azhari

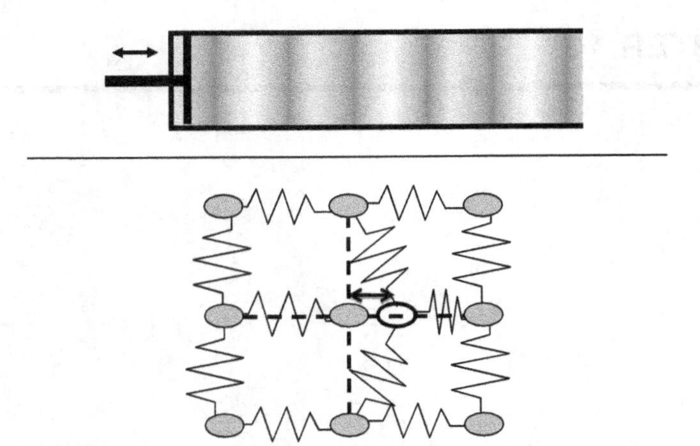

Figure 1.1. Schematic depiction of two possible configuration of mechanical wave propagation. **(Top)** The wave propagation direction is limited to only one direction. **(Bottom)** Although the small masses' motion is horizontal, waves can propagate along the vertical direction as well.

media through which it travels. For clarity, let us observe the two examples shown in Fig. 1.1:

Example 1. A cylinder filled with gas or fluid has a piston on its left side. At a certain time point the piston is pushed back and forth. Consequently, a pressure wave will propagate through the gas or fluid along the cylinder's axis. In this case the wave may be considered as one-dimensional.

Example 2. A matrix consisting of springs and spherical masses is at rest. At a certain time point, one mass is pushed back and forth. As a result, the masses along the line of perturbation will vibrate and a mechanical wave will propagate through the matrix. However, in this case the perturbation will also be associated with deformation of springs from both sides of the moving mass. Hence, one may expect mechanical waves to propagate along directions that are perpendicular to the induced motion as well.

From Example 2, one can conclude that several types of waves, each characterized by different properties, can be generated at the same time and even from the same perturbation. (This issue will be discussed in Chapter 4, where wave propagation in solids is analyzed.) Also, it is important to note that the motion direction of the particles constituting the medium does not have to align with the wave propagation direction. Using this fact, we can apply a preliminary division of wave types based on the relation between the wave propagation direction and the motion direction of the particles constituting the medium (see also Fig. 1.2):

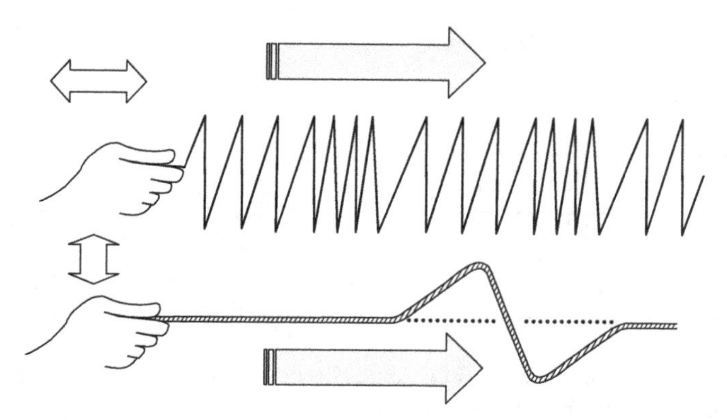

Figure 1.2. Division of waves into two types based on the relation between the wave propagation direction and the motion direction of the particles constituting the medium. **(Top)** In a longitudinal wave the particles motion is parallel to the wave propagation direction. **(Bottom)** In a transverse wave the particles motion is perpendicular to the wave propagation direction.

A Longitudinal Wave—A wave for which the direction of displacement for the medium's particles is *parallel* to the direction of wave propagation

A Transverse Wave—A wave for which the direction of displacement for the medium's particles is *perpendicular* to the direction of wave propagation.

This basic division is frequently used in ultrasound, but it is important to note that there are also wave types for which the direction of motion for the medium's particles is not fixed relative to the wave propagation direction. For example, in surface waves (as in sea waves) the angle between the two directions changes continuously.

A second and important division is based on the wave front geometry. Using this approach we can again divide the waves into two basic types (Fig. 1.3):

A. *A Planar Wave*—A wave for which the wave front is located on a plane that propagates in space.

B. *A Circular Wave*—A wave that propagates symmetrically around a reference point (as a sphere or a ring), or around a reference line (as a cylinder).

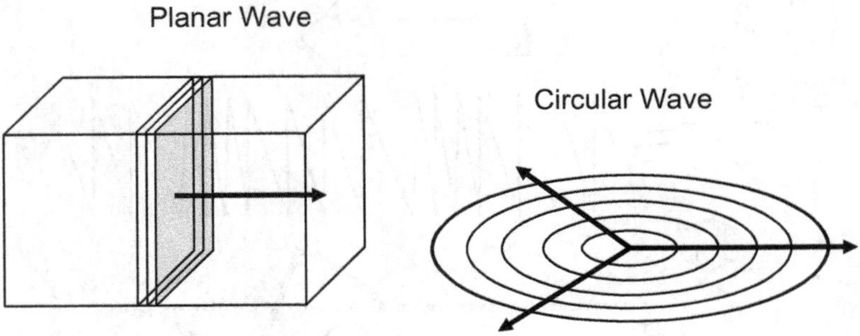

Planar Wave

Circular Wave

Figure 1.3. Division of waves into two types based on the geometry of the wave front. **(Left)** A planar wave. **(Right)** a circular wave.

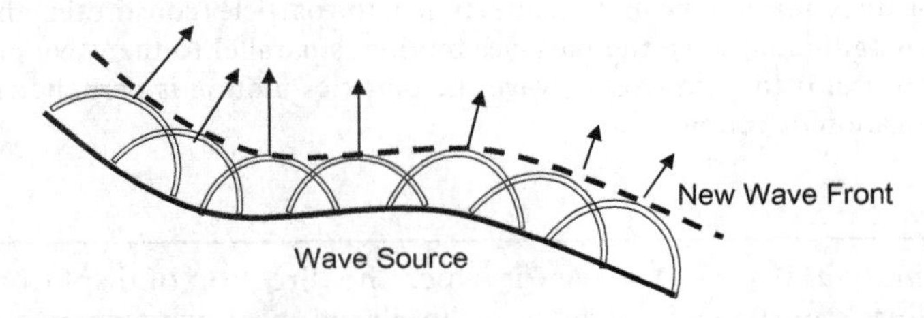

Figure 1.4. Schematic demonstration of the Huygens' Principle implementation to an arbitrarily shaped wave source (or wave front). Each point on the wave front is assumed to emanate a spherical wave (only a few are depicted for clarity). The superposition of all these small spherical waves forms the new wave front.

Of course in nature one can encounter also intermediate type of waves. For example, the acoustic beam transmitted from a disc-shaped transducer (see Chapter 8) has planar wave characteristics (particularly around its center). However, as the ratio between the transducer diameter and the wavelength is decreased, the wave will have more and more spherical wave characteristics.

In the general case, one can use Huygens' Principle to investigate the propagating wave front of an arbitrary shape. Huygens' Principle states that any wave source (or wave front for that matter) can be considered as an infinite collection of spherical wave sources [3]. This is schematically demonstrated in Fig. 1.4.

1.2 GENERAL PROPERTIES OF WAVES—A QUALITATIVE DESCRIPTION

1.2.1 Interference and the Superposition Principle

One interesting property of wave interaction is the interference phenomenon. When two waves collide, they can form "constructive" interference and their

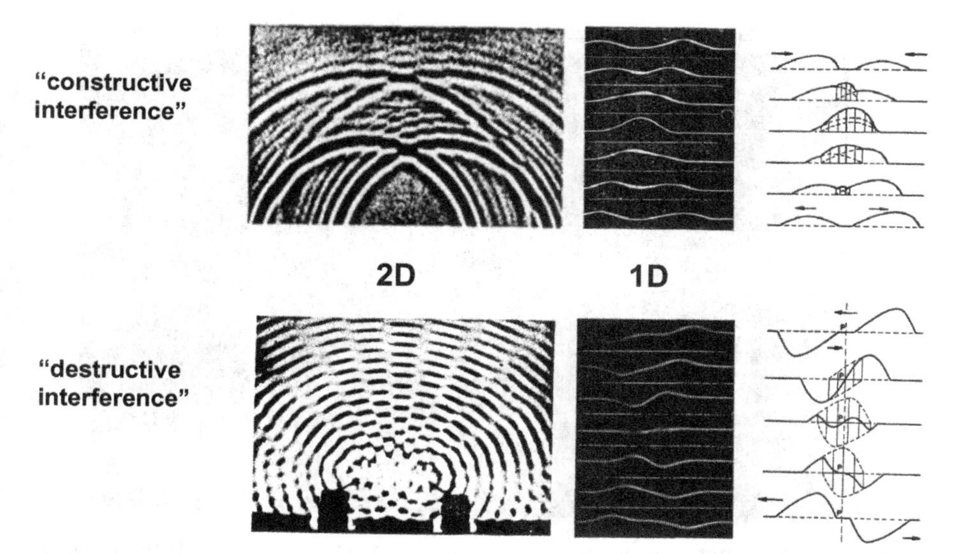

Figure 1.5. (Top) Demonstration of constructive interference in a water bath (2D case—*left*) and for strings (1D case—*middle*). For clarity, the 1D case is also schematically depicted on the right side. **(Bottom)** Demonstration of destructive interference. Reprinted with permission from *PSSC Physics*, 7th edition, by Haber-Schaim, Dodge, Gardner, and Shore, Kendall/Hunt Publishing Company, 1991.

amplitudes seem to enhance each other or "destructive" interference and their amplitudes seem to attenuate each other (see Fig. 1.5) [4]. However, after passing through each other, each wave proceeds as though "nothing has happened" (if nonlinear effects are negligible). The superposition principle states that within an interference zone the net amplitude is the sum of all the interacting wave amplitudes.

1.2.2 Reflection and Transmission of Waves

When a wave passes from one material to another or when it encounters a discontinuity in the medium in which it travels, part of its energy is reflected and part of it is through-transmitted with or without a change in direction. For an acoustic wave the reflected part is usually referred to as an "echo." It is worth noting that in ultrasound we distinguish between the waves reflected along a straight line to the source which are referred to as "backscatter" and which are scatted to other directions. The energy partition and the change in wave properties of the transmitted and scattered waves contain a lot of information that can be utilized for imaging or diagnosis (as indicated in the introduction chapter).

In the one-dimensional (1D) case, when a wave passes from a spring that has certain mechanical properties to another spring that has different

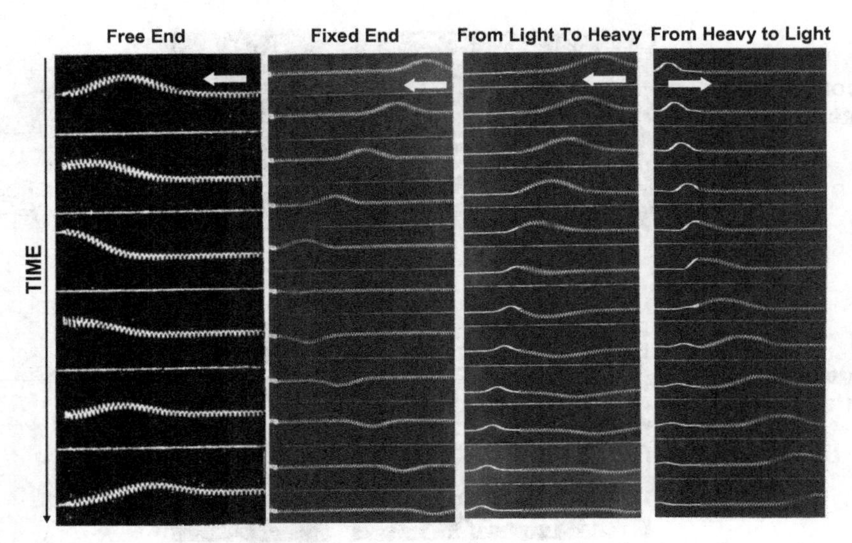

Figure 1.6. A demonstration of wave transmission and reflection phenomenon in springs. The arrow indicates the propagation direction of the impinging wave. Each picture indicates a new time frame. **(Left column)** A wave reaching a free end. **(Second column)** A wave reaching a fixed end. **(Third column)** A wave traveling in a light (low density) spring encounters a heavy (high-density) spring. **(Right column)** A wave traveling in a heavy (high-density) spring encounters a light (low-density) spring. Reprinted with permission from *PSSC Physics*, 7th edition, by Haber-Schaim, Dodge, Gardner, and Shore, Kendall/Hunt Publishing Company, 1991.

properties, or when the wave reaches an end point, reflection occurs and, if possible, transmission occurs as well. In Fig. 1.6, four different scenarios of a wave propagating in a spring and encountering a discontinuity point are demonstrated (from left to right): (a) a wave reaching a free end, (b) a wave reaching a fixed end, (c) a wave traveling in a low-density spring reaching a point connecting the spring with a high-density spring, and (d) a wave traveling in a high-density spring reaching a point connecting the spring with a low-density spring. It is important to note that when the wave encounters a "tough" medium, such as a heavy spring or a fixed end point, the phase of its echo changes (i.e., its amplitude changes signs). Also, it is important to note that when the wave travels from a heavy spring to a light one, its amplitude is *increased* (energy, however, is conserved as will be explained in the next chapter).

In the two-dimensional (2D) case and naturally in the three dimensions (3D), one has to distinguish between circular and planar waves. In the case of a planar wave, the rule is simple: The angle of reflection from a reflector is equal to the angle of incidence. However, for a circular wave (e.g., spherical or cylindrical), things are a little bit more complicated. The reflected wave seems to emanate from a virtual source which is located on the other side of the reflector (see Fig. 1.7).

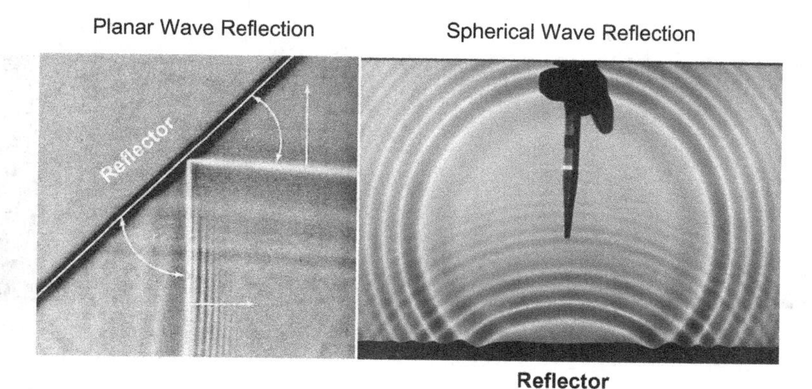

Figure 1.7. A demonstration of the reflection phenomenon for a planar wave **(left)** and for a circular wave **(right)**. Reprinted with permission from *PSSC Physics*, 7th edition, by Haber-Schaim, Dodge, Gardner, and Shore, Kendall/Hunt Publishing Company, 1991.

1.2.3 Diffraction

Another phenomenon that is associated with waves is "diffraction." Waves that impinge upon a corner or pass through a slot in a screen bend their trajectory and propagate into zones that should have been "shadowed." This phenomenon is enhanced for wavelengths that are relatively long relative to the geometry of the obstacle. Diffraction may be significant in medical ultrasound since the typical wavelengths are in the scale of a millimeter (for example, for a frequency of 1 MHz, the wavelength is about 1.5 mm in soft tissues), and so is the needed resolution. The diffraction phenomenon is demonstrated in Fig. 1.8 for clarity.

1.2.4 Standing Waves

Another phenomenon that needs to be noted is the appearance of "standing" waves. When waves rush back and forth within a confined medium (e.g., when strong reflectors are located at each end point of the medium), the waves interfere with each other and with their reflections. Consequently, a motion pattern is formed which is spatially fixed but temporally changes its amplitude (see, for example, Fig. 1.9). This pattern is referred to as a "standing" wave, since an illusion is formed where the wave seems to stand still while its amplitude changes. This type of wave is most probably observed in string-based musical instruments. They may be noticed in vibrating strings and in resonance boxes. In medical applications, this type of wave may be significant when considering the design of instruments and particularly the design of instruments that use constant wave (CW) transmission.

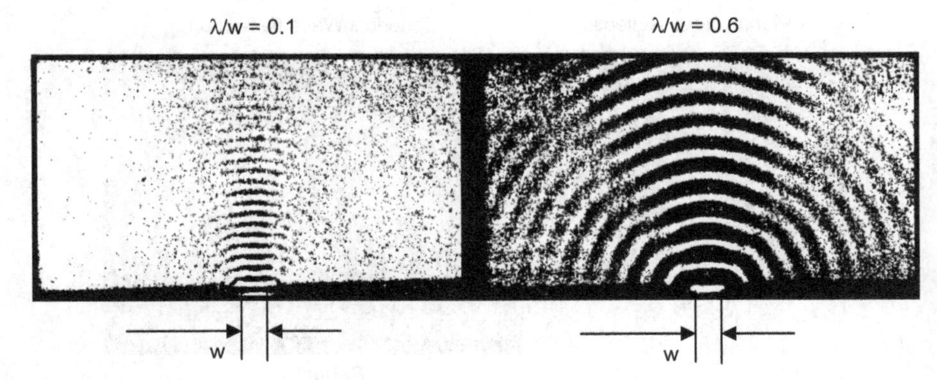

Figure 1.8. Demonstration of the diffraction phenomenon for a planar wave propagating from the bottom toward the top and passing through a slot for which width is W. The ratio between the wavelength λ and the slot width W is 0.1 for the case shown on the left and 0.6 for the case shown on the right. As can be noted, the phenomenon is much more enhanced on the right side. Reprinted with permission from *PSSC Physics*, 7th edition, by Haber-Schaim, Dodge, Gardner, and Shore, Kendall/Hunt Publishing Company, 1991.

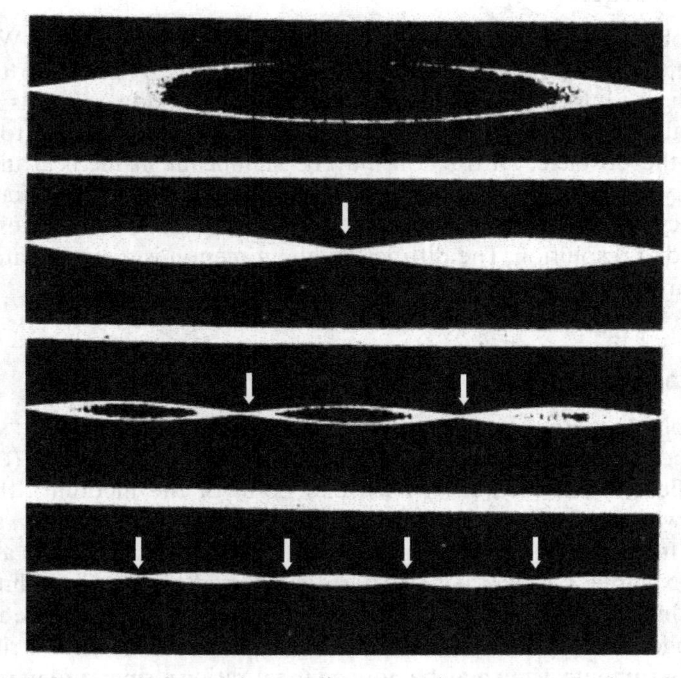

Figure 1.9. A demonstration of the standing wave phenomenon. In these four pictures taken with a long exposure time, four vibration modes for a string tied at both end points are depicted. As can be noted, certain nodal points (marked by the arrows) do not move. Reprinted with permission from *PSSC Physics*, 7th edition, by Haber-Schaim, Dodge, Gardner, and Shore, Kendall/Hunt Publishing Company, 1991.

1.3 MECHANICAL ONE-DIMENSIONAL WAVES

Consider the familiar entertaining device that consists of a set of pendulums made from small metal balls tied in a compact row as shown in Fig. 1.10. At time point $t = -\Delta t_0$ the first pendulum (marked as #0) is tilted by an angle θ_0 and is allowed to swing back. The initial potential energy of its metal ball will be converted into both kinetic and potential energy as the ball swings toward its neighbor. At time $t = 0$ the ball will hit the ball of pendulum #1 and the impact will transfer some of its energy. The ball of pendulum #0 will then swing slightly (if there is a gap between adjacent balls) and resume its original equilibrium position. The ball of pendulum #1 will hit the ball of pendulum #2 and in turn will transfer to it some of its energy and so forth.

If we would take a camera and photograph the set of pendulums at time $t = 0$ and at time $t = t_1$, we would be able to notice that there are "calm" regions (i.e., where the balls are in their equilibrium position and the are "stormy" regions where the balls swing back and forth. The "stormy" region that was created by the initial perturbation propagates along the medium (toward its left side in the case shown in Fig. 1.10). Intuitively, we can relate this experiment to waves. Hence, we shall refer to the propagating perturbation of pendulums from their equilibrium state as a "wave." Comparing our observation to the definition given in the previous sections, we shall indeed note that it suits our criteria. We have in this example a propagating mechanical energy that propagates through matter and that is not associated with mass transfer

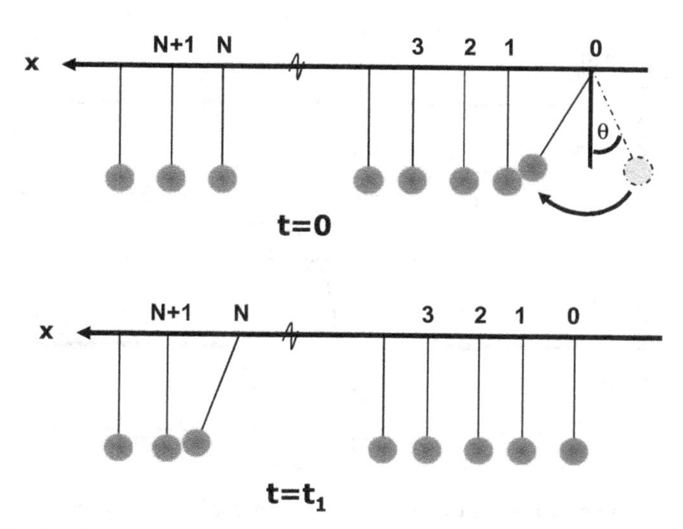

Figure 1.10. A demonstration of a propagating mechanical wave in a "one-dimensional" medium. The perturbation from equilibrium which is induced at time $t = 0$ by swinging the first pendulum will propagate and eventually reach the Nth pendulum at time $t = t_1$.

and that stems from a perturbation induced to the medium. This perturbation, which we refer to as a "wave," varies with time and space and therefore can be designated in the most general case by the term

$$U = U(x, y, z, t) \tag{1.1}$$

where the function U can designate any physical property that characterizes the propagation of the propagating perturbation. In the example given above, U could express the pendulum angle relative to its equilibrium state, the metal ball velocity, or its energy, and so forth. In the one-dimensional case the wave can be represented by

$$U = U(x, t) \tag{1.2}$$

If we set the time to be constant and we plot U as a function of location, we shall obtain a profile of the property described by $U = U(x,t = \text{const})$. On the other hand, we can set the location to be a constant and plot the profile described by $U = U(x = \text{const}, t)$. Comparing the two obtained plots, we shall realize that the two plots characteristics are identical except for the reverse in directions and the natural change in scaling. This plot is called the "wave profile" (see Fig. 1.11).

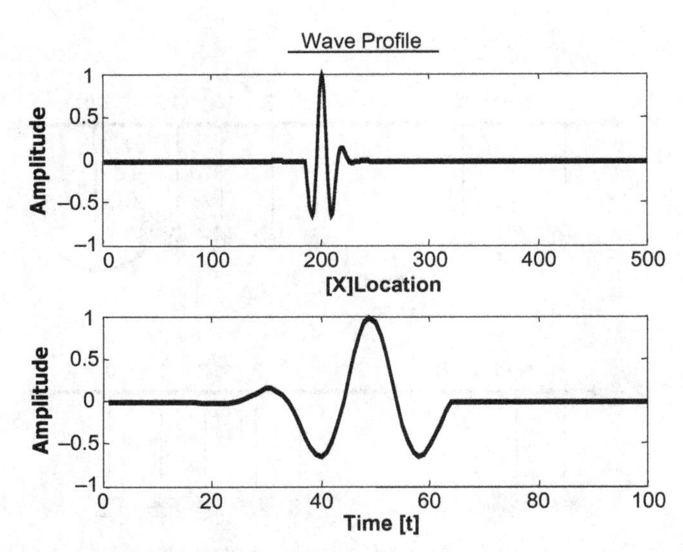

Figure 1.11. A demonstration of a wave profile when plotted as a function of location for a given time point **(top)** or as a function of time for a given spatial location **(bottom)**. As can be noted, the two profile characteristics are identical except for scaling and directions.

1.4 THE WAVE FUNCTION

Generally speaking, a wave propagating along the positive direction of the X axis can be described by the general function [5]

$$U(x,t) = U(x - ct) \tag{1.3}$$

whereas a wave propagating along the negative direction of the X axis can be described by

$$U(x,t) = U(x + ct) \tag{1.4}$$

In order to validate this statement, let us examine a function $U(x,t) = U(x - ct)$ representing a wave that has a profile of maximal amplitude U_0 and a single peak, such as shown for example in the following icon: U_0. At time $t = 0$ the peak of the wave profile is found at some point x_0, that is, $U(x_0, t = 0) = U_0$. Clearly, if for some other point x in time $t = \Delta t$ we can find a location for which $x - c \cdot \Delta t = x_0$, its value must be $U(x - c \cdot \Delta t) = U(x_0, 0) = U_0$. However, since by definition we have $C > 0$, and time is always positive ($t > 0$), it follows that the value of x which fulfills the condition $x - c \cdot \Delta t = x_0$ must continuously increase. Or in other words, the peak of the wave profile must move along the positive direction of x. This can also be demonstrated graphically by plotting the location of the wave profile peak (or any other landmark for that matter) in the plane (x, t). There it will move along a straight line given by $x - x_0 = c \cdot \Delta t$ and whose slope is $1/c = \Delta t/(x - x_0)$ as shown in Fig. 1.12. Similarly, it can be shown that $U(x + ct)$ describes a wave that propagates along the negative direction of x.

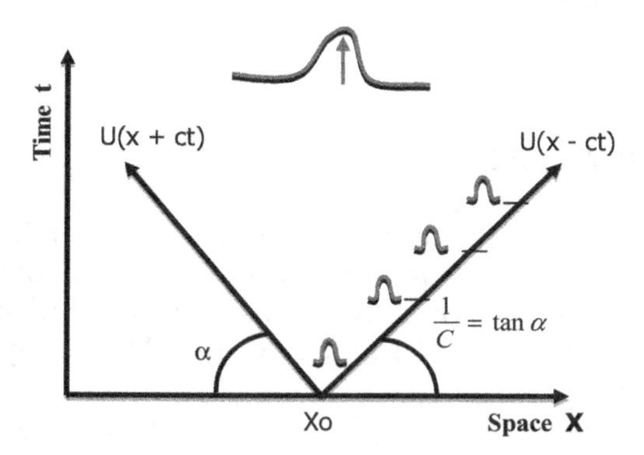

Figure 1.12. Propagation of waves defined by the general function $U(x \mp ct)$ in the space–time (x, t) plane.

1.5 THE WAVE EQUATION

The wave function $U(x \pm ct)$ described above is related to a differential equation for which it can serve as a solution. This equation is defined as the "wave equation" and for the one-dimensional (1D) case in a homogeneous medium is given by [5]

$$\frac{\partial^2 U}{\partial X^2} = \frac{1}{C^2} \cdot \frac{\partial^2 U}{\partial t^2} \quad \text{(one dimension)} \tag{1.5}$$

We can prove this statement for the function $U(x - ct)$. Let us define a new parameter $x - ct \equiv g$ and its corresponding derivative as $\frac{\partial u}{\partial g} = U'$. Using the chain rule, we can write the relation $\frac{\partial u}{\partial x} = \frac{\partial u}{\partial g} \cdot \frac{\partial g}{\partial x}$ and also the relation $\frac{\partial u}{\partial t} = \frac{\partial u}{\partial g} \cdot \frac{\partial g}{\partial t}$. And since $\frac{\partial u}{\partial t} = -C \cdot \frac{\partial u}{\partial g}$, it is easy to see that the relation

$$U'' = \frac{\partial^2 u}{\partial x^2} = \frac{1}{c^2} \cdot \frac{\partial^2 u}{\partial t^2}$$

is valid.

For the three-dimensional (3D) case we can use the term $U = U(\hat{n} \cdot \bar{r} - ct)$ to describe the wave, where $\hat{n}$ is a unit vector along the wave propagation direction and $\bar{r}$ is the spatial location vector. The corresponding wave equation in this case is given by

$$\nabla^2 U = \frac{1}{C^2} \cdot \frac{\partial^2 U}{\partial t^2} \quad \text{(three dimensions)} \tag{1.6}$$

1.6 HARMONIC WAVES

Each wave can be represented using the Fourier transform by a series of harmonic waves. Harmonic waves are waves with a defined periodic profile (such as sine or cosine functions). These types of waves are mathematically convenient to work with and can be described (for the 1D case) by

$$u = Ae^{j[\omega t - kx]} = A[\cos(\omega t - kx)) + j\sin(\omega t - kx))] \tag{1.7}$$

or more generally (for 3D) by

$$u = Ae^{j[\omega t - \bar{k}\cdot\bar{R}]} \tag{1.8}$$

where the wave temporal (also called angular) frequency is given by

$$\omega = 2\pi f = \frac{2\pi}{T} \tag{1.9}$$

and the wave spatial frequency is given by

$$\bar{k} \triangleq k \cdot \hat{n} \tag{1.10}$$

where $\hat{n}$ is a unit vector along the wave propagation direction and k is called the "wave number." The parameter k is required for two reasons:

(a) To cancel the physical dimension of the exponential term.
(b) To ensure the periodicity of the function with a wavelength of λ.

The value of k can be determined by studying a simple harmonic wave, for example:

$$u = A[\cos(\omega t - kx + \varphi)] \tag{1.11}$$

where φ represents the phase of the wave. Naturally, a peak of the wave profile will appear at time $t = 0$ wherever the following relation applies:

$$\cos(\varphi - kx) = 1 \tag{1.12}$$

Thus, for two consecutive peak points x_n and x_{n+1} on the wave profile at time $t = 0$, the following relations must apply:

$$\varphi - kx_n = n \cdot 2\pi \tag{1.13}$$

$$\varphi - kx_{n+1} = (n+1) \cdot 2\pi \tag{1.14}$$

By subtracting the two equations, we shall obtain

$$k(x_{n+1} - x_n) = 2\pi \tag{1.15}$$

And since the distance between two adjacent peak points for a harmonic wave equals the wavelength $\lambda = c/f$, it follows that

$$\boxed{k = \frac{2\pi}{\lambda}} \tag{1.16}$$

It is important to note that the physical units of k are radians/length. Hence, it is actually a parameter defining spatial frequency. The full spatial frequency spectrum of a given wave can be obtained by applying the Fourier transform to the wave profile at a given time point.

1.6.1 Equivalent Presentations

It is also worth noting that harmonic waves can be presented mathematically using alternative arguments. However, the physical interpretation is the same. Three exemplary presentations are given below:

$$U = A\cos\left[2\pi\left(\frac{x}{\lambda} - \frac{t}{T}\right) + \varphi\right], \tag{1.17}$$

$$U = A\cos\left[\frac{2\pi}{\lambda}(x - ct) + \varphi\right], \tag{1.18}$$

$$U = A\cos\left[2\pi\left(\frac{x}{\lambda} - ft\right) + \varphi\right] \tag{1.19}$$

1.7 GROUP WAVES

When a group of waves travel together in a medium, their amplitudes and phases interfere with each other and a pattern is formed which seems to propagate in a speed that differs from the speed of its components. In order to demonstrate this phenomenon, let us consider the simplest case where two waves having the same amplitude, but with different temporal (angular) and spatial frequencies, travel together. For example, if the first wave is given by

$$U_1 = A\cos(k_1 x - \omega_1 t) \tag{1.20}$$

and the second wave that accompanies it is given by

$$U_2 = A\cos(k_2 x - \omega_2 t) \tag{1.21}$$

then the pattern given in Fig. 1.13 is obtained. Marking the envelope of this combination, it would appear as though the envelope travels as a new wave in the medium. This pattern is called a "group" wave and is given mathematically for this specific example by

$$U = 2A\left\{\underbrace{\cos\left[\left(\frac{k_1 + k_2}{2}\right)x - \left(\frac{\omega_1 + \omega_2}{2}\right)t\right]}_{\text{High-frequency components}} \cdot \underbrace{\cos\left[\left(\frac{k_1 - k_2}{2}\right)x - \left(\frac{\omega_1 - \omega_2}{2}\right)t\right]}_{\text{Low-frequency components}}\right\} \tag{1.22}$$

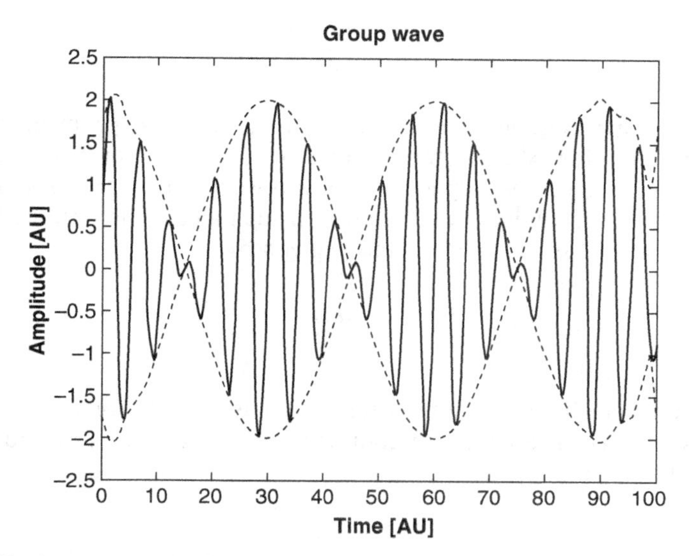

Figure 1.13. A group wave obtained by combining two sinusoidal waves of different frequencies. (The envelope is depicted by the dashed line for clarity.)

As can be noted, the combined pattern is represented by a multiplication between a high-frequency-components (both spatial and temporal) wave (the carrier wave) and a low-frequency-components (modulating) wave.

1.8 WAVE VELOCITY

Wave velocity is not so trivial to define. The more intuitive concept is to track the speed at which the energy carried by the wave propagates. This can be estimated by tracking a specific point of reference on the envelope of a propagating wavepacket (for a group wave). Such point of reference could be, for example, the front edge of the profile envelope or its point of maximal value. In the example given in Fig. 1.10, we can refer to the ball located on the left side of the "stormy" region as the front end of the wave envelope. By measuring the propagation Δx of this landmark at time interval Δt, the wave velocity can be estimated from

$$c = \frac{\Delta x}{\Delta t} \tag{1.23}$$

This type of wave propagation speed estimation by envelope amplitude tracking is called "group velocity." For a group comprised of a set of harmonic waves, the group velocity c_g is given by

$$c_g = \frac{\partial \omega(k)}{\partial k} \tag{1.24}$$

In case of a single-frequency wave, one can track the phase propagation speed. This can be done, for example, by selecting any point of reference on the wave profile. In the case of a wavepacket (a group wave), one can track the phase of one of its high frequency components. In such a case the obtained value is defined as the "phase velocity" c_p and is given by

$$c_p = \frac{\omega(k)}{k} \tag{1.25}$$

If dispersion is negligible or not existent (such as the case for electromagnetic waves in vacuum), the angular (temporal) frequency and the wave number are linearly related, that is,

$$\omega = c \cdot k \tag{1.26}$$

Hence for this case the group velocity and the phase velocity are the same, that is, $c_g = c_p$. If the relation between the angular (temporal) frequency and the wave number is not linear, the speed of sound varies with the wave frequency. This case is discussed in Chapter 5 (Section 5.5).

1.9 STANDING WAVES (A MATHEMATICAL DESCRIPTION)

As was presented in Section 1.2.4, the standing wave phenomenon occurs when a string (or some other type of bounded medium such as a beam or a tube) fixed on both sides vibrates. While most of the particles on the string (or medium) move periodically, there are specific points, called "nodal points," that do not move (as shown in Fig. 1.9). This phenomenon is explained by the existence of two waves of the same amplitude and frequency propagating in the same medium toward opposite directions. Such waves can be generated by the reflections caused by the boundaries (see Fig. 1.6). As a result, the two waves continuously interfere with each other. For a specific length of the medium and for specific frequencies (see Chapter 2), destructive interference occurs constantly at the nodal points and prevent them from moving.

Let us define the wave propagating along the positive x direction as U_1,

$$U_1 = A\cos(kx - \omega t) \tag{1.27}$$

and define the wave propagating along the negative x direction as U_2,

$$U_2 = A\cos(kx + \omega t) \tag{1.28}$$

The resulting displacements within the medium will be given by $U = U_1 + U_2$. And recalling the trigonometric relation,

$$\cos\alpha + \cos\beta = 2\cos\left(\frac{\alpha+\beta}{2}\right)\cdot\cos\left(\frac{\alpha-\beta}{2}\right) \tag{1.29}$$

we obtain

$$U = U_1 + U_2 = \underbrace{2A\cos(kx)}_{X(x)}\cdot\underbrace{\cos(\omega t)}_{\Theta(t)} \tag{1.30}$$

or, in a more general form,

$$U = X(x)\cdot\Theta(t) \tag{1.31}$$

As can be noted, the combined displacement pattern is comprised of a multiplication between two functions: (i) the function $\Theta(t)$, which depends solely on time and which modulates only the wave amplitude, and (ii) the function $X(x)$, which depends solely on location and exhibit a *constant profile* in space. And since at the nodal points we have $X(x) = 0$, no displacement will occur at these points, and hence we have the resulting impression of a "standing wave." (This phenomenon will be discussed again in Chapter 2.)

1.10 SPHERICAL WAVES

In daily life we occasionally encounter waves with circular symmetry (see Fig. 1.14). Most commonly, we associate such waves with the ripples pattern formed when a pebble is thrown into a water pond. In such cases the symmetry is two-dimensional (2D).

However, waves with three-dimensional (3D) symmetry which propagate as a sphere in space (and are therefore referred to as "spherical" waves) are also very common. In fact, such waves are created from almost any noise- or stress-generating event. For example, the noise generated when two objects collide, generated by a ringing bell, or generated by an explosion can all produce spherical waves. If the source for the waves may be considered small enough relative to the wavelength, we can assume that the wave stems from a single-point source and hence may be considered as spherical. Moreover, Huygens' Principle (see above) states that any wave source can be considered as an infinite collection of spherical wave sources.

A spherical wave is characterized by 3D spatial symmetry around its source (in an isotropic medium). Therefore it is natural to describe this type of wave in a spherical coordinate system. Defining the function describing the wave as φ, the three-dimensional wave equation is given by

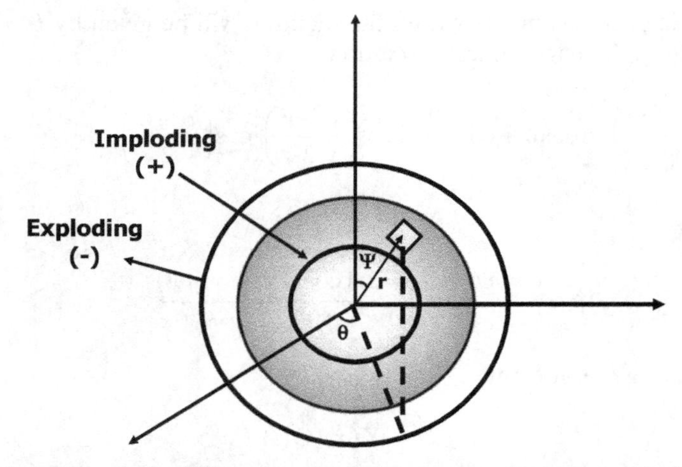

Figure 1.14. Schematic depiction of a spherical wave spreading symmetrically from (or toward) a central point.

$$\nabla^2 \varphi = \frac{1}{c^2} \frac{\partial^2 \varphi}{\partial t^2} \tag{1.32}$$

Rewriting the left-hand side of Eq. (1.32) in a spherical coordinate system r, θ Ψ, , we obtain [6]

$$\nabla^2 \varphi \triangleq \frac{1}{r^2} \frac{\partial}{\partial r}\left(r^2 \frac{\partial \varphi}{\partial r} \right) + \frac{1}{r^2 \sin\theta} \cdot \frac{\partial}{\partial \theta}\left(\sin\theta \frac{\partial \varphi}{\partial \theta} \right) + \frac{1}{r^2 \sin\theta} \cdot \frac{\partial^2 \varphi}{\partial \psi^2} \tag{1.33}$$

And since function φ should be symmetric relative to θ and Ψ, it follows that

$$\frac{\partial^2 \varphi}{\partial \psi^2} = 0 \tag{1.34}$$

$$\frac{\partial \varphi}{\partial \theta} = 0 \tag{1.35}$$

And therefore we shall obtain

$$\nabla^2 \varphi = \frac{\partial^2 \varphi}{\partial r^2} + \frac{2}{r} \frac{\partial \varphi}{\partial r} = \frac{1}{r}\left(\frac{\partial^2 (r\varphi)}{\partial r^2} \right) = \frac{1}{c^2} \frac{\partial^2 \varphi}{\partial t^2} \tag{1.36}$$

It can be easily shown that a solution to this wave equation is given by the generic function:

$$\boxed{\varphi = \frac{A}{r} f\left(t \pm \frac{r}{c}\right)} \tag{1.37}$$

where the "+" sign designates an imploding wave (i.e., propagating toward the central point) and the "–" sign designates an exploding wave. A is the wave amplitude, whose units are the physical units of φ (pressure, for example) multiplied by length.

This can be shown by substitution of Eq. (1.37) into the left-hand side of Eq. (1.36) (for simplicity we shall assume that the amplitude $A = 1$),

$$\nabla^2\varphi = \frac{1}{r}\left(\frac{\partial^2 (r\varphi)}{\partial r^2}\right) = \frac{1}{r}\left(\frac{\partial^2\left[f\left(t \pm \frac{r}{c}\right)\right]}{\partial r^2}\right) = \frac{1}{r}\frac{\partial}{\partial r}\left(\pm\frac{1}{c}f'\right) = \frac{1}{rc^2}f'' \tag{1.38}$$

while from Eq (1.32) we shall obtain the same term,

$$\nabla^2\varphi = \frac{1}{c^2}\frac{\partial^2\varphi}{\partial t^2} = \frac{1}{c^2}\cdot\frac{1}{r}\cdot\frac{\partial^2 f}{\partial t^2} = \frac{1}{c^2}\cdot\frac{1}{r}\cdot f'' \tag{1.39}$$

It should be pointed out that $f\left(t \pm \frac{r}{c}\right)$ is equivalent to $\tilde{f}(ct \pm r)$. But it is very important to note that unlike a planar wave, the amplitude of a spherical wave *decays* with the distance from the source according to $\frac{1}{r}$, even when attenuation is negligible.

1.11 CYLINDRICAL WAVES

Cylindrical waves are those that propagate from a line source and that keep angular symmetry around the source axis. (The source could, for example, be an antenna pole.) A schematic depiction of a cylindrical wave front propagating from or toward a line of symmetry (in this case the Z axis) is shown in Fig. 1.15.

Rewriting the left-hand side of Eq. (1.32) in a cylindrical coordinate system r, z, θ, we obtain [6]

$$\nabla^2\varphi \triangleq \frac{\partial^2\varphi}{\partial r^2} + \frac{1}{r}\frac{\partial\varphi}{\partial r} + \frac{1}{r^2}\frac{\partial^2\varphi}{\partial\theta} + \frac{\partial^2\varphi}{\partial z^2} \tag{1.40}$$

The symmetry around Z and θ yields

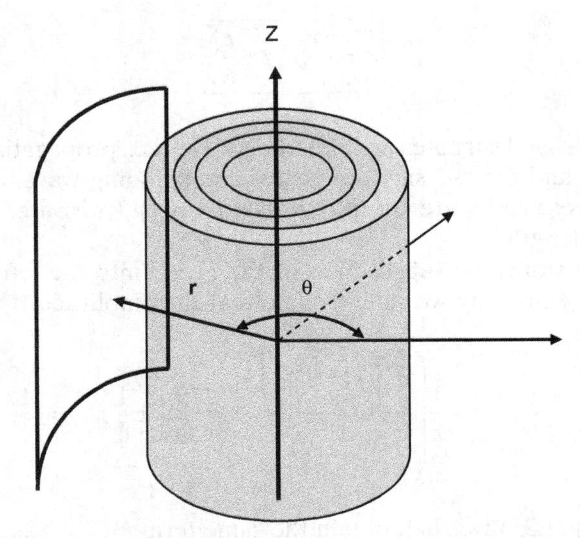

Figure 1.15. Schematic depiction of a cylindrical wave front propagating symmetrically from (or toward) a central line (Z axis in this case).

$$\frac{\partial^2 \varphi}{\partial z^2} = 0, \qquad \frac{\partial^2 \varphi}{\partial \theta^2} = 0 \tag{1.41}$$

And therefore we obtain

$$\nabla^2 \varphi = \frac{\partial^2 \varphi}{\partial r^2} + \frac{1}{r}\frac{\partial \varphi}{\partial r} = \frac{1}{r}\frac{\partial}{\partial r}\left(r\frac{\partial \varphi}{\partial r}\right) \tag{1.42}$$

Substituting Eq. (1.42) into Eq. (1.32), we obtain the wave equation for this type of wave:

$$\frac{\partial^2 \varphi}{\partial r^2} + \frac{1}{r}\frac{\partial \varphi}{\partial r} = \frac{1}{C^2}\frac{\partial^2 \varphi}{\partial t^2} \tag{1.43}$$

Multiplying both sides by r^2 and moving all terms to the left-hand side yields

$$r^2\frac{\partial^2 \varphi}{\partial r^2} + r\frac{\partial \varphi}{\partial r} - \frac{r^2}{C^2}\frac{\partial^2 \varphi}{\partial t^2} = 0 \tag{1.44}$$

To find a solution, we shall assume that the space–time relationship could be presented by two functions multiplied by each other, that is,

$$\varphi = e^{j\omega t} \cdot \psi(r) \tag{1.45}$$

Thus, if follows that

$$\frac{\partial^2 \varphi}{\partial t^2} = -\omega^2 \cdot \varphi \tag{1.46}$$

And if we recall that

$$\frac{\omega^2}{C^2} = \frac{(2\pi f)^2}{(\lambda f)^2} = k^2 \tag{1.47}$$

we shall obtain

$$r^2 \frac{\partial^2 \varphi}{\partial r^2} + r\frac{\partial \varphi}{\partial r} + r^2 k^2 \varphi = 0 \tag{1.48}$$

As can be noted, this equation is actually a Bessel differential equation of order zero. The solution for this equation is given by the Hankel function of the first kind and of order zero $H_0^1(kr)$:

$$\boxed{\varphi = H_0^1(kr) = A \cdot J_0(kr) + B \cdot Y_0(kr)} \tag{1.49}$$

where $J_0(kr)$ and $Y_0(kr)$ are Bessel functions of the first and second types respectively, and A and B are two constants. It should be noted that this solution is a complex function.

For simplicity the following approximation can be used:

$$\boxed{\varphi \approx \frac{D}{\sqrt{r}} \psi(\omega t \pm kr)} \tag{1.50}$$

The exact solution (absolute value) and its approximation are graphically compared in Fig. 1.16. The solutions were calculated for an expanding cylindrical wave propagating in water ($C = 1500$ m/sec) with a frequency of 1 MHz. As can be observed, the approximated solution follows the pattern of the exact solution (particularly for large kr values); however, it has an overestimating offset (which could be partially reduced by subtracting a constant).

1.12 THE WAVE EQUATION IN A NONHOMOGENEOUS MEDIUM

Thus far we have assumed that the medium in which the waves propagate is uniform. However, in the majority of cases this assumption is not true, because the body consists of many complicated structures of tissues and blood vessels.

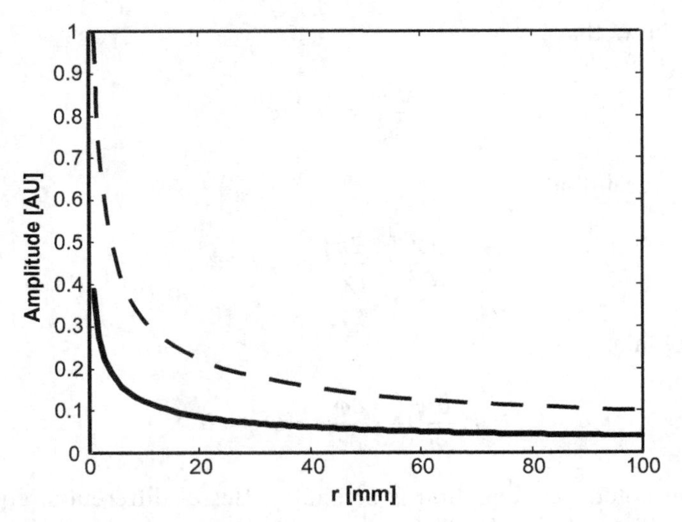

Figure 1.16. A graphic comparison between the absolute value of the exact solution (solid line) and the approximated solution (dashed line) for a cylindrical wave propagating with a frequency of 1 MHz in water. (*Note*: In order to avoid the singularity at the axes origin, the calculated range started at $r = 1$ mm.)

Indeed the assumption of homogeneity is practically convenient; however, in certain applications (e.g., acoustic computed tomography), inhomogeneity has to be accounted for. Inhomogeneity can stem from spatial changes in density or elastic properties; however, since these properties affect the speed of sound, we may consider the speed of sound as the significant variable. Without affecting the generality of the following discussion, we shall consider only one harmonic wave with a given frequency in the equations (the reader should bear in mind that any wave can be expanded into a series of "monochromatic" harmonic waves using the Fourier transform). Consequently, the changes in the speed of sound will be manifested in changes of the wavelength. This spatial variation can thus be presented by the wave number $k(\bar{r})$, where the vector $\bar{r}$ represents the spatial coordinates. Following the derivation presented in Kak and Slaney [7], the wave equation in a nonhomogeneous medium is given by

$$\left[\nabla^2 + k(\bar{r})^2\right]u(\bar{r}) = 0 \tag{1.51}$$

where $u(\bar{r})$ is a function that represents the acoustic field. Using, for example, water (whose acoustic properties are close to that of soft tissues) as a reference medium, this equation can be rewritten as

$$\left[\nabla^2 + k_0^2\right]u(\bar{r}) = -k_0^2\left[n(\bar{r})^2 - 1\right]u(\bar{r}) \tag{1.52}$$

where k_0 is the wave number for that reference medium, and $n(\bar{r})$ is the corresponding refractive index defined by

$$n(\bar{r}) = \frac{C_0}{C(\bar{r})} \tag{1.53}$$

where C_0 is the speed of sound in the reference medium. Commonly, in the context of soft tissues, the changes in the speed of sound are minor. Hence, we may represent the object by a function $F(\bar{r})$ that describes the changes as a perturbation with respect to the reference medium. Thus, the function describing the scanned object is given by

$$F(\bar{r}) = k_0^2\left[n(\bar{r})^2 - 1\right] \tag{1.54}$$

The acoustic field $u(\bar{r})$ within the medium may be presented by two components,

$$u(\bar{r}) = u_0(\bar{r}) + u_S(\bar{r}) \tag{1.55}$$

where $u_0(\bar{r})$ is the (transmitted) incident field, which is the solution for the homogeneous wave equation in the reference medium,

$$[\nabla^2 + k_0^2]u_0(\bar{r}) = 0 \tag{1.56}$$

and the field $u_S(\bar{r})$ describes the scattered field. Substituting into Eq. (1.52) yields

$$\boxed{[\nabla^2 + k_0^2]u_S(\bar{r}) = -F(\bar{r})u(\bar{r})} \tag{1.57}$$

This equation is called the Helmholtz equation. The object function cannot be derived explicitly from this equation; but since $u_0(r^-)$ is assumed to be known and $u_S(r^-)$ can usually be measured at defined locations around the object, we can write an implicit equation relating the measured filed $u_S(r^-)$ to the object function $F(r^-)$. To obtain this relation, we can use the Green function, defined as

$$g(\bar{r} - \bar{r}') = \frac{e^{jK_0|\bar{r}-\bar{r}'|}}{4\pi|\bar{r} - \bar{r}'|} \tag{1.58}$$

Using this function, the following integral equation can be obtained:

$$\boxed{u_S(\bar{r}) = \int g(\bar{r} - \bar{r}')\cdot F(\bar{r}')\cdot u(\bar{r}')\, d\bar{r}'} \tag{1.59}$$

As noted above, this equation cannot be solved directly. However, we can try to approximate $u_S(\bar{r})$ on the right-hand side of Eq. (1.59). There are two known approximations for the scattered field which are given in the following.

1.12.1 The Born Approximation

The idea behind this approximation is to utilize Eq. (1.55) and substitute it into Eq. (1.59). By doing so, we obtain

$$u_S(\bar{r}) = \int g(\bar{r}-\bar{r}')\cdot F(\bar{r}')\cdot u_0(\bar{r}')\,d\bar{r}' + \int g(\bar{r}-\bar{r}')\cdot F(\bar{r}')\cdot u_S(\bar{r}')\,d\bar{r}' \qquad (1.60)$$

But since the scattered field is much weaker than the incident field (which is typically the case for soft tissues), we can omit the second term on the right-hand side of this equation and write the following approximation:

$$\boxed{u_S(\bar{r}) \approx \int g(\bar{r}-\bar{r}')\cdot F(\bar{r}')\cdot u_0(\bar{r}')\,d\bar{r}'} \qquad (1.61)$$

Defining this approximation as the first Born approximation—that is, $u_s(\bar{r}) \triangleq u_B(\bar{r})$—we can go back and substitute its value again into Eq. (1.55) and obtain a better approximation for the scattered field in Eq. (1.60). This approximation is called the second Born approximation and is marked by $u_B^{(2)}(\bar{r})$ and is given by

$$\boxed{u_B^{(2)}(\bar{r}) \approx \int g(\bar{r}-\bar{r}')\cdot F(\bar{r}')\cdot (u_0(\bar{r}') + u_B(\bar{r}'))\,d\bar{r}'} \qquad (1.62)$$

Having this approximation at hand, it doesn't take much to realize that we can further improve it by re-substitution into Eq. (1.55) and Eq. (1.60) and then repeat the procedure over and over again. The general expression for the $u_B^{(i+1)}(\bar{r})$ Born approximation is thus given by

$$\boxed{u_B^{(i+1)}(\bar{r}) \approx \int g(\bar{r}-\bar{r}')\cdot F(\bar{r}')\cdot (u_0(\bar{r}') + u_B^{(i)}(\bar{r}'))\,d\bar{r}'} \qquad (1.63)$$

As noted above, this approximation is based on the assumption that the scattered field is much weaker than the incident field.

1.12.2 The Rytov Approximation

With this approximation the acoustic field is represented by a complex phase that is given by

$$u(\bar{r}) = e^{\phi(\bar{r})} \qquad (1.64)$$

while the incident field is given by

$$u_0(\bar{r}) = e^{\phi_0(\bar{r})} \qquad (1.65)$$

and the total field is given by

$$\phi(\bar{r}) = \phi_0(\bar{r}) + \phi_S(\bar{r}) \tag{1.66}$$

where $\phi_S(\bar{r})$ designates the scattered field. Following some derivations and assumptions, it can be shown that the following relation can be obtained:

$$\boxed{\phi_S(\bar{r}) \approx \frac{1}{u_0(\bar{r})} \int g(\bar{r}-\bar{r}') \cdot F(\bar{r}') \cdot u_0(\bar{r}')\, d\bar{r}'} \tag{1.67}$$

The condition needed for the Rytov approximation to be valid is given by

$$F(\bar{r}) \gg [\nabla \phi_S(\bar{r})]^2 \tag{1.68}$$

As reported in the literature (see, for example, Keller [8]), the Rytov approximation is valid under less restrictive conditions than the Born approximation.

REFERENCES

1. Filippi P, Habault D, Bergassoli A, and Lefebvre JP, *Acoustics: Basic Physics, Theory, and Methods*, Academic Press, New York, 1999.
2. Rose JL, *Ultrasonic Waves in Solid Media*, Cambridge University Press, Cambridge, 1999.
3. Ferwerda HA, *Huygens' Principle 1690–1990: Theory and Applications, Proceedings of an International Symposium; Studies in Mathematical Physics*, Blok HP, Kuiken HK, and Ferweda HA, editors, the Hague/Scheveningen, Holland, November 19–22, 1990.
4. Haber-Schaim U, Dodge JH, Gardner R, Shore EA, *PSSC Physics*, 7th Ed., Kendall/Hunt Publishing Company, Dubuque, IA, 1991.
5. Coulson CA, *Waves: A Mathematical Approach to the Common Types of Wave Motion*, 2nd edition, revised by Alan Jeffrey, Longman, London, 1977.
6. Spiegel MR, *Mathematical Handbook of Formulas and Tables*, Schaum's Outline Series in Mathematics; McGraw-Hill, New York, 1968.
7. Kak AC and Slaney M, *Principles of Computerized Tomographic Imaging*, IEEE Press, New York, 1988.
8. Keller JB, Accuracy and validity of the Born and Rytov approximations, *J Optical Soc Am* **59**:1003–1004, 1969.

CHAPTER 2

WAVES IN A ONE-DIMENSIONAL MEDIUM

Synopsis: In this chapter we shall analyze the behavior of elastic waves in the simplest case of a "one-dimensional" medium such as strings and rods. We shall determine mathematically the relationships between the wave propagation speed and the material properties. We shall study the effect of a discontinuity in the medium and also calculate the corresponding reflection and transmission coefficients.

2.1 THE PROPAGATION SPEED OF TRANSVERSE WAVES IN A STRING

Consider a string such as the one depicted in Fig. 2.1, through which an elastic mechanical wave is propagating. The wave propagates along the horizontal X axis, and its displacements are along the vertical Y axis. Naturally, this is a transverse wave since the displacement direction is perpendicular to the propagation direction (see Chapter 1). The tension in the string is T (Newton) and its mass per unit length is ρ (kg/m). The wave function is designated by $y(x, t)$ which measures the displacement [m] along the vertical direction.

Let us now "zoom-in" and focus on an infinitesimally small element of the string as shown in Fig. 2.1. From Newton's second law of physics, the sum of forces along the y direction acting on this element should equal the multiplication of its mass by its acceleration:

Basics of Biomedical Ultrasound for Engineers, by Haim Azhari

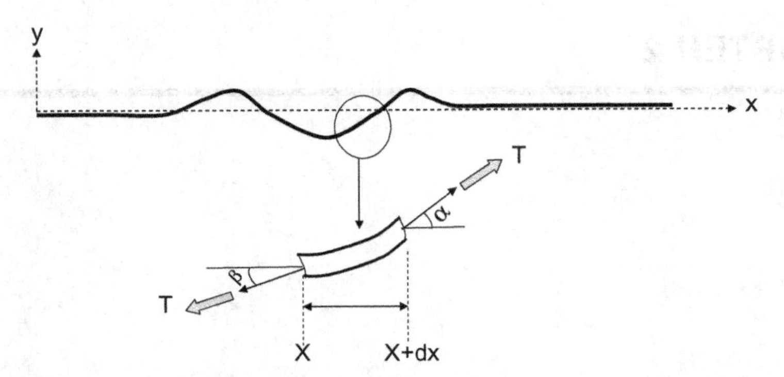

Figure 2.1. Schematic depiction of a transverse wave propagating in a string. A zoomed view of a small element is depicted at the bottom for clarity. The tension in the string is T.

$$\sum F_y = m\frac{\partial^2 y}{\partial t^2} \tag{2.1}$$

The force applied on each side of this element is given by the projection of its tension along the vertical direction. The mass is given by the multiplication of the density by its length. Consequently, we obtain

$$-T\sin(\beta)+T\sin(\alpha)=(\rho\, dx)\frac{\partial^2 y}{\partial t^2} \tag{2.2}$$

Expressing the values of the angles α and β by the spatial derivatives of the wave function at both sides yields

$$\tan\alpha = \left.\frac{\partial y}{\partial x}\right|_{x=x+dx} \tag{2.3}$$

$$\tan\beta = \left.\frac{\partial y}{\partial x}\right|_{x} \tag{2.4}$$

As recalled for small angles, it can be assumed that $\tan(\phi) \approx \sin(\phi)$. Hence, the following approximation can be written

$$-T\left.\frac{\partial y}{\partial x}\right|_{x} + T\left.\frac{\partial y}{\partial x}\right|_{x+\Delta x} = \rho\, dx\frac{\partial^2 y}{\partial t^2} \tag{2.5}$$

Reorganizing Eq. (2.5) yields

$$\underbrace{\frac{\left.\frac{\partial y}{\partial x}\right|_{x+\Delta x}-\left.\frac{\partial y}{\partial x}\right|_{x}}{dx}}_{\frac{\partial^2 y}{\partial x^2}}=\frac{\rho}{T}\cdot\frac{\partial^2 y}{\partial t^2} \tag{2.6}$$

As can be noted, the left-hand side of this equation is actually the second spatial derivative of the wave function, and therefore we can write the following equation:

$$\boxed{\frac{\partial^2 y}{\partial x^2}=\frac{\rho}{T}\cdot\frac{\partial^2 y}{\partial t^2}} \tag{2.7}$$

As can be observed, this is the familiar wave equation! Therefore, it follows that the wave propagation speed fulfills the relation $1/C^2 = \rho/T$ and hence its value is given by

$$\boxed{c=\sqrt{\frac{T}{\rho}}} \tag{2.8}$$

The implication of this equation is that the traveling time for a transverse perturbation from one point to another on a tensed string can be calculated, if its tension and density are known. Furthermore, it also follows that the propagation speed can be controlled by varying the tension T. For example, one can modulate the wave speed in time so as to obtain the relation $C(t)=\sqrt{T(t)/\rho}$.

2.2 VIBRATION FREQUENCIES FOR A BOUNDED STRING

When a mechanical perturbation is applied to a string bounded between two points (such as in musical instruments), it vibrates and standing waves are formed [1]. As recalled from Section 1.9 in Chapter 1, the standing wave can be represented by the function $y(x, t) = X(x)\cdot\Theta(t)$, where $X(x)$ is a function that depends only on the spatial coordinate x and $\Theta(t)$ is a function that depends only on time. Let us first write the corresponding term for the second spatial derivative,

$$\frac{\partial^2 y}{\partial x^2}=\Theta\cdot\frac{\partial^2 X}{\partial x^2} \tag{2.9}$$

and the second temporal derivative,

$$\frac{\partial^2 y}{\partial t^2}=X\cdot\frac{\partial^2 \Theta}{\partial t^2} \tag{2.10}$$

Substituting these terms into the wave equation yields

$$X \cdot \frac{\partial^2 \Theta}{\partial t^2} = c^2 \cdot \Theta \cdot \frac{\partial^2 X}{\partial x^2} \tag{2.11}$$

Separating the terms and reorganizing the equation, we obtain

$$\frac{1}{\Theta} \cdot \frac{\partial^2 \Theta}{\partial t^2} = \frac{c^2}{X} \cdot \frac{\partial^2 X}{\partial x^2} \tag{2.12}$$

And since the two functions are independent of each other, each side should equal a constant. Let us choose arbitrarily the constant $-\omega^2$ and separate the equation into two independent differential equations of the second order,

$$\frac{\partial^2 \Theta}{\partial t^2} + \omega^2 \cdot \Theta = 0 \tag{2.13}$$

$$\frac{\partial^2 X}{\partial x^2} + \frac{\omega^2}{c^2} \cdot X = 0 \tag{2.14}$$

As recalled, the solution for the generic type of equation

$$y'' + \lambda^2 y = 0 \tag{2.15}$$

is given by

$$y = A \cdot \sin(\lambda t) + B \cdot \cos(\lambda t) \tag{2.16}$$

Hence, the solution for the wave equation for a standing wave is given by

$$y(x,t) = X \cdot \Theta = \left[A \cdot \sin\left(\frac{\omega}{c} x\right) + B \cdot \cos\left(\frac{\omega}{c} x\right) \right] \cdot [G \cdot \sin(\omega t) + D \cdot \cos(\omega t)] \tag{2.17}$$

Recalling the fact that the string is bounded at both ends, we can write the following boundary conditions:

$$\begin{aligned} y(0,t) &= 0 \\ y(L,t) &= 0 \end{aligned} \tag{2.18}$$

where L is the length of the string. Substituting for the first boundary condition yields

$$B \cdot [G \sin(\omega t) + D \cos(\omega t)] = 0 \qquad \forall t \Rightarrow B = 0 \tag{2.19}$$

Substituting for the second boundary condition yields

$$A\sin\left(\frac{\omega L}{c}\right)\cdot\Theta(t)=0 \qquad \forall t \tag{2.20}$$

The constant A cannot also be zero because we know that waves do exist in the string. Therefore, it must follow that $\sin(\omega L/c) = 0$, and that the term in the brackets must fulfill the following condition:

$$\Rightarrow \frac{\omega L}{c}=n\cdot\pi, \qquad n=0,1,2,\ldots \tag{2.21}$$

or

$$\Rightarrow \omega_n = 2\pi\cdot f_n = \frac{n\cdot\pi\cdot c}{L}, \qquad n=0,1,2,\ldots \tag{2.22}$$

This implies that only frequencies that fulfill the following condition can exist:

$$\boxed{\Rightarrow f_n = \frac{nc}{2L}, \qquad n=0,1,2,\ldots} \tag{2.23}$$

Or if we wish to define the relation between the wavelength and the string's length, we obtain

$$\boxed{L=\frac{n\cdot C}{2\cdot f_n}=\frac{n\cdot\lambda_n}{2}} \tag{2.24}$$

This indicates that length L is an integer number of half the wavelengths $\lambda/2$; or, more precisely, only wavelengths that fulfill this condition can exist in the string.

Substituting the term for the wave speed from Eq. (2.8) into Eq. (2.23) yields the following relation:

$$\boxed{\Rightarrow f_n = \frac{n}{2L}\cdot\sqrt{\frac{T}{\rho}}, \qquad n=0,1,2,\ldots} \tag{2.25}$$

This equation can teach us several things. First of all we realize that there is an infinite number of vibration frequencies that can be generated in a string. Secondly, all the generated frequencies are multiplications of an integer number of the fundamental frequency. Thirdly, we note that the basic frequency is lower for longer or thicker strings—that is, for which the density ρ is bigger. Finally, we can control the frequency by altering the tension T in the string since the vibration frequency increases as the tension is increased.

Figure 2.2. The guitarist controls the guitar's string frequency by changing its length.

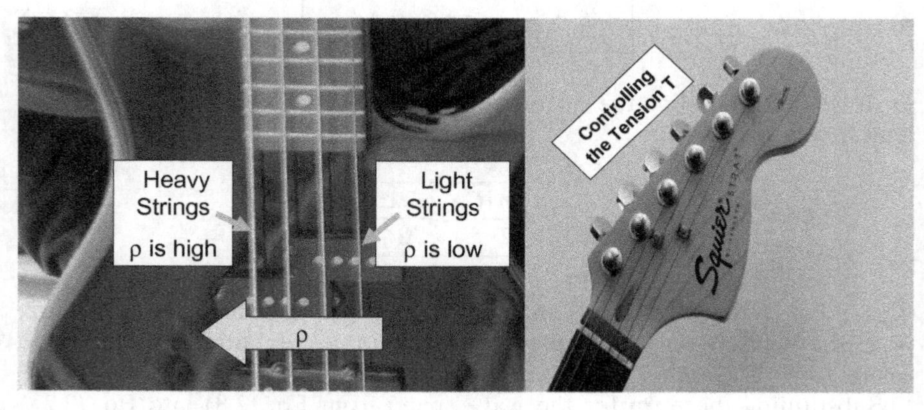

Figure 2.3. (Left) The guitar's strings are thicker for the lower tone (bass) strings. **(Right)** The guitarist tunes the guitar's strings fundamental frequencies by changing their tension using a set of knobs located at the top of its grip handle.

The implications of these relations are very well demonstrated in musical instruments. For example, let us examine the principles by which a guitar is operated. First of all, the tone that is generated by the guitarist is controlled by changing the length of the string L as shown in Fig. 2.2.

Secondly, as shown in Fig. 2.3 (Left), the guitar's strings are arranged in an ascending thicknesses order. Thus, the fundamental frequency for the thicker strings is lower yielding the more bass tones. Thirdly, as shown in Fig. 2.3

(Right), there is a set of knobs at the top of the grip handle. By rotating a knob, the tension in the corresponding string can be increased or decreased. The guitarist uses this fact to tune the guitar's strings fundamental frequencies [see Eq. (2.25)].

2.3 WAVE REFLECTION (ECHO) IN A ONE-DIMENSIONAL MEDIUM

Consider a boundary point connecting two strings, as described in Fig. 2.4. The first string has a density per unit length of ρ_1 (kg/m), and the second string has a density per unit length of ρ_2 (kg/m). The tension in the strings is T (N). A transverse wave U_1 traveling along the first string reaches that boundary point. From experimental observations we know (see Chapter 1) that two new waves will be generated. The first is a reflected wave U_2 (an "echo") that will propagate along the opposite direction. The second is a through transmitted wave U_3 that will continue to propagate along the same direction in the second string.

Making use of the fact that any wave can be decomposed into a set of harmonic waves using the Fourier transform, it is enough to analyze this problem for a single harmonic wave with an angular frequency of ω. The solution will be valid for all other cases. Thus, recalling that the wave propagation speeds are given by Eq. (2.8) ($c_1 = \sqrt{T/\rho_1}$, $c_2 = \sqrt{T/\rho_2}$), we can write the expression for the first wave as

$$U_1 = A_1 \cdot e^{j\omega\left(t - \frac{x}{c_1}\right)}$$

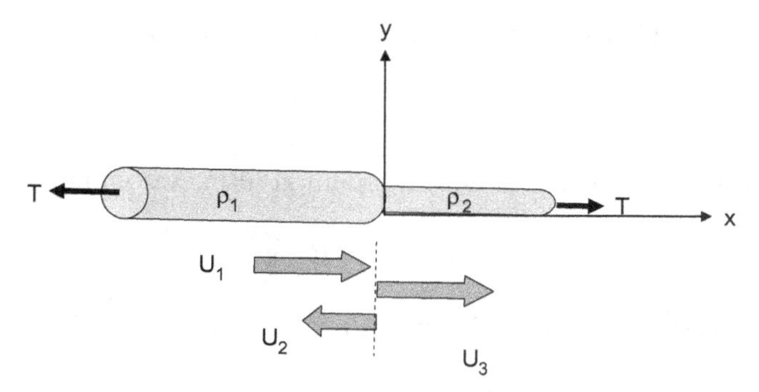

Figure 2.4. A wave U_1 propagating along a string for which the density is ρ_1 and the tension is T encounters a boundary point where the density changes abruptly to ρ_2. Part of the energy is through transmitted as wave U_3, and part of it is reflected (echo) as wave U_2.

and write the expression for the reflected (echo) wave as (note the + sign)

$$U_2 = A_2 \cdot e^{j\omega\left(t + \frac{x}{c_1}\right)}$$

The expression for the third wave is

$$U_3 = B \cdot e^{j\omega\left(t - \frac{x}{c_2}\right)}$$

From the superposition principle we can write the expression for displacements in the first string as the sum of the first two waves:

$$y_1(x,t) = U_1 + U_2 \tag{2.26}$$

And the displacement for the second string is given by

$$y_2(x,t) = U_3 \tag{2.27}$$

At point $x = 0$ where the two strings are attached, we must have continuity in the displacement; hence,

$$y_1\big|_{x=0} = y_2\big|_{x=0} \tag{2.28}$$

In addition, by equating the forces, it follows that the spatial derivatives should also be equal:

$$T \cdot \frac{\partial y_1}{\partial x}\bigg|_{x=0} = T \cdot \frac{\partial y_2}{\partial x}\bigg|_{x=0} \tag{2.29}$$

Substituting Eqs. (2.26) and (2.27) into Eq. (2.28), we obtain

$$A_1 \cdot e^{j\omega t} + A_2 \cdot e^{j\omega t} = B \cdot e^{j\omega t} \tag{2.30}$$

And after division by the common exponential term, we obtain

$$A_1 + A_2 = B \tag{2.31}$$

Substituting Eqs. (2.26) and (2.27) into Eq. (2.29) yields

$$-\frac{j\omega}{c_1} A_1 e^{j\omega t} + \frac{j\omega}{c_1} A_2 e^{j\omega t} = -\frac{j\omega}{c_2} B e^{j\omega t} \tag{2.32}$$

And after rearrangement, we obtain

$$\frac{(A_1 - A_2)}{c_1} = \frac{B}{c_2} \tag{2.33}$$

Solving for Eqs. (2.31) and (2.33) yields

$$\frac{A_2}{A_1} = \frac{c_2 - c_1}{c_2 + c_1} \tag{2.34}$$

But,

$$c_1 = \sqrt{\frac{T}{\rho_1}} \tag{2.35}$$

and

$$c_2 = \sqrt{\frac{T}{\rho_2}} \tag{2.36}$$

thus,

$$\boxed{\frac{A_2}{A_1} = \frac{C_2 - C_1}{C_2 + C_1} = \frac{\sqrt{\rho_1} - \sqrt{\rho_2}}{\sqrt{\rho_1} + \sqrt{\rho_2}}} \tag{2.37}$$

This term expresses the ratio between the reflected and transmitted wave and is defined as the reflection coefficient. Similarly for the transmission coefficient, we obtain

$$\frac{B}{A_1} = \frac{2c_2}{c_1 + c_2} \tag{2.38}$$

or

$$\boxed{\frac{B}{A_1} = \frac{2\sqrt{\rho_1}}{\sqrt{\rho_1} + \sqrt{\rho_2}}} \tag{2.39}$$

2.4 SPECIAL CASES

After we have obtained the mathematical relation between the transmitted and reflected waves at a discontinuity point, let us examine some specific cases.

1. *Case 1.* A point connecting two similar strings, that is, $\rho_1 = \rho_2 = \rho$. Substituting into Eqs. (2.37) and (2.39), we obtain

$$\frac{A_2}{A_1}=\frac{\sqrt{\rho}-\sqrt{\rho}}{\sqrt{\rho}+\sqrt{\rho}}=0$$ There is no echo!

$$\frac{B}{A_1}=\frac{2\sqrt{\rho}}{\sqrt{\rho}+\sqrt{\rho}}=1$$ The transmitted wave is equal to the incident wave!

2. *Case 2.* The string is connected to a wall (a fixed end). This can be represented by $\rho_2 \to \infty$ at point $X = 0$. Consequently, ρ_1 may be considered negligible in comparison (i.e., $\rho_1 \quad \rho_2$). Hence we obtain

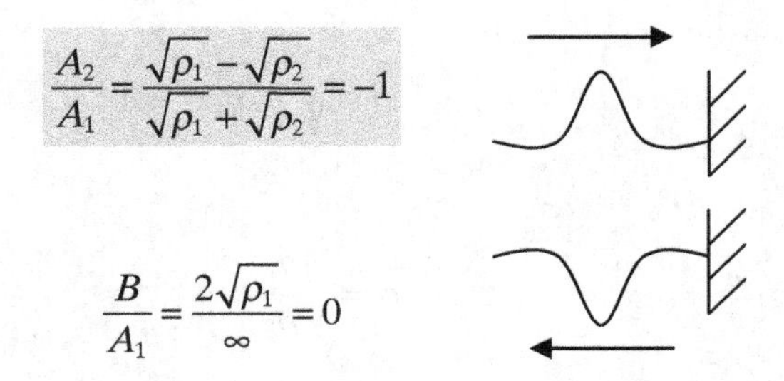

$$\frac{A_2}{A_1}=\frac{\sqrt{\rho_1}-\sqrt{\rho_2}}{\sqrt{\rho_1}+\sqrt{\rho_2}}=-1$$

$$\frac{B}{A_1}=\frac{2\sqrt{\rho_1}}{\infty}=0$$

This implies that the reflected wave has the same amplitude as the impinging wave but an opposite sign (phase), as depicted in the right-hand drawing above.

3. *Case 3.* The string has a free end, that is, $\rho_2 = 0$:

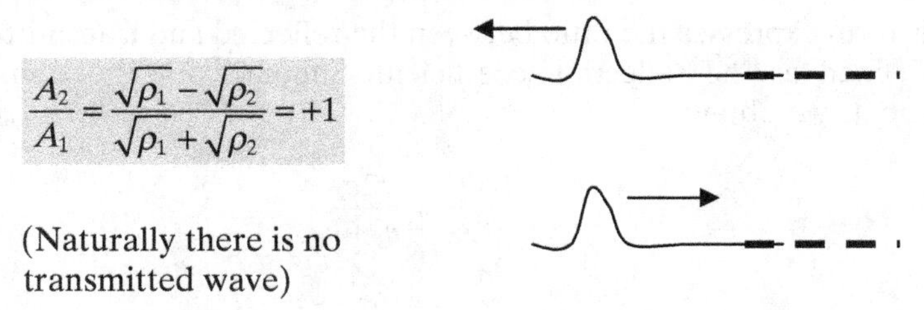

$$\frac{A_2}{A_1}=\frac{\sqrt{\rho_1}-\sqrt{\rho_2}}{\sqrt{\rho_1}+\sqrt{\rho_2}}=+1$$

(Naturally there is no transmitted wave)

As can be noted, the theoretical prediction is that the reflected wave will have the same amplitude and sign as the impinging wave (see drawing above).

Let us compare the theoretical predictions for the last two special cases with the experimental observations [2] shown in Chapter 1 (for clarity they are presented again in Fig. 2.5). As can be noted indeed, the experimental findings are in agreement with the theoretical predictions.

4. *Case 4.* The second string is denser than the first one (i.e., $\rho_1 < \rho_2$). From Eqs. (2.37) and (2.39), it follows that $A_2/A_1 < 0$ and that the through-transmitted wave will be smaller than the impinging wave, that is, $B < A_1$.

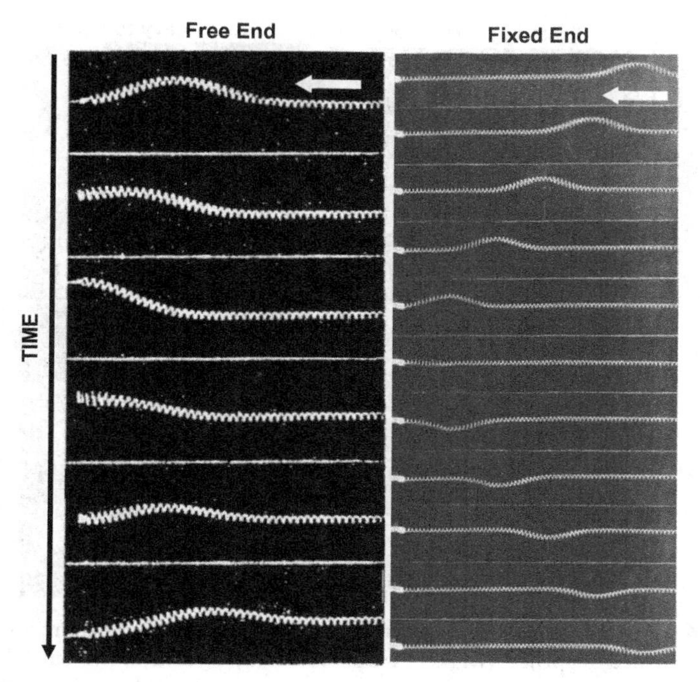

Figure 2.5. **(Right)** A transverse wave traveling (from right to left) along a spring encounters an edge that is tied to a wall. **(Left)** A transverse wave traveling (from right to left) along a spring encounters a free edge point. Reprinted with permission from *PSSC Physics*, 7th edition, by Haber-Schaim, Dodge, Gardner, and Shore, Kendall/Hunt Publishing Company, 1991.

5. *Case 5.* The first string is denser than the second one (i.e., $\rho_1 > \rho_2$). From Eqs. (2.37) and (2.39), it follows that $A_2/A_1 > 0$ and that the through-transmitted wave will be bigger than the impinging wave, that is, $B > A_1$.

Again, when comparing these predictions to the experimental observations shown in Chapter 1 (for clarity they are presented again in Fig. 2.6), it can be noted that they are in agreement.

2.5 WAVE ENERGY IN STRINGS

Let us observe again the element depicted in Fig. 2.1. This element has two energy components: (i) kinetic energy which is related to its velocity and (ii) potential energy stemming from the strings tension. When the element reaches its peak displacement, it has only potential energy. On the other hand, when it is aligned with its equilibrium state (i.e., when its displacement is zero), it has only kinetic energy. Thus by calculating the maximal kinetic energy value,

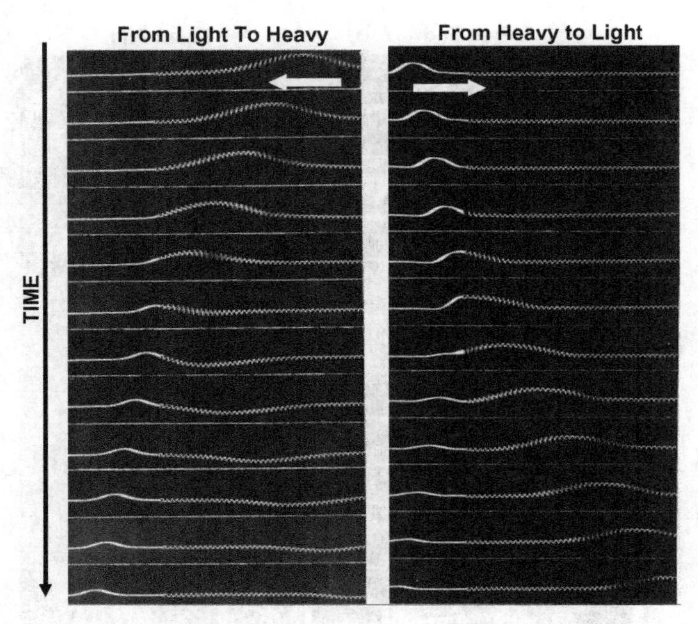

Figure 2.6. (Left) A transverse wave traveling (from right to left) along a light spring encounters a heavy spring. **(Right)** A transverse wave traveling (from left to right) along a heavy spring encounters a light spring. Reprinted with permission from *PSSC Physics*, 7th edition, by Haber-Schaim, Dodge, Gardner, and Shore, Kendall/Hunt Publishing Company, 1991.

we can determine the wave energy. Considering, for example, only the impinging wave, it is easy to show that the element's velocity is

$$\frac{dU_1}{dt} \equiv \dot{U}_1 = j\omega \cdot U_1 \tag{2.40}$$

and its maximal squared value is given by

$$\dot{U}_{1\,\max}^2 = \dot{U}_1 \cdot \dot{U}_1^* = \omega^2 \cdot A_1^2 \tag{2.41}$$

Recalling that the element's mass is given by $\rho \cdot dx$, the following energy terms per unit length can be written:

$E_I = \frac{1}{2}\rho_1 \cdot A_1^2 \cdot \omega^2$ Energy per unit length of the impinging wave

$E_R = \frac{1}{2}\rho_1 \cdot A_2^2 \cdot \omega^2$ Energy per unit length of the reflected wave

$E_T = \frac{1}{2}\rho_2 \cdot B^2 \cdot \omega^2$ Energy per unit length of the through transmitted wave

From energy conservation it follows that the energy "flowing" into the point of discontinuity should equal the energy flowing out of this point—that is, carried by the reflected and through-transmitted waves. Hence, the following equation can be written:

$$E_I \cdot C_1 = E_R \cdot C_1 + E_T \cdot C_2 \tag{2.42}$$

or more explicitly,

$$\frac{1}{2}\rho_1 c_1 \cdot \left(A_1^2 \omega^2\right) = \frac{1}{2}\rho_1 c_1 \cdot \left(A_2^2 \omega^2\right) + \frac{1}{2}\rho_2 c_2 \cdot \left(B^2 \omega^2\right) \tag{2.43}$$

Let us now define the term $Z \triangleq \rho \cdot C$. (This term is called the "acoustic impedance" and is discussed in extent in Chapter 6.) Multiplying by 2 and dividing by ω^2, the following equation is obtained:

$$Z_1 A_1^2 = Z_1 A_2^2 + Z_2 B^2 \tag{2.44}$$

Recalling the ratios obtained above:

$$\begin{aligned} \frac{A_2}{A_1} &= \frac{c_2 - c_1}{c_2 + c_1} \\ \frac{B}{A_1} &= \frac{2c_2}{c_1 + c_2} \end{aligned} \tag{2.45}$$

and substituting them into Eq. (2.44), the following two terms are obtained:

$$\boxed{\begin{aligned} \frac{\text{Reflected energy}}{\text{Impinging energy}} &= \frac{(Z_1 - Z_2)^2}{(Z_1 + Z_2)^2} \\ \frac{\text{Through-transmitted energy}}{\text{Impinging energy}} &= \frac{4Z_1 Z_2}{(Z_1 + Z_2)^2} \end{aligned}} \tag{2.46}$$

From these equations it can be noted that when $Z_1 = Z_2$ all energy is transmitted and when $Z_2 = 0$ all the energy is reflected. It is worth noting that in Chapter 6 it will be shown that identical relations are obtained for a normal incident planar wave in a more general case. It is also worth noting that spring models can be used to analyze more complicated conditions (e.g., reference 3).

2.6 PROPAGATION OF LONGITUDINAL WAVES IN AN ISOTROPIC ROD OR STRING

Thus far we have considered the propagation of transverse waves in a one-dimensional medium. In order to investigate the propagation of longitudinal

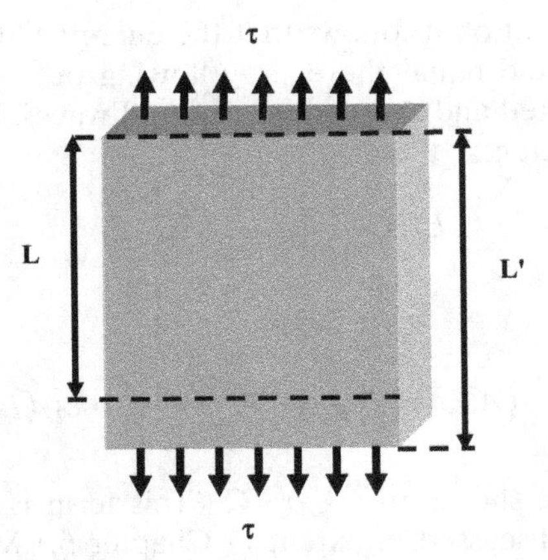

Figure 2.7. An elastic rod of length L subjected to stress τ from both sides will be stretched to a new length L'.

waves in such a medium, we must first be acquainted with basic terms in solid mechanics [4]. (This field is discussed in length in Chapter 4.) Consider an isotropic rod of length L such as the one depicted in Fig. 2.7. The rod is subjected to stress τ along its axial direction, from both sides. As a result, the elastic rod will be elongated to the new length L'.

Defining the strain ε as the ratio between the elongation ΔL and the initial length L, we obtain

$$\varepsilon = \frac{L' - L}{L} = \frac{\Delta L}{L} \tag{2.47}$$

For elastic (small) strains in isotropic materials, such as steel, aluminum, and so on, a linear relation was found between the magnitude of the stress applied to the rod and its strain. This linear relation is called *Hooke's Law* and is defined as

$$\varepsilon = \frac{\tau}{E} \tag{2.48}$$

where E is an elastic coefficient that is called the *modulus of elasticity*, also sometimes referred to as Young's modulus. The magnitude of E is characteristic to each material. For a nonisotropic material, additional elastic coefficients are required to relate the applied stress and the resulting deformations (see Chapter 4).

Now consider an isotropic rod of cross-sectional area A and a modulus of elasticity E as the one shown in Fig. 2.8. At time $t = 0$, a big hammer hits the rod's left side. Consequently, it can be assumed that its left face will be abruptly displaced along the horizontal direction, leading to the generation of

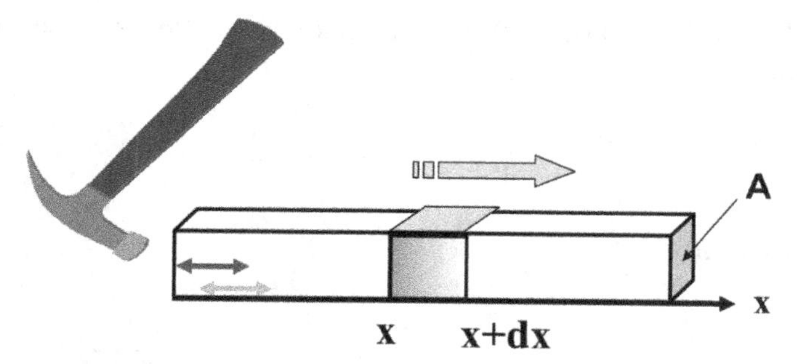

Figure 2.8. Longitudinal planar wave in an elastic rod can be generated by displacing abruptly one of its sides (for example, by hitting it with a big hammer).

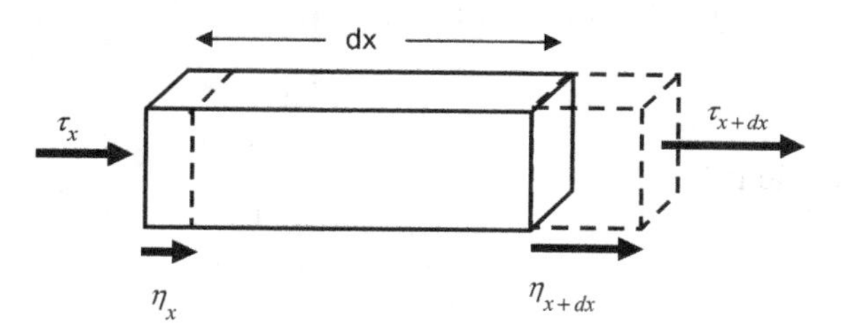

Figure 2.9. When a longitudinal wave passes through a rod's element located at x, its left side will be displace by η_x while its right side will be displaced by η_{x+dx}.

a planar longitudinal wave that will travel along the X axis toward the rod's right side.

If we examine closely a small element of the rod located at point x (see Fig. 2.9) when the wave passes through it, we shall find out that its side faces are not displaced exactly the same. Because of the time needed for the wave to travel and because of the rod's elasticity, its left side will be displaced by η_x while its right side will be displaced by η_{x+dx}.

Using Hooke's relation for stress and strain, we can write

$$\tau = E \cdot \varepsilon \tag{2.49}$$

Consequently, the force applied to the left side is given by

$$F_{\text{left}} = A\tau_x = A \cdot E \cdot \varepsilon = A \cdot E \cdot \left.\frac{\partial \eta}{\partial x}\right|_x \tag{2.50}$$

Similarly, the force applied to the right side is given by

$$F_{\text{right}} = A \cdot E \cdot \left.\frac{\partial \eta}{\partial x}\right|_{x+dx} \tag{2.51}$$

If the rod's density is ρ (kg/m³), we can make use of Newton's second law of physics and write

$$\Delta F = F_{\text{right}} - F_{\text{left}} = \frac{\partial^2 \eta}{\partial t^2} \tag{2.52}$$

or, more explicitly,

$$\Rightarrow AE \cdot \left(\left.\frac{\partial \eta}{\partial X}\right|_{x+dx} - \left.\frac{\partial \eta}{\partial X}\right|_{x} \right) = \underbrace{\rho \cdot A \cdot dx}_{m} \cdot \ddot{\eta} \left| \cdot \frac{1}{A\,dx} \right.$$

$$\Rightarrow E \cdot \left(\frac{\left.\frac{\partial \eta}{\partial X}\right|_{x+dx} - \left.\frac{\partial \eta}{\partial X}\right|_{x}}{dx} \right) = \rho \cdot \ddot{\eta} \tag{2.53}$$

which is equivalent to

$$\boxed{E \cdot \frac{\partial^2 \eta}{\partial x^2} = \rho \cdot \ddot{\eta}} \tag{2.54}$$

Noting that this is in fact the wave equation for the rod, it follows that

$$\frac{\partial^2 \eta}{\partial x^2} = \frac{1}{\frac{E}{\rho}} \cdot \ddot{\eta} \Rightarrow \frac{1}{C^2} = \frac{\rho}{E} \tag{2.55}$$

The wave propagation speed is therefore given by

$$\boxed{C = \sqrt{\frac{E}{\rho}}} \tag{2.56}$$

As can be noted, the reciprocal ratio of the rod's density (note that the units are mass per volume in this case) plays again a major role in determining the wave's propagation speed. However, one should not hastily conclude that the speed of sound for denser materials is lower than for lighter materials. The elasticity of the material is also important and can counterbalance the effect of density. It is only the ratio between the two that sets the value of the speed of sound. Importantly, it should also be emphasized that this equation is valid only for rods and strings! In the 3D case when waves travel within a solid medium, other elastic relations must be accounted for as will be explained in Chapter 4. (Reflections and transmissions for longitudinal waves will be discussed in Chapter 6.)

2.7 A CLINICAL APPLICATION OF LONGITUDINAL WAVES IN A STRING

We shall conclude this chapter with a practical example that has made use of longitudinal waves in a one-dimensional medium. A company called Angiosonics has developed a catheter that is actually a metal string, for delivering longitudinal ultrasonic waves into occluded blood vessels [5].

The clinical problem for which this company has provided a solution is the occlusion of a blood vessel by a blood clot. When a blood vessel is partially occluded (a state of stenosis) due to sclerosis, its clear cross-sectional area through which blood can flow is reduced. A big blood clot that is carried by the bloodstream can reach this narrow passage and occlude it (as depicted schematically in Fig. 2.10). This occlusion may cause severe ischemia and can lead to damage of the tissues located downstream.

Ultrasonic waves can disintegrate blood clots. The idea of Angiosonics was to build a device that will deliver the ultrasonic waves to the clot. Their device is comprised of a metal catheter that is attached to a wave generator. The free end of the catheter is inserted through a small cut into the blood vessel. The catheter is guided by X rays until its tip reaches the blood clot. Then, by pressing a pedal the physician activates the wave generator. The wave generator causes the proximal catheter side to vibrate (in a manner similar to the hammer depicted in Fig. 2.8). Longitudinal intensive ultrasonic planar waves were then generated at the catheter side that was located outside the body. The waves had a frequency of 400 kHz and traveled through the catheter until they have reached its tip located near the blood clot. The vibrations at the tip generated cavitation bubbles (see Chapter 12). These vibrations and the micro shock waves resulting from the bubbles collapse, disintegrated the clot. Luckily as a result of acoustic streaming (see Chapter 12), the clot debris remained near the catheter tip until complete disintegration. The reopening of the passage to blood could salvage the tissue located downstream with reference to that blood vessel. Pictures of the system and the catheter tip taken from Angiosonics publications are shown in Figures 2.11 and 2.12.

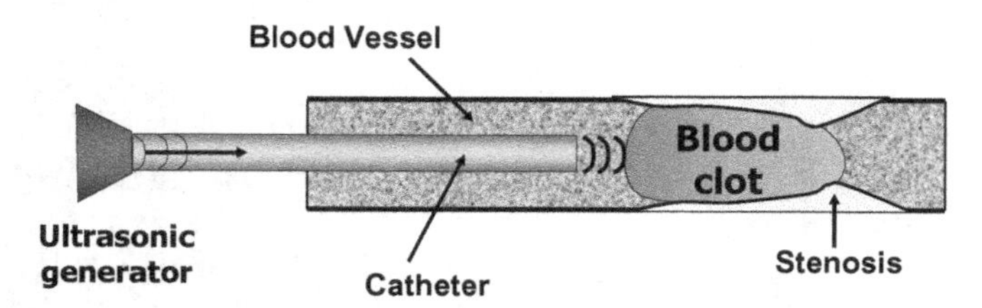

Figure 2.10. When a blood clot reaches a stenosis in a blood vessel, it can occlude it. A catheter (made by Angiosonics) can be used to deliver ultrasonic waves to the clot, causing it to disintegrate.

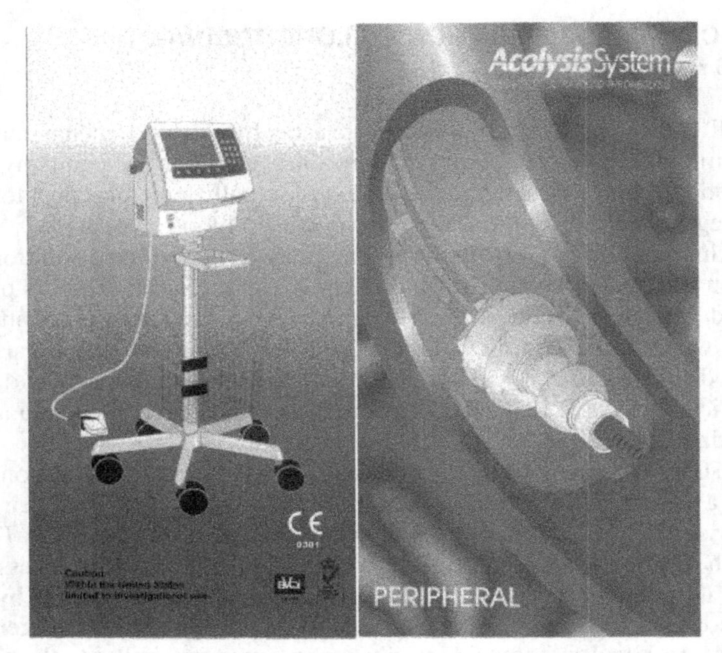

Figure 2.11. (Left) Picture of the Acolysis system suggested by Angiosonics company. (Right) Catheter tip shown schematically "digging" through a thrombus. Copyright © Angiosonics, all rights reserved.

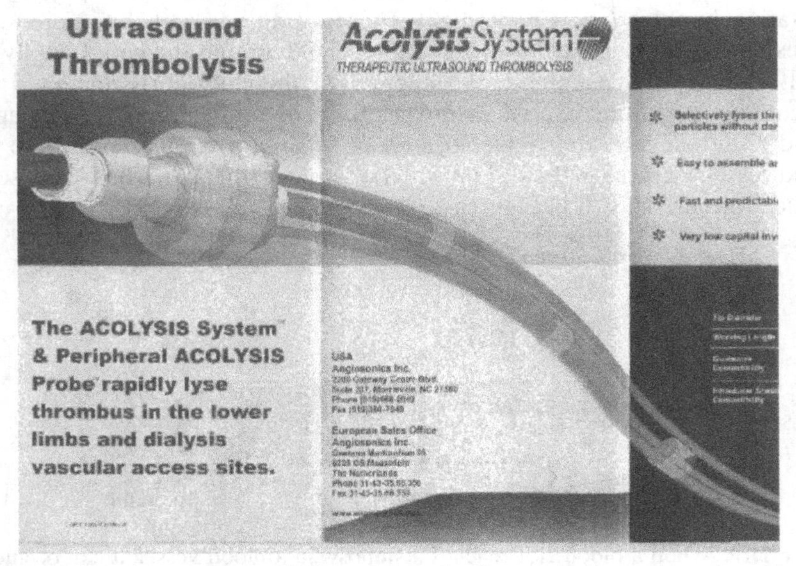

Figure 2.12. Picture of the Acolysis catheter tip. Copyright © Angiosonics, all rights reserved.

REFERENCES

1. Seto WW, *Theory and Problems of: Mechanical Vibrations*, Schaum's Outline Series in Engineering, McGraw-Hill, New York, 1964.
2. Haber-Schaim U, Dodge JH, Gardner R, Shore EA, *PSSC Physics*, 7th Ed., Kendall/Hunt Publishing Company, Dubuque, IA, 1991.
3. Delsanto PP and Scalerandi M, A spring model for the simulation of the propagation of ultrasonic pulses through imperfect contact interfaces, *J Acoust Soc Am* 104:2584–2591, 1998.
4. Ross CTF, *Mechanics of Solids*, Horwood Engineering Science Series, Chichester, 1999.
5. Levin PS, Saltonstall J, Nguyen L, and Rosenschein U, Ultrasound transmission apparatus and method of using same, US Patent 5,971,949, Angiosonics Inc., 1999.

CHAPTER 3

ULTRASONIC WAVES IN FLUIDS

Synopsis: In this chapter we shall analyze the behavior of ultrasonic waves in a fluidic material. We shall describe the relation between the fluid properties and the speed of sound. We shall define the wave "intensity." We shall learn about acoustic "radiation pressure" and show how it could be used for measuring the intensity.

(It is worth noting that most of the soft tissues in the body can be well described, in the context of ultrasound, as a fluidic material and hence the importance of this chapter.)

3.1 WAVES IN FLUIDS

A fluidic material is characterized in the context of ultrasonic waves by two significant properties. First of all, fluidic materials are isotropic in nature. Hence, the wave's properties are not affected by the propagation direction. Secondly, if the fluid is nonviscous, shear (transverse) waves can hardly exist or at least decay very rapidly in such a medium.

In order to understand why shear waves can hardly propagate in fluids, we have to remember that these waves are characterized by particle motion that is *perpendicular* to the wave's propagation direction. This type of wave propagation requires a strong enough mechanical bonding between the material's layers. This bonding is needed in order to transfer through shear forces the mechanical energy from one layer to another. However, in nonviscous fluids

Basics of Biomedical Ultrasound for Engineers, by Haim Azhari

(and naturally in gases) the bonding between the layers is relatively weak. Consequently, material layers can slide relative to each other.

The above argument does not, obviously, preclude the existence of *surface* waves in fluids. The most familiar example is the sea wave. This type of wave can also occur between two fluidic material layers that have different physical properties. But this type of wave will not be discussed here.

By studying the acoustic properties of various soft tissues (see Appendix A), it can be shown that their acoustic properties are very similar to those of water; that is, the mean speed of sound in soft tissue [1–3] is $c_{\text{soft tissue}} \approx 1540$ (m/sec) and the mean density is $\rho_{\text{soft tissue}} \approx 1.06$ (g/cm^3) as compared to a speed of $c_{\text{water}} \approx 1480$ (m/sec) and a density of $\rho_{\text{water}} \approx 1$ (g/cm^3) in water. This indicates that as a first approximation we can treat soft tissues as a fluidic material and assume that mainly longitudinal waves propagate there. (The propagation of ultrasonic waves in bones is discussed in Chapter 4.) Nevertheless, it is important to note that low-frequency shear waves can exist in soft tissues. In fact, these waves are used in a field called "Elastography," which utilizes special ultrasonic imaging methods (see Chapter 10) or MRI (e.g., reference 4) to characterize the elastic properties of tissues.

3.2 COMPRESSIBILITY

In order to relate the wave equation with the fluid properties, we first need to define a physical property called "compressibility." Consider the cylinder depicted in Fig. 3.1. The cylinder is filled with a fluid (or gas) with an initial pressure P_0 and its initial volume V. If the piston that forms its left side is pushed so that the volume is decreased by ΔV, the pressure will increase as a result by ΔP. If the cylinder is filled with gas, the change in volume ΔV is commonly large enough for us to notice even if the force applied to the piston is relatively small. One can appreciate this when inflating the tires of a bicycle.

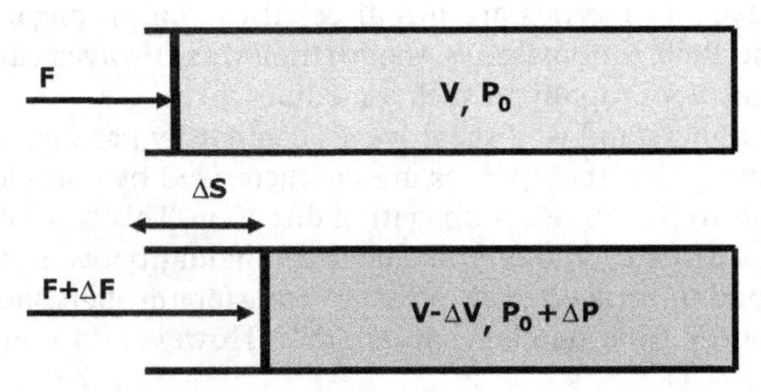

Figure 3.1. A piston compressing a cylinder filled with gas or a fluid with an initial pressure P_0 and its initial volume V **(top)** will cause its volume to decrease by ΔV and cause its pressure to increase by ΔP **(bottom)**.

On the other hand, we tend to assume that fluids such as water or oil may be considered "incompressible." This assumption may be practical enough in many applications. However, it is not true. Fluids can be compressed when subjected to pressure. Nevertheless, the change in volume is very minor even when the amount of pressure applied is large. In the context of ultrasonic wave propagation, this compressibility plays a major role and hence needs to be accounted for.

Under isothermal conditions, the coefficient of compressibility β for a fluid is defined as [5]

$$\beta = -\frac{1}{V}\left(\frac{\Delta V}{\Delta P}\right)_T \tag{3.1}$$

where ΔV and ΔP are the corresponding changes in volume and pressure, respectively, V is the initial volume, and the suffix T indicates that the temperature is constant. And since in the analysis that follows we are interested in the wave pressure, we shall rewrite Eq. (3.1) as

$$\Delta P = -\frac{1}{\beta}\left(\frac{\Delta V}{V}\right) \tag{3.2}$$

3.3 LONGITUDINAL WAVES IN FLUIDS

Referring again to Fig. 3.1, let us assume that the piston on the left side of the cylinder is moved rapidly and periodically to the left and to the right. Resulting from the rapid changes in the local density, the regional pressure changes and a longitudinal planar wave is generated. This wave will propagate in the fluid (as explained in Chapter 1) toward the right-hand side. The wave's profile can be obtained by plotting the pressure in the fluid as a function of location at a given time point. Assuming an harmonic motion of the piston, we can expect to see a profile such as the one depicted in Fig. 3.2, where P_0 is the equilibrium pressure (i.e., before the appearance of the wave) and $\Delta P(z)$ is the change in the pressure resulting from the wave at a given time point.

By studying the wave profile, one can realize that every point within the fluid is subjected to a pressure gradient when the ultrasonic wave passes through it. Consequently, in accordance with Newton's second law of physics, the fluid particles will be accelerated and displaced. This motion of the particles is actually what makes the wave propagate. However, there is a slight time delay between the motions of two adjacent material points. The time elapse between the moment that a small material element "feels" a change in the pressure at one side and the moment that it transfers that change in pressure to the other side is determined by the speed of sound.

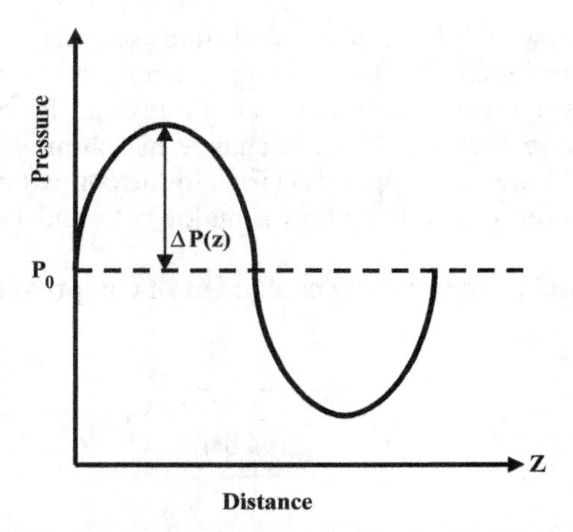

Figure 3.2. A longitudinal plane wave profile can be depicted by plotting the pressure as a function of position at a given time point.

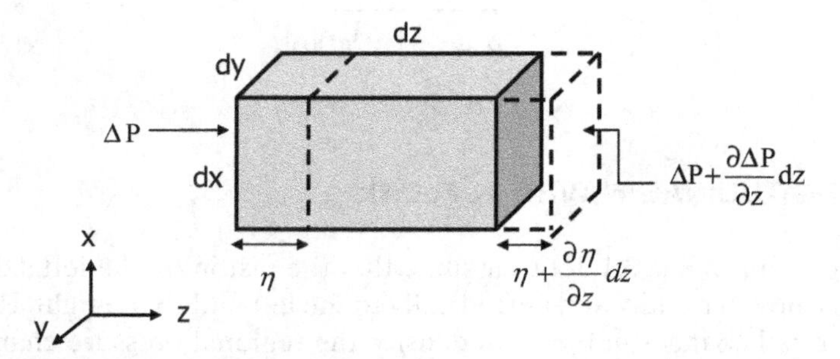

Figure 3.3. A longitudinal plan wave propagating toward the positive side of the Z axis is passing through a fluid element. Consequently, there will be a gradient in the pressure between the two sides of the elements. In addition, the displacement of both sides would not be the same, as depicted.

Consider a wave passing through an infinitesimally small fluid element such as the one depicted in Fig. 3.3. The left-hand side "senses" a change in the pressure by an amount of ΔP while the pressure at the right-hand side is approximated, using the first terms of the Taylor's series expansion, by $\Delta P + \partial(\Delta P)/\partial z\, dz$. Resulting from this change in pressure, the left-hand face of the element is displaced by W to the right side. Resulting from the compressibility of the fluid, the displacement on the right-hand face will not be the same, but it will be displaced by $\eta + \partial\eta/\partial z\, dz$ as shown in Fig. 3.3.

The force applied on each face of the element is given by the multiplication of its area and the pressure applied on it. Therefore, the force applied on the left side is given by

$$\rightarrow left\ F = P \cdot dx \cdot dy \tag{3.3}$$

and the force applied on the right side is given by

$$\leftarrow right\ F = dxdy\left(P + \frac{\partial P}{\partial z}dz\right) = P \cdot dx \cdot dy + \frac{\partial P}{\partial z}dx \cdot dy \cdot dz \tag{3.4}$$

Consequently, the net force pushing this element is given by

$$\begin{aligned}\Delta F_z = left\ F - right\ F &= \underbrace{P \cdot dx \cdot dy - P \cdot dx \cdot dy}_{=0} - \frac{\partial P}{\partial z}dx \cdot dy \cdot dz \\ &= -\frac{\partial P}{\partial z}dx \cdot dy \cdot dz\end{aligned} \tag{3.5}$$

In accordance with Newton's second law of physics, this force should equal the mass times the acceleration:

$$\Delta F_z = m \cdot \ddot{\eta} \tag{3.6}$$

where the mass is given by

$$m = dx \cdot dy \cdot dz \cdot \rho \tag{3.7}$$

Therefore,

$$\Delta F_z = dx \cdot dy \cdot dz \cdot \rho \cdot \ddot{\eta} \tag{3.8}$$

Substituting Eq. (3.5) into Eq. (3.8) yields

$$-\frac{\partial P}{\partial z}dx \cdot dy \cdot dz = \rho \cdot dx \cdot dy \cdot dz \cdot \frac{\partial^2 \eta}{\partial t^2} \tag{3.9}$$

After dividing by the volume, we obtain

$$-\frac{\partial P}{\partial z} = \rho \cdot \frac{\partial^2 \eta}{\partial t^2} \tag{3.10}$$

The change in the element's volume divided by its original volume is given by

$$\frac{\Delta V}{V}=\frac{dxdy\left(\eta+\frac{\partial \eta}{\partial z}dz-\eta\right)}{dxdydz}=\frac{\partial \eta}{\partial z} \tag{3.11}$$

Substituting this ratio into Eq. (3.2) yields

$$\Delta P=-\frac{1}{\beta}\left(\frac{\partial \eta}{\partial z}\right) \tag{3.12}$$

It should be noted that ΔP may be large and that the instantaneous pressure is given by

$$P\equiv \underbrace{P_0}_{\text{Rest}}+\Delta P \tag{3.13}$$

Hence,

$$-\frac{\partial P}{\partial z}=-\frac{\partial (P_0+\Delta P)}{\partial z}=-\frac{\partial (\Delta P)}{\partial z} \tag{3.14}$$

Substituting Eq. (3.12) into Eq. (3.14) yields

$$-\frac{\partial P}{\partial z}=-\frac{\partial}{\partial z}\left(-\frac{1}{\beta}\frac{\partial \eta}{\partial z}\right)=\frac{1}{\beta}\cdot\frac{\partial^2 \eta}{\partial z^2} \tag{3.15}$$

Equating Eq. (3.15) and Eq. (3.10) yields

$$\boxed{\frac{1}{\beta}\cdot\frac{\partial^2 \eta}{\partial z^2}=\rho\cdot\frac{\partial^2 \eta}{\partial t^2}} \tag{3.16}$$

or, after rearrangement,

$$\boxed{\frac{\partial^2 \eta}{\partial z^2}=(\beta\cdot\rho)\frac{\partial^2 \eta}{\partial t^2}} \tag{3.17}$$

As can be observed, this is the wave equation for the fluid material and therefore the following relation applies:

$$\frac{1}{C^2}=(\beta\cdot\rho) \tag{3.18}$$

Consequently the speed of sound in the fluid is given by

$$\boxed{C=\frac{1}{\sqrt{\beta\cdot\rho}}} \tag{3.19}$$

3.4 THE WAVE ENERGY

As stated in the first chapter, the acoustic wave is actually a flow of mechanical energy. In order to assess the wave energy, let us use a simple model in which the material molecules are represented by small masses and the intermolecular forces by a set of springs connecting these masses, as depicted in Fig. 3.4.

When the wave passes through the medium, it displaces the molecule layers (as shown schematically in Fig. 3.4). And when the distance between the layers changes relative to the equilibrium state, the intermolecular forces try to resist this motion and bring the displaced layer back to its original position. Consequently, the material energy is comprised of two components: (a) potential energy E_p which stems from the intermolecular forces, and (b) kinetic energy E_k, resulting from the molecules velocity V, that is,

$$\boxed{E_{\text{Total}} = E_k + E_P} \tag{3.20}$$

The kinetic part of the energy is given by

$$\boxed{E_K = \frac{m \cdot V^2}{2} = dx \cdot dy \cdot dz \cdot \rho \cdot \frac{1}{2}\left(\frac{\partial \eta(z,t)}{\partial t}\right)^2} \tag{3.21}$$

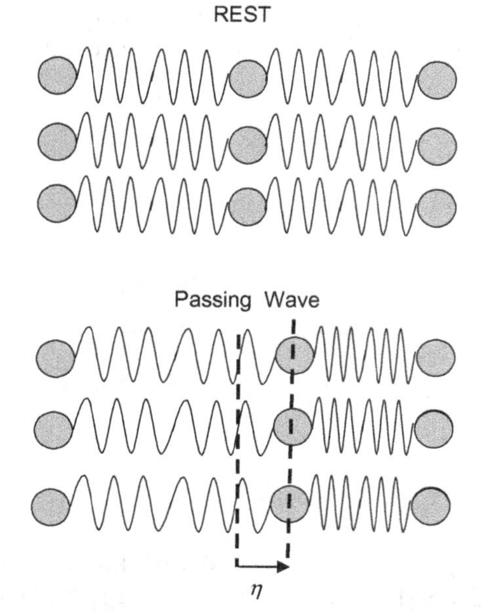

Figure 3.4. A simple model comprising of small masses and springs represents the material at rest **(top)** and when a longitudinal waves passes through it **(bottom)**.

where $\eta(z, t)$ is the displacement induced by the wave. For simplicity, let us assume that the displacement function of the wave is real and harmonic, for example:

$$\eta(z,t) = \eta_0 \cos(\Omega t - Kz) \tag{3.22}$$

where η_0 is the wave amplitude. (*Note*: The symbol Ω was utilized here to represent the angular frequency in order to visually emphasize its role in the following equations.) In such case the particles velocity is given by

$$\frac{\partial \eta}{\partial t} = -\eta_0 \Omega \sin(\Omega t - Kz) \tag{3.23}$$

and the kinetic energy per unit volume is given by

$$E'_K = \frac{E_K}{\text{unit volume}} = \frac{E_K}{dxdydz} = \rho \cdot \frac{1}{2} \eta_0^2 \Omega^2 \sin^2(\Omega t - Kz) \tag{3.24}$$

As can be understood from this model, the relative portion of each energy component (kinetic or potential) varies according to the particles location (as in a pendulum). There are two extreme points at which only one component exists. The first is when the particles reach their peak displacement; in this case their velocity and kinetic energy will be equal to zero. And the second point is when the particles pass through their equilibrium position; in this case their velocity will be maximal but their potential energy will be equal to zero. Referring to Eq. (3.24), it can be noted that the second case occurs when the following relation is met:

$$\sin(\Omega t - Kz) = 1 \tag{3.25}$$

And since at this point $E' = E'_K$, we can write

$$\boxed{E' = E'_K = \frac{\rho \cdot \eta_0^2 \Omega^2}{2}} \tag{3.26}$$

This is the wave energy per unit volume when the wave passes through the material.

3.5 INTENSITY

Another important parameter associated with the wave is its intensity. The wave intensity is defined as the amount of energy "flowing" through a unit area per time unit [6]. (The physical units commonly used are watts per squared centimeter.) By definition the intensity is therefore given by

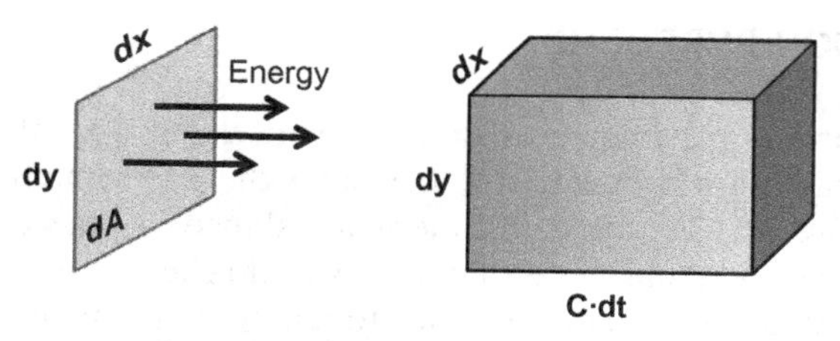

Figure 3.5. The amount of energy "flowing" through a unit area **(left)** per time dt is equal to the energy enclosed within the volume of the virtual cube shown on the right.

$$I = \frac{E}{dA \cdot dt} \tag{3.27}$$

As recalled, the amount of energy carried by the wave per unit volume is given by Eq. (3.26). This energy propagates at the speed of sound c. Thus, after time dt it will travel a distance of $c \cdot dt$, covering a volume of $dxdy \cdot c \cdot dt$ (see Fig. 3.5). The energy enclosed by this virtual cube is therefore given by

$$E = E' \cdot dv = E' \cdot dxdy \cdot c \cdot dt \tag{3.28}$$

The area through which the energy has flown through is given by $dA = dxdy$. The intensity I is therefore given by

$$I = \frac{E' \cdot dxdy \cdot c \cdot dt}{dxdy \cdot dt} = E' \cdot c \tag{3.29}$$

or, after substitution of Eq. (3.26),

$$I = \frac{1}{2}\rho \cdot \eta_0^2 \cdot \Omega^2 \cdot c \tag{3.30}$$

The intensity I is an important metric used for assessing the amount of energy reaching the tissue. For example, if the ultrasonic beam width at position x is $A(x)$ and the intensity is $I(x, t)$, the energy that has entered an insonified tissue element after time Δt is given by

$$E = A(x) \cdot \int_0^{\Delta t} I(x,t)\, dt \tag{3.31}$$

The combination of irradiation time and intensity is an important safety factor that has to be accounted for when designing an ultrasonic instrument (see Chapter 12). It is also an important factor when designing a therapeutic device (e.g., reference 7).

3.6 RADIATION PRESSURE

Another important parameter associated with the propagation of acoustic waves is the "radiation pressure" [8]. Consider the experiment depicted schematically in Fig. 3.6. In this experiment a balance is placed within a bath containing a fluid. A metal plate that serves as a reflector is placed on one of its sides. Underneath this plate (and not touching it) an ultrasonic transducer is placed. At the beginning of the experiment the transducer is turned off. A small mass is placed on the other side so that the balance is horizontally set at an equilibrium state (marked by the dashed line). The ultrasonic transducer is then turned on and transmits ultrasonic waves toward the reflecting plate. Although the transmitted wave may be purely sinusoidal (i.e., with a mean amplitude of zero), the experimental observation is that the plate will move upwards—that is, along the wave propagation direction. This implies that a mean positive pressure is applied on the plate as a result of the impinging ultrasonic waves. Let us try to understand where this pressure stems from and what its magnitude is.

When a fluid element changes its volume as a result of a propagating acoustic wave, its density ρ changes by the following ratio:

$$\frac{\Delta\rho}{\rho} = -\frac{\Delta V}{V} \tag{3.32}$$

The negative sign on the right-hand side indicates that the density increases as the volume is decreased. Substituting Eq. (3.1)—that is, $\beta = -\frac{1}{\Delta P}\left(\frac{\Delta V}{V}\right)_T$ —into Eq. (3.32) yields

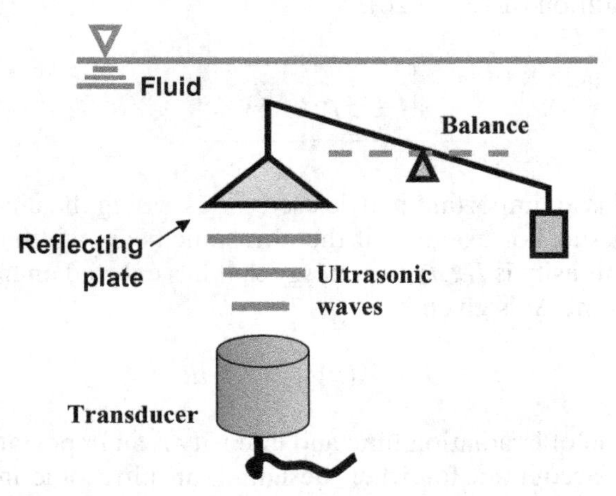

Figure 3.6. Ultrasonic waves transmitted toward a reflecting plate will cause it to move along the wave propagation direction.

$$\beta = \frac{\Delta\rho}{\rho} \cdot \frac{1}{\Delta P} \tag{3.33}$$

Recalling Eq. (3.12)—that is, $\Delta P = -\frac{1}{\beta}\left(\frac{\partial \eta}{\partial z}\right)$—we shall assume again that the wave is a real harmonic one that is given by the following equation:

$$\eta(z,t) = \eta_0 \cos(\Omega t - Kz) \tag{3.34}$$

Taking its spatial derivative, we obtain

$$\frac{\partial \eta}{\partial z} = K \cdot \eta_0 \cdot \sin(\Omega t - Kz) \tag{3.35}$$

and we also obtain its corresponding temporal derivative:

$$\frac{\partial \eta}{\partial t} = -\Omega \cdot \eta_0 \cdot \sin(\Omega t - Kz) \tag{3.36}$$

Substituting the above two equations into Eq. (3.12) yields

$$\Delta P = -\frac{1}{\beta}\left(\frac{\partial \eta}{\partial z}\right) = -\frac{K \cdot \eta_0}{\beta} \sin(\Omega t - Kz) \quad \left| \cdot \frac{\Omega}{\Omega} \right. \tag{3.37}$$

$$\Delta P = -\left[\frac{K}{\Omega}\right] \cdot \frac{1}{\beta} \cdot \eta_0 \cdot \Omega \cdot \sin(\Omega t - Kz) \tag{3.38}$$

Studying the term in the squared brackets, we can write

$$\frac{K}{\Omega} = \frac{2\pi}{\lambda} \cdot \frac{1}{2\pi f} = \frac{1}{\lambda \cdot f} = \frac{1}{c} \tag{3.39}$$

and

$$c = \frac{1}{\sqrt{\rho\beta}} \Rightarrow \beta = \frac{1}{\rho c^2} \tag{3.40}$$

Also, it can be noted that the following relation is by definition the particles velocity:

$$-\eta_0 \cdot \Omega \cdot \sin(\Omega t - Kz) = \frac{\partial \eta}{\partial t} \equiv U \qquad \text{(wave particles velocity)} \tag{3.41}$$

Substituting the above relations into Eq. (3.38) yields

$$\Delta P = \frac{1}{c} \cdot \rho c^2 \cdot \frac{\partial \eta}{\partial t} = \rho \cdot c \cdot U \tag{3.42}$$

This is by first approximation the acoustic pressure generated by the wave. In Eq. (3.33), the relation between ΔP and ρ is also given. Rearranging this equation, we obtain

$$\Delta P \cdot \beta = \frac{\Delta \rho}{\rho} \tag{3.43}$$

And the change in density is therefore given by

$$\Rightarrow \Delta \rho = \rho_0 \cdot \Delta P \cdot \beta \tag{3.44}$$

where ρ_0 is the density at rest (i.e., before the wave passes through the element). Expressing the instantaneous density as $\rho = \rho_0 + \Delta\rho$, we can use Eq. (3.44) to obtain a more accurate approximation by

$$\rho = \rho_0 + \Delta \rho = \rho_0 + \rho_0 \cdot \Delta P \cdot \beta \tag{3.45}$$

or, equivalently,

$$\rho = \rho_0 \cdot (1 + \Delta P \cdot \beta) \tag{3.46}$$

Substituting Eq. (3.42) into this equation yields

$$\Delta P = \rho \cdot c \cdot U = \rho_0 \cdot (1 + \Delta P \cdot \beta) \cdot c \cdot U \tag{3.47}$$

As can be noted, Eq. (3.47) is a recursive equation for the change in pressure ΔP. Assuming that the change in density is very small so that we can approximate it by $\rho \approx \rho_0$, the following approximation for the pressure change is obtained:

$$\Delta P = \rho \cdot c \cdot U = \rho_0 \cdot c \cdot U \tag{3.48}$$

Therefore,

$$\Delta P = \rho \cdot c \cdot U = \rho_0 \cdot \left(1 + \underbrace{\rho_0 \cdot c \cdot U}_{\Delta P} \cdot \beta \right) \cdot c \cdot U \tag{3.49}$$

Comment. An alternative way to obtain this term is to extract Δ_P from Eq. (3.47),

$$\overset{\downarrow}{\Delta P} = \rho \cdot c \cdot U = \rho_0 \cdot (1 + \Delta P \cdot \beta) \cdot c \cdot U = \rho_0 \cdot c \cdot U + \rho_0 \cdot \overset{\downarrow}{\Delta P} \cdot \beta \cdot c \cdot U \tag{3.50}$$

which can be rearranged to yield

$$\Delta P \cdot (1 - \rho_0 \cdot \beta \cdot c \cdot U) = \rho_0 \cdot c \cdot U \tag{3.51}$$

After further rearrangement and multiplying the nominator and denominator as follows:

$$\Delta P = \frac{\rho_0 \cdot c \cdot U}{(1 - \rho_0 \cdot \beta \cdot c \cdot U)} \quad \Bigg| \cdot \frac{1 + \rho_0 \cdot \beta \cdot c \cdot U}{1 + \rho_0 \cdot \beta \cdot c \cdot U} \tag{3.52}$$

we obtain

$$\Delta P = \frac{\rho_0 \cdot c \cdot U \cdot (1 + \rho_0 \cdot \beta \cdot c \cdot U)}{1 - \underbrace{(\rho_0 \cdot \beta \cdot c \cdot U)^2}_{\text{negligible}}} = \rho_0 \cdot c \cdot U \cdot (1 + \rho_0 \cdot \beta \cdot c \cdot U) \tag{3.53}$$

As can be noted, this expression is identical to Eq. (3.49).

Let us now substitute the value of U from Eq. (3.41) into Eq. (3.49); this will yield

$$\begin{aligned} \Delta P &= \rho \cdot c \cdot U = \rho_0 \cdot (1 + \rho_0 \cdot c \cdot U \cdot \beta) \cdot c \cdot U = \rho_0 \cdot c \cdot U + \rho_0^2 \cdot c^2 \cdot U^2 \cdot \beta \\ &= \rho_0 \cdot c \cdot [-\eta_0 \Omega \cdot \sin(\Omega t - Kz)] + \rho_0^2 \cdot c^2 \cdot \beta \cdot \Omega^2 \cdot \eta_0^2 \cdot \sin^2(\Omega t - Kz) \\ &= \underbrace{-\rho_0 \cdot c \cdot \Omega \cdot \eta_0 \cdot \sin(\Omega t - Kz)}_{} + \underbrace{\rho_0^2 \cdot c^2 \cdot \beta \cdot \Omega^2 \cdot \eta_0^2 \cdot \sin^2(\Omega t - Kz)}_{} \end{aligned} \tag{3.54}$$

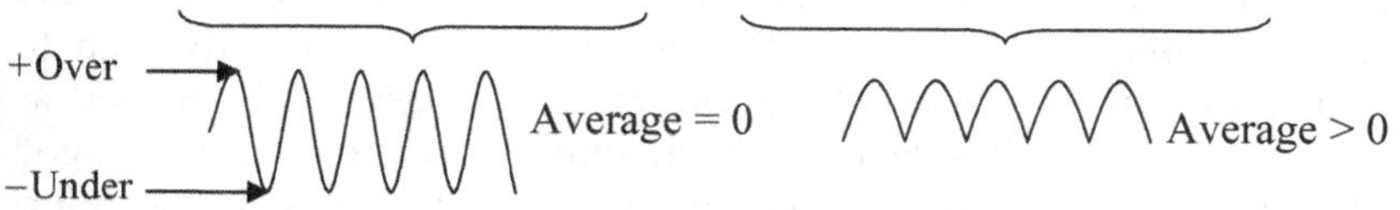

+Over

−Under

Average = 0

Average > 0

As can be observed, the right-hand side of this equation is comprised of two terms. The first term is a pure sinusoidal oscillating pressure. The average change in the pressure contributed by this term is therefore zero. However, the second term, which stems from the changes in the density (the nonlinear term), is always positive. Therefore, the average change in the pressure contributed by this term is bigger than zero. This implies that a "constant" positive average pressure is induced by the wave along its propagation direction. In order to find the value of this average pressure, we need to integrate ΔP over one period and divide by its duration,

$$\Gamma = \frac{1}{T} \int_0^T \Delta P \, dt \tag{3.55}$$

Substituting Eq. (3.54) into the integral yields

$$
\begin{aligned}
\Gamma &= \frac{1}{T}\int_0^T [\rho_0^2 \cdot c^2 \cdot \beta \cdot \Omega^2 \cdot \eta_0^2 \cdot \sin^2(\Omega t - Kz)]\, dt \\
&= \rho_0^2 \cdot c^2 \cdot \beta \cdot \Omega^2 \cdot \eta_0^2 \cdot \frac{1}{T}\int_0^T [\sin^2(\Omega t - Kz)]\, dt \\
&= \frac{1}{2} \cdot \rho_0^2 \cdot c^2 \cdot \beta \cdot \Omega^2 \cdot \eta_0^2 \Big|_{z=0}
\end{aligned} \tag{3.56}
$$

Substituting for $C = \frac{1}{\sqrt{\beta\rho}} \approx \frac{1}{\sqrt{\beta\rho_0}}$ [see Eq. (3.19)], we obtain

$$
\Gamma = \frac{1}{2} \cdot \rho_0^2 \cdot c^2 \cdot \beta \cdot \Omega^2 \cdot \eta_0^2 = \frac{1}{2} \cdot \rho_0 \cdot \Omega^2 \cdot \eta_0^2 \tag{3.57}
$$

Recalling the relation given by Eq. (3.30)—that is, $I = \frac{1}{2}\rho \cdot \eta_0^2 \cdot \Omega^2 \cdot c$—we can write down the expression for the average pressure as

$$
\boxed{\Gamma = \frac{I}{c}} \tag{3.58}
$$

This is the radiation pressure where no reflection occurs.

When encountering a reflective surface, this radiation pressure will induce a "radiation force." This radiation force can be used to "push" tissue noninvasively. By focusing the acoustic beam (as explained in Chapter 7), the obtained intensity, and consequently the radiation force, can be sufficiently high to induce local strain in the tissue. This strain can be ultrasonically imaged as explained in Chapter 9. This fact has been utilized in a recently emerged imaging technique called acoustic radiation force impulse or ARFI (e.g., references 9 and 10).

3.7 A PERFECT REFLECTOR

When a wave reaches a free surface or encounters an extremely stiff surface, the wave will be fully reflected (as was demonstrated in the figures shown in the previous chapter for strings). The reflected wave intensity will be equal to that of the incident wave. Consequently, the momentum of the material particles will be changed and a pressure will be applied on the surface. The pressure will be applied along the wave propagation direction. Let us try to describe the situation using the schematic illustration depicted in Fig. 3.7. In this illustration the wave momentum is described as a virtual cube propagating toward the reflecting surface. The cube has the same area as the plate. The momentum of the reflected wave is described as another virtual cube propagating away

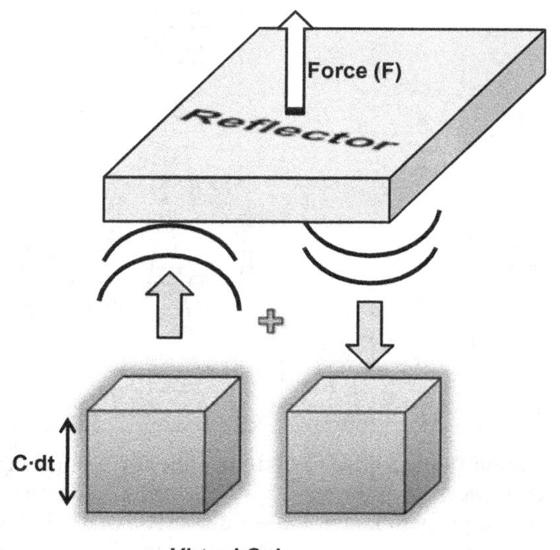

Figure 3.7. Schematic depiction of the momentum flow toward and away from a perfect reflector. Each wave is represented by a virtual cube. The incident and reflected waves were separated just for clarity.

from this surface. Although the incident wave and the reflected wave occupy the same volume, they were separated in this illustration for clarity.

The momentum of the incident wave that propagates upward ↑ (along the z axis) is given by

$$M_z^{(1)} = \overbrace{\rho \cdot A \cdot c \cdot dt}^{\text{mass}} \cdot U \tag{3.59}$$

The momentum of the reflected wave ↓ is given by

$$M_z^{(2)} = -\rho \cdot A \cdot c \cdot dt \cdot U \tag{3.60}$$

The net change in the momentum is therefore given by

$$dM_z = M_z^{(1)} - M_z^{(2)} = 2M_z^{(1)} \tag{3.61}$$

Referring to Newton's law of physics, it follows that the force applied on the plate is given by

$$F_z = \frac{dM_z}{dt} = 2 \cdot \rho \cdot A \cdot c \cdot U \tag{3.62}$$

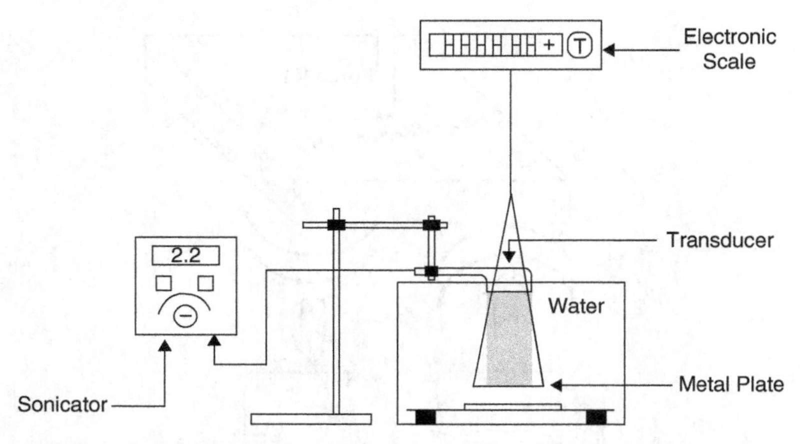

Figure 3.8. A schematic depiction of a system built for evaluating the intensity of ultrasonic transducers used for physiotherapy. The transducer was placed in a water tank so that its beam was perpendicular to a stainless steel plate that was held by strings and attached to an electronic scale. The intensity of the beam was controlled by the sonicator. (Drawing was prepared by Victor Frenkel.)

Substituting the term for ΔP given by Eq. (3.49), we obtain

$$F_z = 2 \cdot (\rho \cdot c \cdot U) \cdot A = 2\Delta P \cdot A \tag{3.63}$$

The average pressure applied on the reflector is given by

$$\Gamma_R = \frac{1}{T}\int_0^T \frac{F_z}{A}\, dt = \frac{2}{T}\int_0^T \Delta P\, dt \tag{3.64}$$

which is simply [see Eq. (3.55) and Eq. (3.58)] equal to

$$\boxed{\Gamma_R = \frac{2I}{C}} \tag{3.65}$$

This equation can be used when the medium contains highly reflective substances such as the gas bubbles used for contrast enhancement [11]. It can also be used for evaluating the intensity I of ultrasonic probes. To demonstrate this application, the system shown schematically in Fig. 3.8 was built. It was used for evaluating the intensity of ultrasonic transducers used for physiotherapy. A transducer was placed in a water tank so that its beam was pointing downward and perpendicular to a stainless steel plate also immersed in the water. The plate was held by strings and attached to an electronic scale. The intensity of the beam was controlled by a sonicator.

When the transducer was turned on—that is, transmitting a sinusoidal wave—the pressure applied by the beam on the plate has changed the reading of the scale. The scale would read a higher value as though the mass of the plate has increased. Designating the increase in the mass reading of the scale by m and using Eq. (3.65), the following relation can be written:

$$\Gamma = \frac{2I}{C} = \frac{F}{A} = \frac{m \cdot g}{A} \tag{3.66}$$

where F is the radiation force applied by the wave on the plate, g is the gravitational acceleration, A is the beam cross-sectional area, and C is the speed of sound in the water.

The beam intensity can thus be calculated from

$$\boxed{I = \frac{m \cdot g \cdot C}{2 \cdot A}} \tag{3.67}$$

Assuming that the gravitational acceleration is equal to 10 m/sec² and that the speed of sound in water is 1500 m/sec, along with the fact that the change in mass is measured in grams and that the beam area is measured in squared centimeters, we can write the following practical relation:

$$I\left(\mathrm{W/cm^2}\right) = \frac{m(\mathrm{g})}{0.1333 \cdot A\left(\mathrm{cm^2}\right)} \tag{3.68}$$

The experimental results obtained with this system are shown in Fig. 3.9 and Fig. 3.10 for 3-MHz and 1-MHz transducers, respectively. The horizontal axis

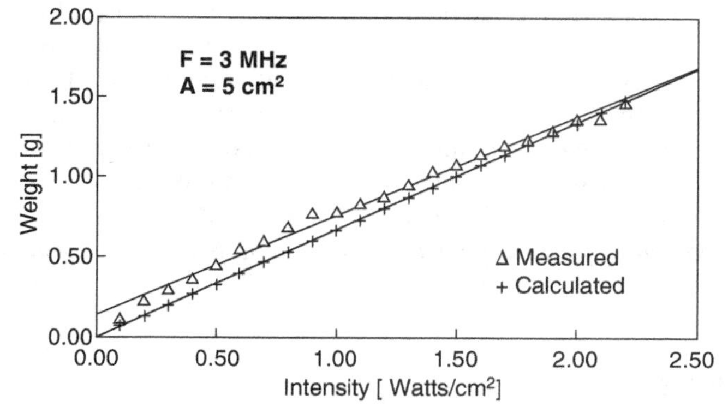

Figure 3.9. Results obtained for a 3-MHz transducer with a beam cross-sectional area of A = 5 cm². The calculated mass change readings are marked by a plus sign (+), and the measured values are indicated by a triangle (Δ). The solid lines designate the linear regression lines. (Prepared by Victor Frenkel.)

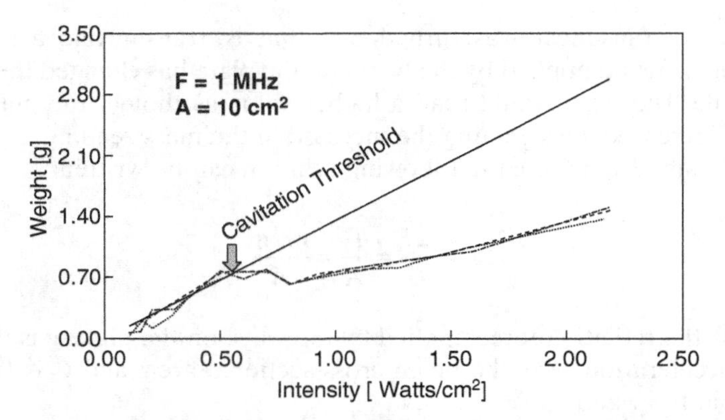

Figure 3.10. Results obtained for a 1-MHz transducer with a beam cross-sectional area of $A = 10\,\text{cm}^2$. The calculated mass change readings are marked by the solid line. The dashed lines correspond to the readings obtained in three experiments. As can be noted, the agreement between the theoretical and experimental values is good as long as the cavitation threshold was not exceeded. (Prepared by Victor Frenkel.)

depicts the intensity readings provided by the sonicator (assumed as our gold standard). The vertical axis depicts the change in mass readings. The theoretical change in mass readings was calculated from

$$m = \frac{2I \cdot A}{C \cdot g} \tag{3.69}$$

As can be noted from the results obtained for the 3-MHz transducer (Fig. 3.9), the calculated and measured values are in very good agreement. On the other hand, the results obtained for the 1-MHz transducer (Fig. 3.10) demonstrate clearly how important it is to degas the water when using ultrasound. The water in this experiment was not degassed. At an intensity of about $0.5\,\text{W/cm}^2$, the cavitation threshold was exceeded and air bubbles appeared. These bubbles disrupted the beam and the linear relation. Nevertheless, as can be observed, the calculated and measured values were correlated very well as long as the cavitation threshold was not exceeded. The cavitation effect is discussed in Chapter 12. *In any case, one is highly advised to make sure that the medium is properly degassed (if possible) when working at high intensities and particularly at low frequencies!*

REFERENCES

1. Duck FA, *Physical Properties of Tissue*, Academic Press, London, 1990.
2. Duck FA, Propagation of sound through tissue, in *The Safe Use of Ultrasound in Medical Diagnosis*, ter Haar G and Duck FA, editors, British Institute of Radiology, London, 2000, Chapter 2, pp. 4–15.

3. Wells PNT, *Biomedical Ultrasonics*, Academic Press, New York, 1977.
4. Lopez O, Amrami KK, Manduca A, and Ehman RL, Characterization of the dynamic shear properties of hyaline cartilage using high-frequency dynamic MR elastography, *Magn Reson Med* **59**(2):356–364, 2008.
5. *CRC Handbook of Chemistry and Physics*, 61st edition, Weast RC and Astle MJ, editors, CRC Press, Boca Raton, FL, 1980–1981.
6. Beyer RT and Letcher SV, *Physical Ultrasonic*, Academic Press, New York, 1969.
7. Ginter S, Liebler M, Steiger E, Dreyer T, and Riedlinger RE, Full-wave modeling of therapeutic ultrasound: Nonlinear ultrasound propagation in ideal fluids, *J Acoust Soc Am* **111**(5, Pt. 1):2049–2059, 2002.
8. Duck FA, Baker AC, and Starritt HC, *Ultrasound in Medicine*, Bristol, Institute In Physics Pub., 1998.
9. Dumont D, Behler R, Nichols T, Merricks E, and Gallippi C, ARFI imaging for noninvasive material characterization of atherosclerosis. *Ultrasound Med Biol* **32**(11):1703–1711, 2006.
10. Palmeri ML, Frinkley KD, Zhai L, Gottfried M, Bentley RC, Ludwig K, and Nightingale KR, Acoustic radiation force impulse (ARFI) imaging of the gastrointestinal tract, *Ultrason Imaging* **27**(2):75–88, 2005.
11. Dayton PA, Allen JS, and Ferrara KW, The magnitude of radiation force on ultrasound contrast agents, *J Acoust Soc Am* **112**(5):2183–2192, 2002.

CHAPTER 4

PROPAGATION OF ACOUSTIC WAVES IN SOLID MATERIALS

Synopsis: In this chapter we shall try to understand the relations between the properties of a homogeneous and nonhomogeneous solid material serving as a medium and the acoustic waves propagating though it. The main characteristics of solid materials in this context are the connection between the wave propagation directions and speed of sound and the coexistence of longitudinal and transverse (shear) waves. This chapter will start with a brief revision of basic definitions taken from the theory of elasticity—that is, stresses and strains. Then the relationships between the material properties and the speed of sound will be presented. (A reader who is familiar with the theory of elasticity for solids may skip the first part.)

4.1 INTRODUCTION TO THE MECHANICS OF SOLIDS

In order to analyze the propagation phenomenon of acoustic waves in a solid material, we must first become familiarized with several basic terms taken from the mechanics of solids. Then we should incorporate these definitions with the various material properties and find their relations to the speed of sound.

4.1.1 Stress

Stress is a measure of the "density" (or flux) of internal forces acting within the body as a result of external forces and body forces (such as gravity), acting on an object. This measure relates the specific geometry at every point within the studied object and the specific configuration by which external and body

Basics of Biomedical Ultrasound for Engineers, by Haim Azhari

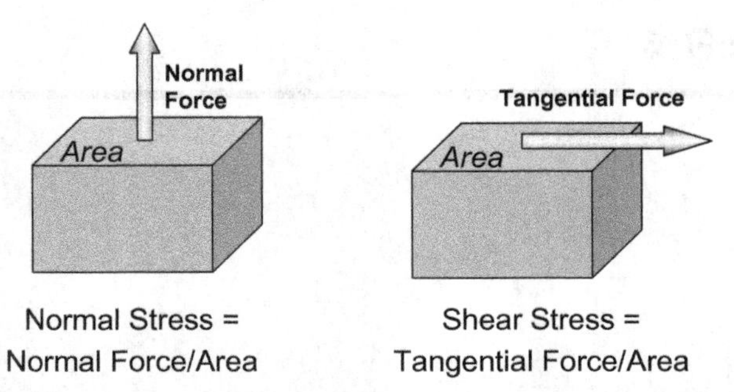

Figure 4.1. Schematic depiction of the two basic types of stress applied to a solid cube. **(Left)** Normal stress. **(Right)** Shear stress.

forces are applied to it. The generic definition of stress is the ratio between the magnitude of a force acting upon a plan and the area upon which this force is applied. As can be noted, stress has the same physical dimensions as pressure. However, when calculating a stress value, the relative direction between the applied force and the normal to the plan upon which it is applied has to be accounted for. In Fig. 4.1, for instance, two simple examples are given: (a) a normal compression or tension stress, where the direction of the applied external force and the normal to the plan upon which it is applied are parallel to each other, and (b) a shear stress, where the direction of the applied external force and the normal to the plan upon which it is applied are perpendicular to each other. From these two basic examples, one can realize that for a more general 3D case we can obtain a different combination of normal and shear stress at each material point. In fact, as will be shown in the following, we shall need nine different stress components to describe the load at a given point.

4.1.2 Strain

Strain is a measure of the deformation of a body subjected to stress. This measure also depends on the spatial directions of the applied stress and the corresponding elastic properties of the matter. It should be pointed out that in certain solids (anisotropic materials) the elastic properties may be different along different directions. Strain has also nine components in the 3D general case.

4.1.3 Special Issues to Be Noted when Investigating Wave Propagation in Solids

- An isotropic material has physical properties that are the same along every measured direction.
- There are many solids that are anisotropic. Consequently, speed of sound can be different along different propagation directions. Thus, the orienta-

tion of the natural coordinate system of the matter relative to the wave has to be known and accounted for.

- In the general case of a solid material, all elastic coefficients may affect the value of the wave propagation speed.
- Solid materials can support the simultaneous propagation of longitudinal and shear waves.
- When impinging upon a boundary of a solid material, longitudinal waves can be transformed or lead to the generation of shear waves, and vice versa (see Chapter 6).

In order to relate the speed of sound and material properties, we first need to be acquainted with the basic relations between stresses and strains. This field is known as "elasticity theory," which will be briefly presented in the following sections.

4.2 THE ELASTIC STRAIN

When a solid material is subjected to stress along any direction (regardless of the matter's natural coordinate system), changes in the matter's dimensions (deformations) occur.

Let us consider two material points P and Q within a solid material (Fig. 4.2). The distance between these two points is dl, and their corresponding projections on the X axis are X_P and X_Q, respectively. When the matter is subjected to stress, the two points will move to a new location P' and Q', respectively, the distance between them will change to dl' and their corresponding projections on the X axis will change to $X_{p'}$ and $Q_{p'}$, respectively. Hence, the displacement of point P along the X axis will be given by

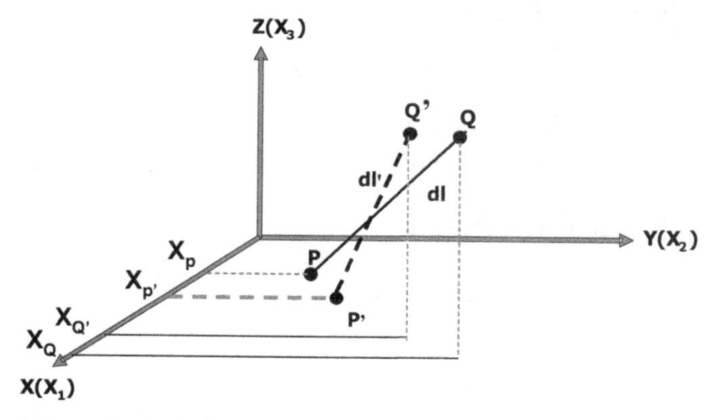

Figure 4.2. Schematic depiction of the displacement of two points in a solid material resulting from its deformation.

$$\eta^{(p)} = X'_P - X_P \tag{4.1}$$

and the displacement of point Q,

$$\eta^{(Q)} = X'_Q - X_Q \tag{4.2}$$

The corresponding change in length along the X axis will thus be given by

$$d\eta = \eta^{(Q)} - \eta^{(P)} = (X'_Q - X'_P) - (X_Q - X_P) \equiv dX' - dX \tag{4.3}$$

or, after rearrangement,

$$dX' = d\eta + dX \tag{4.4}$$

It should be noted at this point that it is common in elasticity to denote the axes as $X_1 = X$, $X_2 = Y$, $X_3 = Z$. Thus, we can write the general term for the new projection along the ith direction as

$$dx'_i = d\eta_i + dx_i \tag{4.5}$$

The length of the line connecting points P and Q before the deformation can now be written as

$$dl^2 = dx^2 + dy^2 + dz^2 = dx_1^2 + dx_2^2 + dx_3^2 \tag{4.6}$$

and the length after deformation can be written as

$$\begin{aligned} dl'^2 &= dx'^2 + dy'^2 + dz'^2 = dx_1'^2 + dx_2'^2 + dx_3'^2 = \sum_i dx_i'^2 \\ &= \sum_i (d\eta_i + dx_i)^2 = \sum_i d\eta_i^2 + dx_i^2 + 2d\eta_i \cdot dx_i \end{aligned} \tag{4.7}$$

Using an abbreviated format [1] where

$$\begin{aligned} &\sum_i A_i B_i \triangleq A_i B_i \\ &\frac{\partial A_i}{\partial X_j} \triangleq A_{i,j} \end{aligned} \tag{4.8}$$

the two lengths can be rewritten as

$$dl^2 = \sum_i dx_i^2 = dx_i \cdot dx_i \tag{4.9}$$

and

$$dl'^2 = d\eta_i \cdot d\eta_i + \underbrace{dx_i \cdot dx_i}_{dl^2} + 2d\eta_i \cdot dx_i \tag{4.10}$$

The term for $d\eta$ can also be compactly written as

$$d\eta_i = \left(\frac{\partial \eta_i}{\partial x}\right)dx + \left(\frac{\partial \eta_i}{\partial y}\right)dy + \left(\frac{\partial \eta_i}{\partial z}\right)dz \stackrel{\Delta}{=} \left(\frac{\partial \eta_i}{\partial x_k}\right)dx_k \tag{4.11}$$

Substituting this term in Eq. (4.10), we obtain

$$dl'^2 = dl^2 + \left(\frac{\partial \eta_l}{\partial x_j}\right)dx_j\left(\frac{\partial \eta_l}{\partial x_k}\right)dx_k + 2\left(\frac{\partial \eta_i}{\partial x_k}\right)dx_i dx_k \tag{4.12}$$

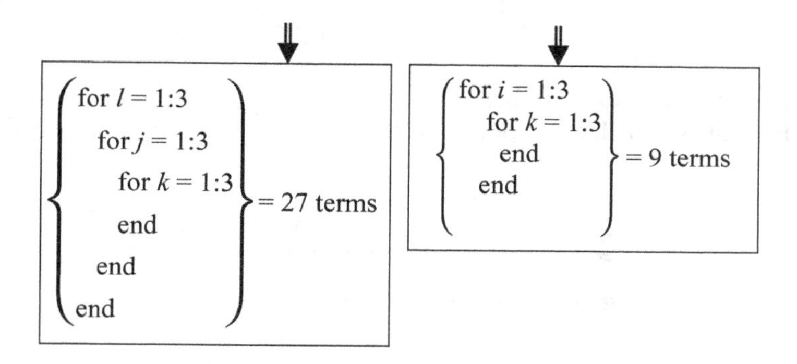

This equation has 36 terms on its right-hand side! For clarity, computer-programming-style loops are written below the equation to show the corresponding needed changes in variables.

Let us now define the term "strain" as [2]

$$U_{ik} \stackrel{\Delta}{=} \frac{1}{2}\left[\frac{\partial \eta_i}{\partial x_k} + \frac{\partial \eta_k}{\partial x_i} + \underbrace{\left(\frac{\partial \eta_l}{\partial x_i}\right)\left(\frac{\partial \eta_l}{\partial x_k}\right)}_{\text{3 Terms}}\right] \tag{4.13}$$

Using this definition, the new length can be written as

$$\Rightarrow \underbrace{dl'^2}_{\text{New length}} = \underbrace{dl^2}_{\text{Old length}} + \underbrace{2U_{ik}dx_i dx_k}_{\text{9 Deformation-induced terms}} \tag{4.14}$$

Because the deformations associated with ultrasound (and in fact in most other applications) are very small (i.e., $\partial\eta/\partial x << 1$), the three higher-order terms in Eq. (4.13) can be neglected. The accepted definition of the strain is thus given by

$$\boxed{U_{ik} = \frac{1}{2}\left[\frac{\partial \eta_i}{\partial x_k} + \frac{\partial \eta_k}{\partial x_i}\right]} \tag{4.15}$$

4.2.1 Strain Properties

A. Strain is nondimensional.
B. Strain is symmetrical, that is $U_{ik} = U_{ki}$.
C. The strain for $i = k$ measures the relative amount of stretching/shrinking of an element—for example, for an element AB shown in Fig. 4.3.

The length of this element after it has been stretched is $dx_i' = dx_i + d\eta_i$. Hence, its corresponding strain is given by

$$U_{11} = \frac{1}{2}\left(\frac{2\partial\eta_1}{\partial x_1}\right) = \frac{d\eta_1}{dx_1} = \frac{\text{Change in length}}{\text{Original length}} \tag{4.16}$$

D. For the $i \neq k$ case, things are slightly more complicated. Referring to Fig. 4.4, we mark three points AOB which define a right angle. Following

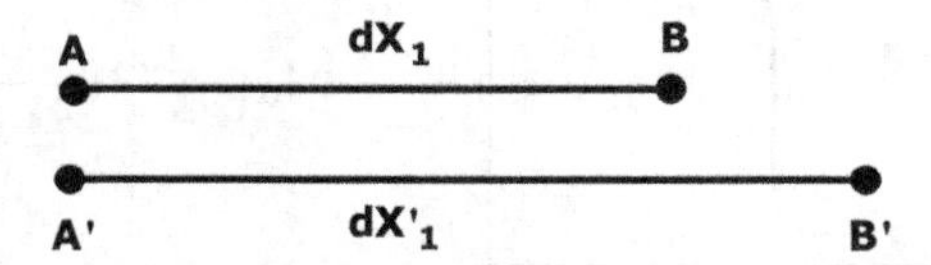

Figure 4.3. A line element connecting material points A and B, before **(top)** and after **(bottom)** deformation along the X_1 direction.

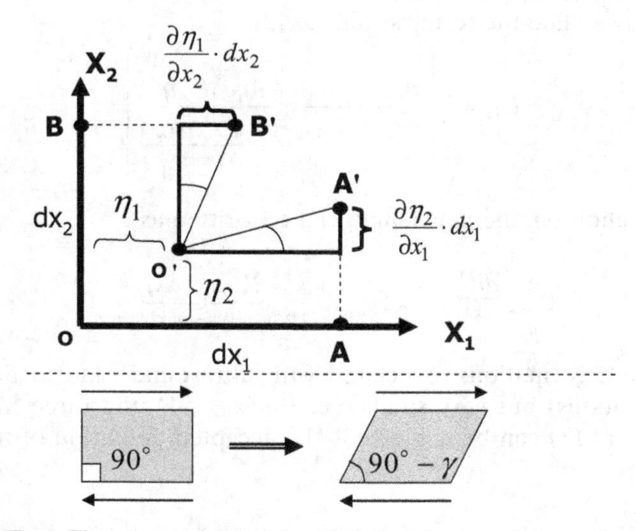

Figure 4.4. (Top) Three material points AOB that form a right angle are displaced due to shear deformation into new locations $A'O'B'$. Consequently, the angle they form is different from 90°. **(Bottom)** A right angle is deformed by an angle γ.

material deformation, these three points have been displaced to new locations $A'O'B'$. The angle defined by these three points is no longer a right angle. This kind of deformation is commonly associated with shear stresses, hence it is referred to as shear deformation.

For small deformations, the angles α and β are given by

$$\begin{aligned}\angle\alpha &= \left[\left(\frac{\partial\eta_2}{\partial x_1}\right)dx_1\right]\Big/dx_1 = \frac{\partial\eta_2}{\partial x_1}\\ \angle\beta &= \left[\left(\frac{\partial\eta_1}{\partial x_2}\right)dx_2\right]\Big/dx_2 = \frac{\partial\eta_1}{\partial x_2}\end{aligned} \tag{4.17}$$

The calculated strain $U_{12} = U_{21}$ therefore measures their average value:

$$U_{12} = U_{21} = \frac{1}{2}\left(\frac{\partial\eta_2}{\partial x_1} + \frac{\partial\eta_1}{\partial x_2}\right) = \frac{1}{2}\left(\underbrace{\alpha+\beta}_{=\gamma}\right) = \frac{\gamma}{2} \tag{4.18}$$

The strain $U_{12} = U_{21}$ also represents (half) the angular change as presented at the bottom of Fig. 4.4.

4.3 STRESS

As explained above, stress represents the force applied per unit area. Or, using a more precise mathematical definition [3], we obtain

$$\tau \triangleq \lim_{\Delta A\to 0}\frac{\Delta F}{\Delta A} \quad \left[\frac{\text{N}}{\text{m}^2} = \text{pascal}\right] \tag{4.19}$$

The accepted notation for stress (as depicted in Fig. 4.5) is τ_{ij}, where the first index (i) indicates the direction of the normal to the surface upon which the

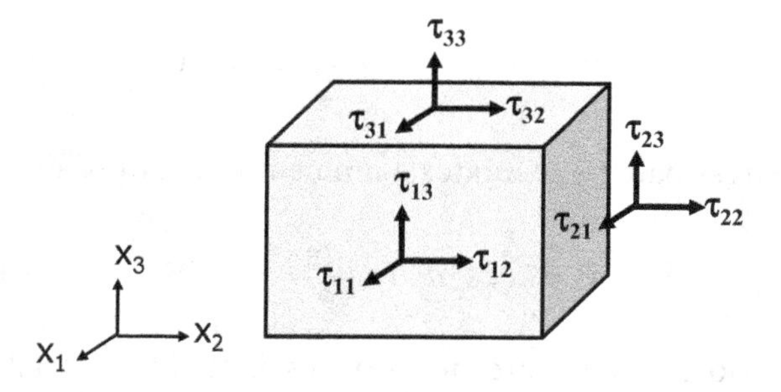

Figure 4.5. Schematic depiction of the stresses applied on a cubic element. It should be pointed out that similar stresses are also applied on the other three surfaces of the element.

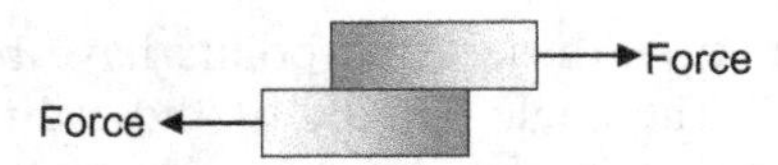

Figure 4.6. Shear stress applied on an element is generated when two opposite forces (marked by the arrows) try to cause material layers to slide relative to each other, as schematically depicted here.

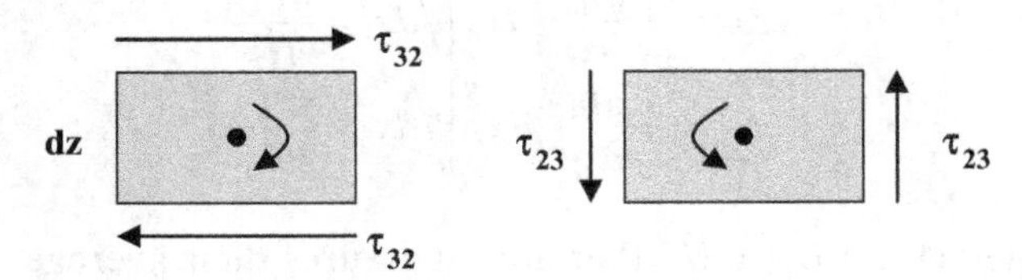

Figure 4.7. Each pair of shear stresses creates a moment on the material element which attempts to rotate it.

stress is applied. The second index (j) indicates the direction of the applied force. In Fig. 4.5 the arrows designate the force direction. When the two indices are identical—that is, τ_{ij}, where $i = j$—the stress corresponds to tension or pressure applied on the element. This type of stress is referred to as "normal" stress. When the two indices are not identical, the stress is referred to as "shear" stress. This type of stress acts like scissors and attempts to separate the material layers as schematically depicted in Fig. 4.6.

As can be understood, stresses must be applied in pairs under static conditions. In the case of shear stresses—that is, for $i, j\ i \neq j$—there must be also equality between τ_{ij} and τ_{ji}. Observing Fig. 4.7, we realize that every stress generates a force of $\tau \cdot dA$ on the material element. Consequently, every pair of shear stresses will attempt to rotate the material element. Considering, for example, the stress pair τ_{32}, it can be easily derived that the moment it generates is given by

$$M = 2(\tau_{32} \cdot dxdy) \cdot \frac{dz}{2} = \tau_{23} \cdot dV \tag{4.20}$$

whereas the stress pair τ_{23} generates a moment in the opposite direction,

$$M = 2(\tau_{32} \cdot dxdz) \cdot \frac{dy}{2} = \tau_{23} \cdot dV \tag{4.21}$$

And since obviously there are no vortexes in solid material, the element should not rotate under static conditions. Thus, it follows that

$$\Rightarrow \quad \tau_{32} = \tau_{23} \tag{4.22}$$

or, for the general case,

$$\tau_{ij} = \tau_{ji} \tag{4.23}$$

4.4 HOOKE'S LAW AND ELASTIC COEFFICIENTS

When the strains are very small, as is the case for a propagating ultrasonic wave, one can assume linear relations between stresses and strains [2]. These relations are dictated by the elastic properties of the material. In the general case the strain–stress relation for any solid elastic material can be defined using the generalized Hooke's law (the simplest case of Hooke's law was presented in Chapter 2, Section 2.6):

$$\boxed{\tau_{ij} = C_{ijKL} U_{KL}} \tag{4.24}$$

This implies that when all the elastic coefficients C_{ijKL} and the stresses τ_{ij} are known, the corresponding strains U_{KL} can be determined. And of course when two of the three parameters are known, the third one could potentially be calculated (the solution is not always simple though). Recalling the fact that each of the indices i, j, K, L can obtain three different values in space, it follows that the stress tensor τ_{ij} is comprised of nine elements, and so is the strain tensor U_{KL}. It follows therefore that in order to use the strain–stress relationship defined by the generalized Hooke's law, 81 (=9 × 9) coefficients should be known! Fortunately, using the symmetry that stems from the relation $U_{ij} = U_{ji}$ and $\tau_{ij} = \tau_{ji}$, the number of independent coefficients can be reduced to 21. Furthermore, considering the crystalloid structure of the matter and including other considerations, the number of independent coefficients can substantially be reduced, as shown in Table 4.1. It should be pointed out that metals have typically a cubic structure, thus only three coefficients are needed to define their stress–strain relations. In cortical bones, the structure is hexagonal, thus

Table 4.1. Crystalloid Structure and the Number of Elastic Coefficients

Crystalloid Structure	Number of Elastic Coefficients
Cubic	3
Hexagonal	5
Tetragonal	6 or 7
Trigonal	6 or 7
Orthorhombic	9
Monoclinic	13
Triclinic	21

five elastic coefficients are needed. For more information about crystalloid structure and the number of elastic coefficients, see, for example, Koerber [4].

Isotropic materials have only three elastic coefficients, and since they are dependent, the number can further be reduced to only 2! These are called the Lamé coefficients and are commonly marked by the symbols λ and μ. (*Importantly*: The reader should be aware of the fact that the symbol λ may sometimes designate an elastic coefficient and not a wavelength and avoid confusion!)

Using the two Lamé coefficients, other practical coefficients may be determined for elastic isotropic materials, as shown herein:

$$\begin{cases} \text{Young's modulus} & E = \dfrac{\mu(3\lambda + 2\mu)}{(\lambda + \mu)} \\ \text{Poisson's ratio} & \upsilon = \dfrac{\lambda}{2(\lambda + \mu)} \\ \text{Bulk modulus} & B = -\dfrac{P}{\Delta} = \lambda + \dfrac{2}{3}\mu \\ \text{Shear modulus} & \mu \equiv G = \dfrac{E}{2(1 + \upsilon)} \end{cases} \tag{4.25}$$

In Eq. (4.25), Δ designates a volumetric strain. It is important to note that in some textbooks the Poisson's ratio is given in its absolute (i.e., positive) value and sometimes in its physical (negative) value. To learn more about the above coefficients, the reader is referred to reference 5 for example.

4.5 THE WAVE EQUATION FOR AN ELASTIC SOLID MATERIAL

As recalled, in a solid material normal and shear stresses can be applied simultaneously and along various directions. Hence, resulting from the application of these stresses, one can expect that longitudinal and shear waves will be generated. Moreover, the shear waves may have different polarity; that is, the resulting particles displacements may be along different planes perpendicular to the wave propagation direction.

Consider a solid material element through which several elastic waves travel concurrently, causing material displacements long the X_2 direction, as depicted in Fig. 4.8. (It is important to note that in order to relate the elastic coefficients C_{ijKL} to the strains and stresses, the chosen directions X_1, X_2, X_3 must be aligned with the natural directions of the material.) The propagating longitudinal wave is associated with unbalanced normal stress τ_{ii} applied to the front and back surfaces of the solid element. The propagating shear waves, on the other hand, are associated with unbalanced shear stresses τ_{ij} applied on each pair of surfaces of the solid element. Resulting from this imbalance, the material particles will move relatively to each other.

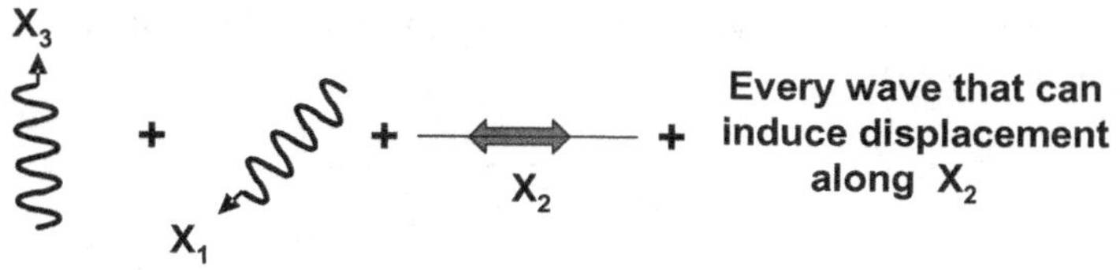

X_3 + + X_2 + Every wave that can induce displacement along X_2

X_1

Figure 4.8. Several shear (wiggly lines) and longitudinal waves (straight line) can travel through the solid material concurrently and induce displacements along X_2.

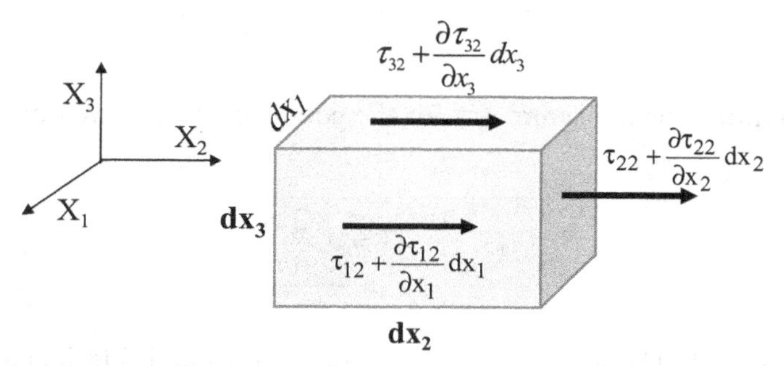

Figure 4.9. The stresses applied on the surfaces of a solid element cube during the propagation of elastic waves that induce displacements along the X_2 direction. The stresses applied on the three opposite walls of the element are without the differential addition.

Using linear approximation, we may consider the imbalance to be represented by the differential of the stresses. This is shown schematically in Fig. 4.9 where for clarity only three surfaces are shown. The stresses applied on the three opposite walls of the element are without the differential addition. From Newton's second law of physics, the sum of all the forces applied along the X_2 direction should equal the mass times the acceleration along the same direction:

$$\sum F_{x_2} = m\ddot{\eta}_2 \tag{4.26}$$

And since the force equals the multiplication of the stress times the surface area upon which it is applied, it follows that

$$\frac{\partial \tau_{22}}{\partial x_2} dx_2 (dx_1 dx_3) + \frac{\partial \tau_{32}}{\partial x_3} dx_3 (dx_1 dx_2) + \frac{\partial \tau_{12}}{\partial x_1} dx_1 (dx_2 dx_3) = \rho_0 \ddot{\eta}_2 dx_1 dx_2 dx_3 \tag{4.27}$$

or, after simplification,

$$\frac{\partial \tau_{12}}{\partial x_1} + \frac{\partial \tau_{22}}{\partial x_2} + \frac{\partial \tau_{32}}{\partial x_3} = \rho_0 \frac{\partial^2 \eta_2}{\partial t^2} \tag{4.28}$$

where ρ_0 is the rest (i.e., no wave) density. Using the abbreviated format, this equation can be written as

$$\boxed{\tau_{ij,i} = \rho_0 \ddot{\eta}_j} \tag{4.29}$$

Substituting $\tau_{i,j}$ with strains and elastic coefficients using Eq. (4.24) yields

$$\frac{\partial}{\partial x_i}(C_{ijKL}U_{KL}) = \frac{1}{2}C_{ijKL}\frac{\partial}{\partial x_i}\left(\frac{\partial \eta_K}{\partial x_L} + \frac{\partial \eta_L}{\partial x_K}\right) = \rho_0 \frac{\partial^2 \eta_j}{\partial t^2} \tag{4.30}$$

And accounting for the symmetry for the coefficients C_{ijKL}, we obtain [6]

$$\boxed{C_{ijKL}\frac{\partial^2 \eta_K}{\partial x_i \partial x_L} = \rho_0 \frac{\partial^2 \eta_j}{\partial t^2}} \tag{4.31}$$

Although Eq. (4.31) contains many terms on its left-hand side, it can be noted that all these terms are spatial second derivatives for the displacement. The right-hand side, on the other hand, is a simple temporal second derivative for the displacement. This corresponds to the following template:

$$\boxed{\frac{1}{\rho_0}\{\text{Const}\}\cdot \eta'' = \ddot{\eta}} \tag{4.32}$$

This is actually the most general form of the wave equation in the material. By solving this equation, the speed of sound can be calculated. It is important to note that since there is no limitation on the direction of the propagating wave relative to the natural coordinate system of the material, different waves with propagation velocity that is direction-dependent can be obtained. Also, it should be noted that Eq. (4.31) has summation over the three indices i, k, L; thus in the most general case we may need to account for 27 terms on the left-hand side. For further elaboration on this topic the interested reader is referred to references 6 and 7.

4.6 PROPAGATION OF A HARMONIC PLANAR WAVE IN A SOLID MATERIAL

In order to solve Eq. (4.31) for a propagating plane wave, we shall first make two assumptions that simplify its form but do not reduce its generality. The

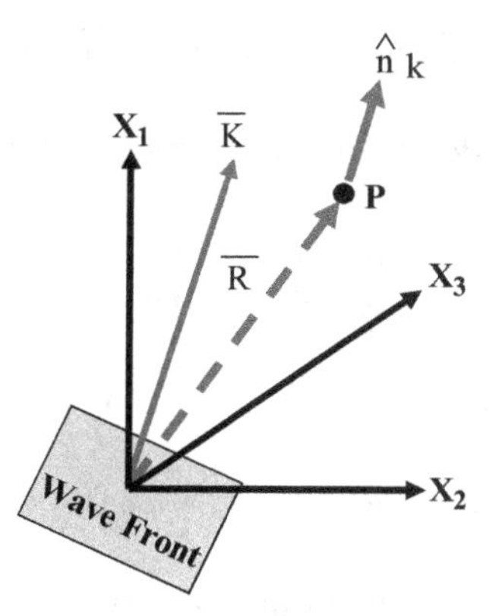

Figure 4.10. A planar wave propagates along $\overline{K}$ will reach point P located at $\overline{R}$ and will cause it to be displaced.

first assumption is that the propagating wave is a harmonic wave. The second assumption is that the coordinate system aligns with the "natural" coordinate system of the material (as noted above). At this point we make reference to Fig. 4.10. As shown there, a planar wave front is assumed to be aligned with the coordinate system's point of origin at time 0. The wave will naturally propagate further in space along vector $\overline{K}$, reaching later a point P whose location is given by vector $\overline{R}$. The unit vector $\hat{n}_k$ designates the wave propagation direction. The wave number vector $\overline{K}$ is given by

$$\overline{K} = K_1\hat{x}_1 + K_2\hat{x}_2 + K_3\hat{x}_3 \tag{4.33}$$

The location of point P is given by vector $\overline{K}$,

$$\overline{R} = x_1\hat{x}_1 + x_2\hat{x}_2 + x_3\hat{x}_3 \tag{4.34}$$

When the wave (which is a harmonic wave in our case) passes through point P, it will cause it to be displaced from its equilibrium state. The displacement projection along direction i is defined as η_i and is given at time t by

$$\eta_i = \eta_{i0}\exp[j(wt - (k_1x_1 + k_2x_2 + k_3x_3))] = \eta_{i0}\exp\left[j\left(wt - \overline{K}\cdot\overline{R}\right)\right] \tag{4.35}$$

where η_{i0} is the amplitude of the displacement projection along direction i, and w is the wave angular frequency. Substituting this expression into Eq. (4.31) and canceling the exponential terms yields

$$-C_{ijKL}\cdot K_jK_L\eta_{K0} = -\rho_0 w^2\eta_{i0} \tag{4.36}$$

In order to bring the equation into a more convenient format, we shall replace η_{i0} by

$$\eta_{i0} = \delta_{iK}\eta_{K0} = \delta_{i1}\eta_{10} + \delta_{i2}\eta_{20} + \delta_{i3}\eta_{30} \tag{4.37}$$

where δ_{iK} is Kronecker's delta function, which is defined as follows:

$$\begin{array}{l} \delta_{ij} = 1 \quad \text{if} \quad i = j, \\ \delta_{ij} = 0 \quad \text{if} \quad i \neq j \end{array} \tag{4.38}$$

This substitution leads to the general equation:

$$\left(C_{ijKL} \cdot K_j K_L - \delta_{iK} w^2 \rho_0\right)\eta_{i0} = 0 \tag{4.39}$$

And since we have three directions in space, we can write it explicitly as a set of three equations given by [6]

$$\boxed{\begin{array}{l} \left(C_{1j1L}k_j k_L - w^2\rho_0\right)\eta_{10} + C_{1j2L}k_j k_L \eta_{20} + C_{1j3L}k_j k_L \eta_{30} = 0 \\ C_{2j1L}k_j k_L \eta_{10} + \left(C_{2j2L}k_j k_L - w^2\rho_0\right)\eta_{20} + C_{2j3L}k_j k_L \eta_{30} = 0 \\ C_{3j1L}k_j k_L \eta_{10} + C_{3j2L}k_j k_L \eta_{20} + \left(C_{3j3L}k_j k_L - w^2\rho_0\right)\eta_{30} = 0 \end{array}} \tag{4.40}$$

This set of three equations must be valid for every value of η_{i0}, the wave amplitude, which is set arbitrarily by the transmitting device or wave source. And since w (the wave angular frequency) is known and so are (usually) the elastic coefficients of the material C_{ijKL}, it follows that the three unknowns are k_1, k_2, k_3, which define the wave number along each of the three directions. Solving Eq. (4.40) for these values will enable us to determine the magnitude of $\bar{K}$. From the relation $k = 2\pi/\lambda$ the wavelength λ can be determined. Finally, since the wave angular frequency w is also known, the speed of sound along the general direction K can be determined from $c = \lambda \cdot \frac{w}{2\pi}$.

The mathematical "trick" to solve this set of equation is to ignore for the time being the fact the unknowns are k_1, k_2, k_3 and pretend that the three values of η_{i0} are the unknowns. Thus, we can rewrite it as

$$\begin{array}{l} A_{11}\eta_{10} + A_{12}\eta_{20} + A_{13}\eta_{30} = 0 \\ A_{21}\eta_{10} + A_{22}\eta_{20} + A_{23}\eta_{30} = 0 \\ A_{31}\eta_{10} + A_{32}\eta_{20} + A_{33}\eta_{30} = 0 \end{array} \tag{4.41}$$

where A_{ij} are the elements of the corresponding matrix of constants $\bar{\bar{A}}$. In order for the set of equations (4.41) to have a solution for every arbitrary amplitude η_{i0}, the determinant of the constants matrix should be equal to zero, that is, $\det\left\{\bar{\bar{A}}\right\} = 0$. (This stems from the fact that the right-hand side is equal to zero.) And since the matrix of constants $\bar{\bar{A}}$ is a function of k_1, k_2, k_3, the corresponding wave numbers can be determined.

4.6.1 Special Case #1

Assume, for example, the special case where the wave propagates along the Z axis, so that $K_3 = K$ and $K_1 = K_2 = 0$. This could be easily achieved by selecting the coordinate system according to the material properties. Writing the explicit equations for $\det\left(\bar{\bar{A}}\right)=0$ in this case yields

$$\begin{vmatrix} C_{1313}K^2 - w^2\rho_0 & C_{1323}K^2 & C_{1333}K^2 \\ C_{2313}K^2 & C_{2323}K^2 - w^2\rho_0 & C_{2333}K^2 \\ C_{3313}K^2 & C_{3323}K^2 & C_{3333}K^2 - w^2\rho_0 \end{vmatrix} = 0 \tag{4.42}$$

As can be noted, the only unknown here is K. However, this is a sixth-order equation—that is, with K^6. Hence, we may expect to have several alternative solutions. This implies that several wave types can coexist in this general solid material and propagate along the same direction with different velocities.

4.6.2 Special Case #2

Let us simplify further the problem and assume that the material is homogeneous. In this case of course the solution is identical for all directions. Furthermore, we can make use of the two Lamé coeffients and substitute the elastic coefficients with their values (kindly note that in order to distinguish between wavelengths and the Lamé coefficient, the symbol λ' is used):

$$\begin{aligned} C_{1313} &= C_{2323} = \mu \\ C_{3333} &= \lambda' + 2\mu \end{aligned} \tag{4.43}$$

And since all other coefficients $C_{ijkl} = 0$, it follows that the following equation is obtained:

$$\begin{gathered} \begin{vmatrix} \mu K^2 - w^2\rho_0 & 0 & 0 \\ 0 & \mu K^2 - w^2\rho_0 & 0 \\ 0 & 0 & (\lambda'+2\mu)K^2 - w^2\rho_0 \end{vmatrix} = 0 \\ \Rightarrow \left(\mu K^2 - w^2\rho_0\right)^2 \cdot \left[(\lambda'+2\mu)K^2 - w^2\rho_0\right] = 0 \end{gathered} \tag{4.44}$$

As can be noted, Eq. (4.44) has only two independent solutions for K. Recalling the relations

$$\begin{aligned} &\text{(a)}\ K = \frac{2\pi}{\lambda} \\ &\left.\begin{aligned} &\text{(b)}\ w = 2\pi f \\ &\text{(c)}\ f = \frac{c}{\lambda} \end{aligned}\right\} \quad w = \frac{2\pi c}{\lambda} \Rightarrow w = K \cdot c \Rightarrow c = \frac{w}{K} \end{aligned} \tag{4.45}$$

and solving for the term in the left brackets of Eq. (4.44), we obtain

$$(\mu K^2 - w^2 \rho_0) = 0 \quad \Rightarrow \quad \frac{w^2}{K^2} = \frac{\mu}{\rho_0} = c^2 \tag{4.46}$$

From this relation we derive the propagation speed for a shear (transverse) wave in a homogeneous solid material:

$$\boxed{C_{\text{shear}} = \sqrt{\frac{\mu}{\rho_0}}} \tag{4.47}$$

Similarly, by opening the right-hand side brackets, we obtain

$$[(\lambda' + 2\mu) K^2 - w^2 \rho_0] = 0 \quad \Rightarrow \quad \frac{w^2}{K^2} = \frac{\lambda' + 2\mu}{\rho_0} = c^2 \tag{4.48}$$

From this equation the propagation speed for longitudinal waves in an isotropic material is obtained:

$$\boxed{C_{\text{Long}} = \sqrt{\frac{(\lambda' + 2\mu)}{\rho_0}}} \tag{4.49}$$

These are the main two wave speeds for an isotropic solid material, and they are the same for all directions. Nevertheless, it should be pointed out that other types of waves can also exist in a solid material—for example, Rayleigh surface waves where the particles have both longitudinal and transverse motion. The particles move in elliptical paths, with the major axis of the ellipse perpendicular to the surface of the solid. As the depth into the solid increases, the elliptical path narrows; and after 1/5th of a wavelength, it changes its rotation direction. There are also plate waves called Lamb or Love types [8, 9]. These types of waves will not be discussed here.

4.6.3 Special Case #3

The most prominent solid (natural) objects in our bodies are bones. Bones form a complicated structure. A bone is basically a composite material with two main phases containing organic and inorganic substances. The organic phase is comprised mainly of a collagen fiber matrix, but also contain proteins and lipids. The inorganic phase is comprised of a mineral called hydroxyaptite. The spatial structure of a bone changes according to its specific anatomical function. Consequently—and this point is important in our context—its mechanical properties and elasticity differ according to its loading directions.

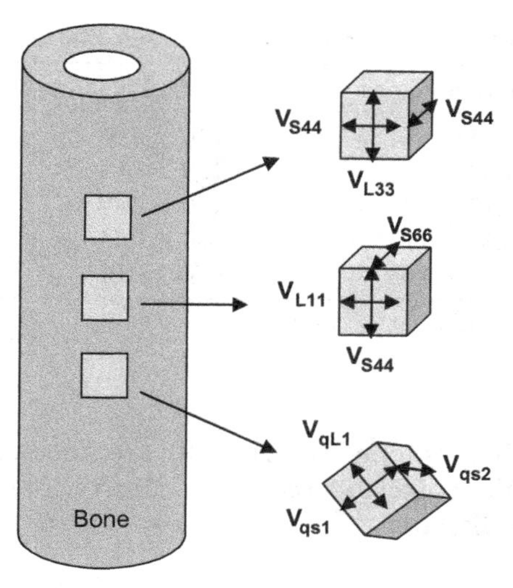

Figure 4.11. Schematic description of the bone samples S. B. Lang [10] has taken for measuring the speed of sound along the directions marked by the arrows. The indices L and S correspond to longitudinal and shear waves, respectively.

Generally, it can be stated that the stiffness is maximal along the direction for which the stresses are maximal. Thus, for example, a cortical bone in the leg will have maximal stiffness along its axial direction.

The properties of the long bones in the limbs can be characterized by five independent elastic coefficients. (The teeth, by the way, have a different structure and different acoustic properties than bones.) A method for measuring these coefficients with ultrasonic waves have been suggested by Lang [10].

Lang has taken bone samples as described in Fig. 4.11, and he has measured the speed of sound (indices are marked in the drawing) along the directions described in this figure. Given the samples' density ρ, he has calculated the bones' five elastic coefficients from the following set of equations:

$$\begin{cases} C_{11} \equiv C_{1111} = \rho\cdot(V_{L11})^2 \\ C_{33} \equiv C_{3333} = \rho\cdot(V_{L33})^2 \\ C_{44} \equiv C_{2323} = \rho\cdot(V_{L11a})^2 \\ C_{44} \equiv C_{2323} = \rho\cdot(V_{L11b})^2 \\ C_{66} \equiv C_{1212} = \frac{1}{2}(C_{11} - C_{12}) = \rho\cdot(V_{s66})^2 \\ C_{13} \equiv C_{1133} = [(C_{11} - C_{44} - 2\rho\cdot V^2)\cdot(C_{11} - C_{44} - 2\rho\cdot V^2)]^{1/2} - C_{44} \\ \text{where } V \text{ is either } V_{qL} \text{ or } V_{qs1} \\ \frac{1}{2}(C_{66} + C_{44}) = \rho\cdot(V_{qs2})^2 \end{cases}$$

Table 4.2. Summary of Lang's Findings [10] on Bones Elastic Coefficients

10^{10} N/m^2	Dried Femur	Fresh Phalanx	Dried Phalanx
C_{11}	2.38 ± 0.14	1.97 ± 0.05	2.12 ± 0.07
C_{12}	1.02 ± 0.06	1.21 ± 0.04	0.95 ± 0.03
C_{13}	1.12 ± 0.21	1.26 ± 0.12	1.02 ± 0.14
C_{33}	3.34 ± 0.12	3.2 ± 0.11	3.74 ± 0.16
C_{44}	0.82 ± 0.02	0.54 ± 0.01	0.75 ± 0.02

The elastic coefficients found by Lang [10] are summarized in Table 4.2.

Using these elastic coefficients, one can use Eq. (4.40) to estimate the speed of sound along any direction.

REFERENCES

1. Malvern LE, *Introduction to the Mechanics of a Continuous Medium*, Prentice-Hall, Englewood Cliffs, NJ, 1969.
2. Ross CTF, *Mechanics of Solids*, Horwood Engineering Science Series, Chichester, 1999.
3. Shigley JE, *Mechanical Engineering Design*, 3rd edition, McGraw-Hill, New York, 1977.
4. Koerber GG, *Properties of Solids*, Prentice-Hall, Englewood Cliffs, NJ, 1962.
5. Fung YC, *Foundations of Solid Mechanics*, Prentice-Hall, Englewood Cliffs, NJ, 1965.
6. Beyer RT and Lechter SV, *Physical Ultrasonics*, Academic Press, New York, 1969.
7. Beltzer AI, *Acoustics of Solids*, Springer-Verlag, Berlin, 1988.
8. Abramov OV, *Ultrasound in Liquid and Solid Metals*, Published by CRC Press, 1994.
9. Rose JL, *Ultrasonic Waves in Solid Media*, Cambridge University Press, Cambridge, 1999.
10. Lang SB, Ultrasonic Method for measuring elastic coefficients of bovine bones. *IEEE Trans Biomed Eng* **17**:101–105, 1970.

CHAPTER 5

ATTENUATION AND DISPERSION

Synopsis: In this chapter we shall introduce the wave attenuation phenomenon in a homogeneous medium. We shall try to understand the mechanism of this phenomenon and its characteristics. We shall also learn about the relation between frequency, attenuation and speed of sound. And finally we shall introduce the nonlinear parameter B/A.

5.1 THE ATTENUATION PHENOMENON

From experimental observations we learn that the amplitude of a planar acoustic wave propagating in a homogeneous medium decreases with the wave traveling distance, as shown in Fig. 5.1. This phenomenon stems from energy absorption by the medium through which the wave travels. Consequently, the energy carried by the traveling wave decreases and so does its amplitude. This reduction in amplitude versus distance is exponential in nature; and if dependence on frequency is neglected (see the following for frequency dependence), it can be approximated by

$$P(x) = P_0 e^{-\alpha x} \tag{5.1}$$

where P_0 is the wave pressure at some reference point (e.g., at the transmitting transducer's surface) and $P(x)$ is the wave pressure at distance x from that

Basics of Biomedical Ultrasound for Engineers, by Haim Azhari

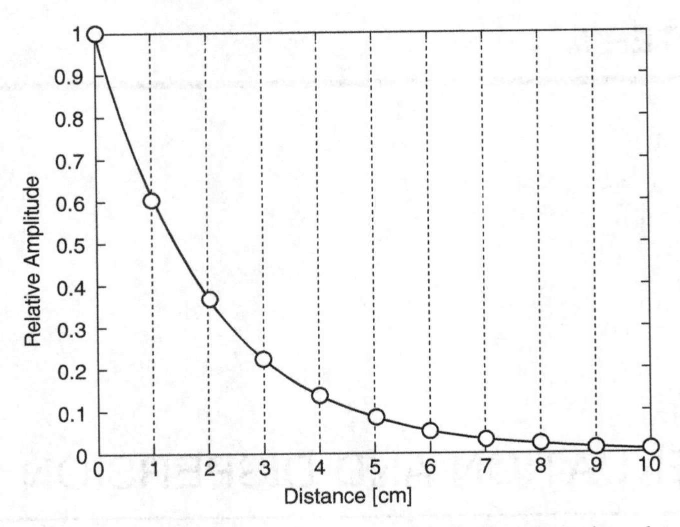

Figure 5.1. Amplitude attenuation of a propagating plan wave in a homogeneous medium as a function of distance.

reference point. The constant α is characteristic to the medium through which the wave travels (see Table A.2 in Appendix A). In many applications, one may be interested in the wave intensity and not its amplitude. Thus, recalling the relation $I = P^2/2z$, which implies that intensity is proportional to the square of the pressure, one can use the following relationship:

$$I(x) = I_0 e^{-2\alpha x} = I_0 e^{-\mu x} \tag{5.2}$$

where I_0 is the wave intensity at some reference point and $I(x)$ is the wave intensity at distance x from that reference point. The constant μ is simply $\mu = 2\alpha$. Alternatively, one can use this relation using the following ratio form:

$$\frac{I(x)}{I_0} = e^{-\mu x} \tag{5.3}$$

This ratio is conveniently measured in decibel units,

$$\text{db} = 10 \log_{10} \frac{I_2}{I_1} \tag{5.4}$$

where I_1 and I_2 are two values measured at two defined points.

Recalling again the relation between pressure and intensity (i.e., $I = P^2/2z$), one can assess the attenuation in wave *power* using the relation

$$db = 10\log_{10}\left(\frac{P_2}{P_1}\right)^2 = 20\log_{10}\left(\frac{P_2}{P_1}\right) \tag{5.5}$$

However, if one would like to measure the attenuation in wave *amplitude*, the relation $db = 10\log_{10}\left(\frac{P_2}{P_1}\right)$ should be used.

It is important to note the above definitions in order to avoid confusion. Also, one should realize that when a wave is attenuated, its decibel value will be negative since the ratio in the parentheses is smaller than 1. On the other hand, when using an amplifier to amplify a signal, the gain is given in positive decibel units. Finally, it is worth pointing out that for an attenuation of 50% in wave *amplitude* the corresponding decibel value is: –6 db, whereas for an attenuation of 50% in wave *intensity* the corresponding decibel value is –3 db.

In some textbooks and other publications the logarithmic relation is measured using the natural logarithm. In such case the decibel units are replaced by units that are called "nepers," where

$$\text{neper} = \ln\frac{I_2}{I_1} \tag{5.6}$$

and

$$1\text{ db} = 8.686\text{ nepers} \tag{5.7}$$

5.2 EXPLAINING ATTENUATION WITH A SIMPLE MODEL

Thus far, we have treated the medium through which the waves travel as a perfectly elastic medium. This implies that we have assumed a simple linear relationship between stress (or pressure) applied on the matter and its resulting deformations and strains. In a perfectly elastic medium the energy inserted in order to reach a certain amount of deformation is equal to the energy released when the applied stress (or pressure) is unloaded (as is the case for an ideal spring). Therefore, when plotting the strain–stress relationship for an ideal elastic matter, a straight line is obtained (see Fig. 5.2, left). However, in many cases when plotting the strain–stress curve while loading and unloading the matter, we commonly observe a discrepancy between the loading (ascending) curve and the unloading (descending) curve, as depicted in Fig. 5.2, right). This phenomenon is called "hysteresis," and it stems from the gap between the amount of mechanical energy inserted (the area under the ascending curve) and the amount of energy released (the area under the descending curve). The area enclosed between the two curves is the energy absorbed by the matter and most of it is transformed into heat.

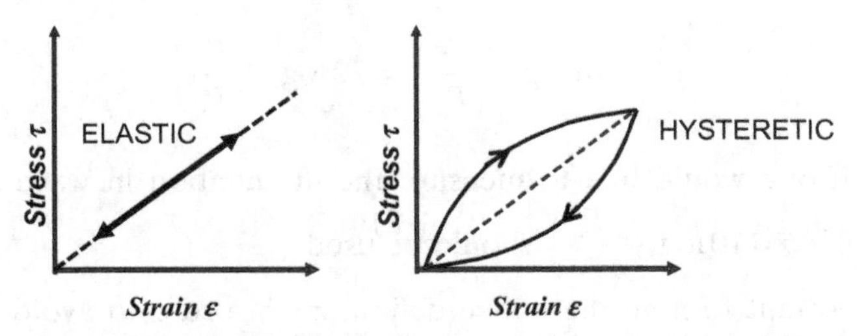

Figure 5.2. **(Left)** Strain–stress relationship for an ideal elastic matter. **(Right)** Strain–stress relationship for a viscoelastic matter depicting hysteresis.

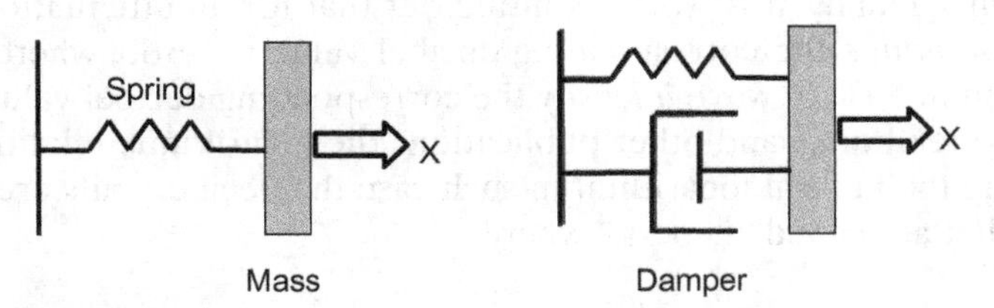

Figure 5.3. **(Left)** A model consisting of a mass and a spring used for representing a particle of a perfectly elastic matter. **(Right)** A damper added to the mass and a spring can be used for representing a particle of a viscoelastic matter.

Using a simple analogy, one can refer to each particle of a perfectly elastic matter as a miniature assembly consisting of a small mass and a spring (see Fig. 5.3, left). The particles of a matter with hysteretic behavior (viscoelastic matter), on the other hand, can be represented by an assembly that has an additional "damper" unit (see Fig 5.3, right). This damper as the shock absorber in cars applies a force that resists deformation and that is proportional to the rate of deformation; that is, the higher the temporal derivative of the forced deformation, the stronger the resisting force. While the force that the spring generates is linearly proportional to the induced deformation, that is, $F_{\text{spring}} = kx$ (where k is the spring elastic coefficient), the force produced by the damper is $F_{\text{damper}} = D \cdot dx/dt$ (where D is the damper coefficient). The direction of the force produced by the damper is always opposite to the deformation direction. Therefore, the damper always absorbs energy.

Using the second analogy, the deformation temporal derivative dx/dt represents the local particle velocity induced by the acoustic wave traveling through the matter. For a harmonic wave, the particles oscillate periodically around their equilibrium position. As the wave frequency is increased, the number of oscillations per time unit which the particles are forced to do increases, and so does the corresponding particle velocity (if the amplitude is kept the same). As a result, the force opposing deformation increases and so does the amount of energy absorbed. Hence, we can expect to observe higher attenuation for higher frequencies.

5.3 ATTENUATION DEPENDENCY ON FREQUENCY

After defining the above model that represents matter more realistically, we realize that the attenuation coefficient depends of the wave frequency through some relation $\alpha = H(f)$. In order to further explore this relation, let us use the model and study its response to a sudden (step function type) loading. Using a first-order approximation, we estimate the rate of deformation by the relation

$$\frac{dx}{dt} = \frac{1}{\tau_r}(\bar{x} - x) \tag{5.8}$$

where $\bar{x}$ is the final (asymptotic) state of deformation and τ_r is the characteristic time constant. A schematic depiction of such loading and the expected deformation as a function of time are shown in Fig. 5.4.

Typically, in our context of a harmonic acoustic wave propagation, the loading is periodical with a defined frequency. Hence, the characteristic time constant τ_r can be associated with a "characteristic relaxation frequency" f_r, which is defined by

$$f_r = \frac{1}{\tau_r} \tag{5.9}$$

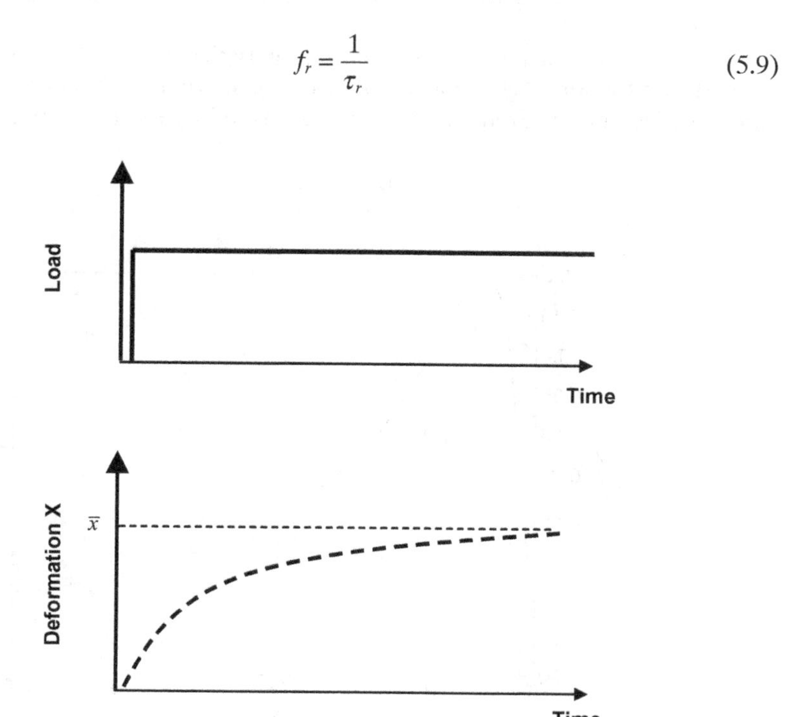

Figure 5.4. **(Bottom)** Schematic depiction of the expected response to a step-function-type loading **(top)**.

(To learn more about relaxation models of propagating acoustic waves, see, for example, references 1–3).

And since the described model has a characteristic relaxation frequency, one can relate it to the attenuation coefficient. In the literature it was shown that the attenuation coefficient α, the characteristic relaxation frequency f_r and the wave frequency f are related through the following relation:

$$\frac{\alpha}{f^2} = \frac{A_1 \cdot f_r^2}{f_r^2 + f^2} + B \tag{5.10}$$

where A_1 and B are two empirical constants. Alternatively, when incorporating the wavelength λ the following relation may be used:

$$\boxed{\alpha \cdot \lambda = \frac{A_2 \cdot f_r \cdot f}{f_r^2 + f^2}} \tag{5.11}$$

where A_2 is another constant (which actually depends on A_1). It is worth noting that this function has a characteristic peak around f_r where its value is maximal (this fact can be used to determine f_r). This is demonstrated graphically in Fig. 5.5.

For many biological tissues and within the typical range of frequencies used in medical ultrasound (low megahertz range), the relation between the attenuation coefficient and frequency can be described by the following equation:

$$\boxed{\alpha(f) = af^b} \tag{5.12}$$

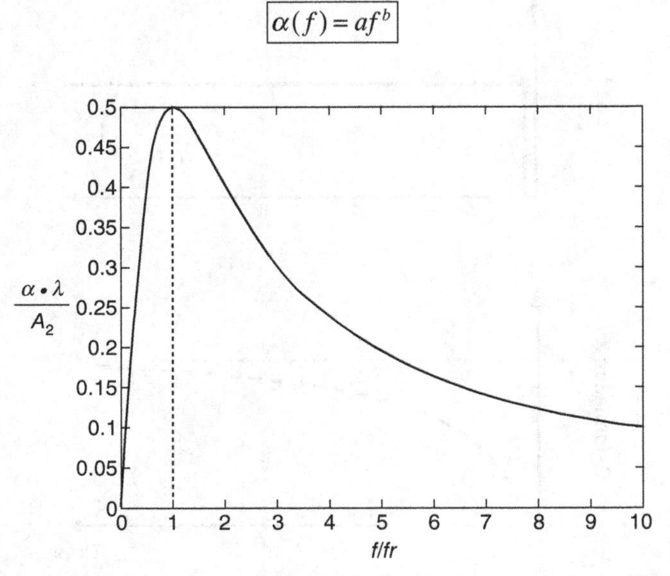

Figure 5.5. Graphical depiction of Eq. (5.11). As can be noted, the maximal value is obtained for f_r.

where a and b are empirical constants whose values differ from one type of tissue to another. Observing the values outlined in Table A.2 in Appendix A, for soft tissues one can note that $b \approx 1$. Hence, we can approximate this relation by

$$\boxed{\alpha(f) = a \cdot f} \tag{5.13}$$

Or if we wish to obtain a better approximation (or in order to cover a wide range of frequencies), one can assume the following linear relation:

$$\boxed{\alpha(f) = \alpha_0 + \alpha_1 \cdot f} \tag{5.14}$$

where α_0 and α_1 are empirical constants, for the specific tissue or material studied.

The above relation has interesting implications on the spectral distribution of broadband signals. If, for example, a medical imaging setup is considered, a short and therefore a broadband pulse is required (see Chapter 9). In many applications it is common to use or mathematically approximate the transmitted pulse by a "Gaussian" pulse. This pulse has a Gaussian envelope in both temporal and spectral domains (see Fig. 5.6), and it can conveniently be described by the following mathematical expression:

$$p(t) = e^{(-\beta \cdot t^2)} \cdot \cos(2\pi \cdot f_0 \cdot t) \tag{5.15}$$

where f_0 is the central frequency of the pulse and β defines its width. The corresponding spectral distribution of this pulse is given by

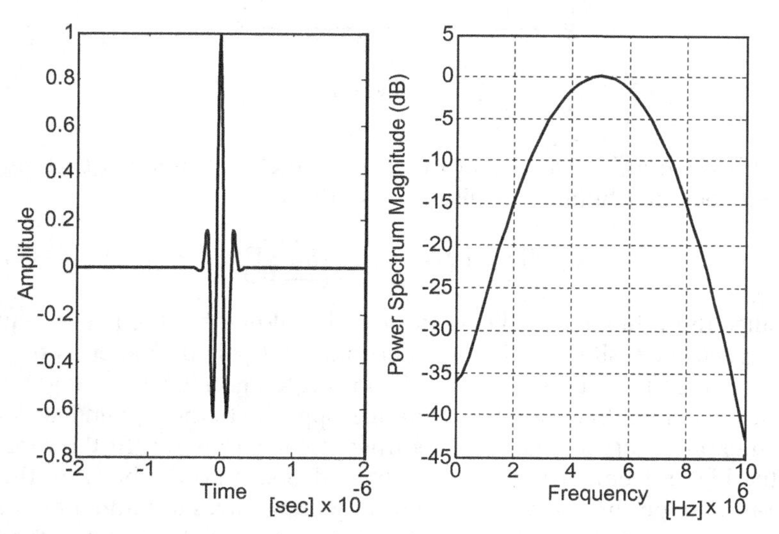

Figure 5.6. Graphical depiction of a "Gaussian" pulse with central frequency of 5 MHz in the time domain **(left)** and in the frequency domain **(right)**.

$$P(f) = \frac{1}{2}\sqrt{\frac{\pi}{\beta}} \cdot e^{-\frac{(f-f_0)^2}{4\beta}} \tag{5.16}$$

If dependency on frequency which is represented by Eq. (5.14) is combined with the dependence on distance represented by Eq. (5.1), the frequency–distance attenuation transfer function can be written as

$$H(x,f) = e^{-(\alpha_0 \cdot x + \alpha_1 \cdot f \cdot x)} \tag{5.17}$$

The spectral density function for a signal $S(x,f)$ measured for a Gaussian pulse at location x is therefore given by

$$\begin{aligned} S(x,f) &= P(f) \cdot H(x,f) \\ &= \frac{1}{2}\sqrt{\frac{\pi}{\beta}} \cdot e^{-\frac{(f-f_0)^2}{4\beta}} \cdot e^{-(\alpha_0 \cdot x + \alpha_1 \cdot f \cdot x)} \\ &= \frac{1}{2}\sqrt{\frac{\pi}{\beta}} \cdot e^{-\frac{(f-f_0)^2}{4\beta} - (\alpha_0 \cdot x + \alpha_1 \cdot f \cdot x)} \end{aligned} \tag{5.18}$$

Studying this function, it can be noticed that the peak of the spectral distribution function will occur when the exponential power term will be minimized. In order to find the frequency for which this function is maximal, let us take the derivative of this term and require that it would equal zero,

$$\frac{\partial \left[\frac{-f^2 - f_0^2 + 2f \cdot f_0 - 4\beta \cdot \alpha_1 \cdot x \cdot f}{4\beta} - \alpha_0 \cdot x \right]}{\partial f} = 0$$

$$\Rightarrow -2f + 2f_0 - 4\beta \cdot \alpha_1 \cdot x = 0 \tag{5.19}$$

From this relation it follows that the peak of the spectrum will be found for the frequency that fulfills the following condition:

$$\boxed{f_{\max}(x) = f_0 - 2\beta \cdot \alpha_1 \cdot x} \tag{5.20}$$

And since the value of x is always positive, it follows that the peak frequency of the spectral density function will become *smaller* and smaller as the wave travels through the medium. This is demonstrated graphically in Fig. 5.7. It is interesting to note that this phenomenon is opposite to the "Beam Hardening" phenomenon observed when X rays travel through matter. In the case of X rays, the higher-frequency waves penetrate deeper into the body. In the case of ultrasound, on the other hand, high-frequency waves are attenuated faster and it is the low-frequency waves that can penetrate deeper into the body. Thus, we can describe this phenomenon as "beam softening."

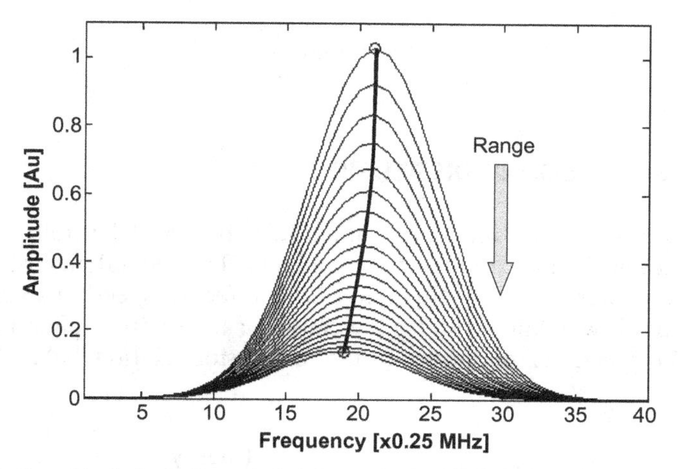

Figure 5.7. Graphical depiction of the change in the spectral density function for a Gaussian pulse as a function of range. Note that the amplitude decreases and the peak of the spectrum shifts toward the lower frequencies as the wave travels deeper into the medium. Also note the asymmetry of the spectral profile for the higher range.

5.4 THE COMPLEX WAVE NUMBER

Let us study the mathematical term describing the pressure amplitude of a propagating plane wave in a homogeneous medium and incorporate the term describing its attenuation:

$$P = P_0 e^{-\alpha x} e^{j(\omega t - kx)} = P_0 e^{j[\omega t - x(k - j\alpha)]} \tag{5.21}$$

As can be observed, the right-hand side of the obtained exponential power term $k - j\alpha$ has the physical dimensions of a wave number (spatial frequency). Thus, it can be stated that it represents a "complex wave number." And since the wave number and the speed of sound are related, we can argue that the speed of sound c is also a complex term defined by

$$c = \frac{\omega}{k - j\alpha} \tag{5.22}$$

or

$$c = \frac{\omega}{\sqrt{k^2 + \alpha^2}} e^{j\theta} \tag{5.23}$$

where

$$\theta \equiv \arctan\left(\frac{\alpha}{k}\right) \tag{5.24}$$

5.5 SPEED OF SOUND DISPERSION

Since both α and k are frequency-related [e.g., Eq. (5.14)], it follows that the speed of sound is also frequency-dependent. This dependency of the wave properties is referred to as "dispersion." The effect of speed of sound dispersion in optics is well known and is the cause for the rainbow phenomena. The amount of dispersion is quantified by the "dispersion relation," which is defined as

$$\text{Dispersion relation} = \frac{\text{Energy}}{\text{Momentum}} \tag{5.25}$$

As explained in Chapter 1, the wave propagation speed can be measured by tracking some marker on the wave profile envelope. If the propagation speed is measured by amplitude tracking, it is called "group velocity." Typically, in ultrasonic imaging, broadband pulses are used (see Chapter 9). Such pulses actually consist of a group of harmonic waves propagating together at different frequencies (a wavepacket). As stated in Chapter 1, the group velocity c_g is given by

$$c_g = \frac{\partial \omega(k)}{\partial k} \tag{5.26}$$

Alternatively, one can track the phase propagation speed. In such a case the obtained value is defined as the "phase velocity" c_p and is given by

$$c_p = \frac{\omega(k)}{k} \tag{5.27}$$

If dispersion is negligible or nonexistent, the angular (temporal) frequency and the wave number are linearly related, that is,

$$\omega = c \cdot k \tag{5.28}$$

Hence for this case, the group velocity and the phase velocity are the same, that is, $c_g = c_p$.

If the relation between the angular (temporal) frequency and the wave number is nonlinear, there will be a discrepancy between the two speed values. This discrepancy may vary with the frequency f. The group velocity and the phase velocity for frequency f are related by the following equation [4]:

$$c_g = \frac{c_P}{1 - \frac{f}{c_P}\frac{dc_P}{df}} \tag{5.29}$$

Hence, the greater the speed of sound dispersion, the larger the value of dc_p/df and the larger the discrepancy between c_g and c_p. While the dependence of the attenuation coefficient on frequency is very significant, the speed of sound dispersion as assessed by dc_p/df in tissues is very low and considered negligible for most practical applications. For example, Levy et al. [5] has reported that this dispersion index in soft tissues (in vitro) ranged for a bovine heart from 0.63 ± 0.24 (m/sec·MHz) at 1.5 MHz to 0.27 ± 0.05 (m/sec·MHz) at 4.5 MHz and for a turkey breast from 1.3 ± 0.28 (m/sec·MHz) at 1.75 MHz to 0.73 ± 0.1 (m/sec·MHz) at 3.8 MHz.

Speed of sound dispersion has also been reported for hemoglobin solutions by Carstensen and Schwan [6], for lung tissues by Pedersen and Ozcan [7], for normal brain by Kremkau et al. [8], and for bones by Wear [9], Strelitzki and Evans [4], and Droin et al. [10]. Although it is a relatively weak phenomenon, speed of sound dispersion does differ from one tissue type to another. Thus, its potential as a tool for tissue characterization still needs to be explored.

5.6 THE NONLINEAR PARAMETER B/A

When an acoustic wave propagates with high intensity through a medium, nonlinear effects occur. These effects may be manifested by the appearance of high harmonic waves (i.e., waves whose frequency is an integer multiplication of the transmitted wave). Or in the case where micro-bubbles are generated or injected into the body (see Chapters 9 and 10), subharmonic waves (i.e., at half the basic frequency) may be generated. A characteristic measure of the medium's nonlinearity is a ratio called B/A.

As explained in Chapter 3, the changes in the local pressure induced by the propagating wave is associated with changes in the local medium density. The relation between these two properties can be presented using a Taylor series expansion. As shown by Beyer [11], if the change in density is defined by

$$\Upsilon = \frac{\rho - \rho_0}{\rho_0} \tag{5.30}$$

where ρ_0, ρ are the rest and instantaneous densities, respectively, the instantaneous pressure p can be approximated by

$$p = p_0 + A \cdot \Upsilon + B \cdot \frac{\Upsilon^2}{2!} + C \cdot \frac{\Upsilon^3}{3!} + \ldots \tag{5.31}$$

where p_0 is the pressure at rest and the three constants above are given under isentropic conditions by

$$A = \rho_0 \left(\frac{\partial p}{\partial \rho}\right)_{S,\rho=\rho_0} = \rho_0 \cdot c_0^2$$
$$B = \rho_0^2 \left(\frac{\partial^2 p}{\partial \rho^2}\right)_{S,\rho=\rho_0} \tag{5.32}$$
$$C = \rho_0^3 \left(\frac{\partial^3 p}{\partial \rho^3}\right)_{S,\rho=\rho_0}$$

Using this relation, the instantaneous speed of sound can be assessed from

$$c^2 = c_0^2 \left[1 + \left(\frac{B}{A}\right) \cdot \Upsilon + \left(\frac{C}{A}\right) \cdot \frac{\Upsilon^2}{2!} + \cdots\right] \tag{5.33}$$

where c_0 is the speed of sound at rest and the ratio B/A is given by

$$\frac{B}{A} = 2\rho_0 c_0 \left(\frac{\partial c}{\partial p}\right)_T + \frac{2\beta T c_0}{\mathbb{C}_p}\left(\frac{\partial c}{\partial T}\right)_p, \quad \text{where} \quad \beta = \frac{1}{V}\left(\frac{\partial V}{\partial T}\right)_p \tag{5.34}$$

where T is the temperature and $\mathbb{C}_p$ is the specific heat at constant pressure coefficient. The C/A ratio can also be calculated from

$$\frac{C}{A} = \frac{3}{2}\left(\frac{B}{A}\right)^2 + 2\rho_0^2 c_0^3 \left(\frac{\partial^2 c}{\partial p^2}\right)_S \tag{5.35}$$

There are several methods for measuring the nonlinear parameter B/A. See for example Gong et al. [12], and Zhang and Dunn [13], and Fatemi and Greenleaf [14]. Furthermore, methods for imaging this ratio have been suggested by Kim et al. [15] and Cain and Houshmand [16]. And even the application of a computed tomography (see Chapter 10) has been developed by Zhang et al. [17]. The B/A ratio differs for different type of tissues as shown in Table A.2 of Appendix A. Hence, it can be used for tissue characterization.

REFERENCES

1. Herzfeld KF and Litovitz TA, *Absorption and Dispersion of Ultrasonic Waves*, Academic Press, New York, 1959.
2. Markham JJ, Beyer RT, and Lindsay RB, Absorption of sound in fluids, *Rev Mod Phys* **23**:353–411, 1951.

3. Szabo TL and Wu J, A model for longitudinal and shear wave propagation in viscoelastic media, *J Acoust Soc Am* **107**(5):2437–2446, 2000.
4. Strelitzki R and Evans JA, On the measurements of the velocity of ultrasound on the os calcis using short pulse, *Eur J Ultrasound* **4**:205–213, 1996.
5. Levy Y, Agnon Y, and Azhari H, Measurement of speed of sound dispersion in soft tissues using a double frequency continuous wave method, *Ultrasound Med Biol* **32**(7):1065–1071, 2006.
6. Carstensen EL and Schwan HP, Acoustic properties of hemoglobin solutions, *J Accoust Soc Am* **31**:305–311, 1959.
7. Pedersen PC and Ozcan HS, Ultrasound properties of lung tissue and their measurements, *Ultrasound Med Biol* **12**(6):483–499, 1986.
8. Kremkau FW, Barnes RW, and McGraw CP, Ultrasonic attenuation and propagation speed in normal human brain, *J Acoust Soc Am* **70**:29–38, 1981.
9. Wear KA, Measurments of phase velocity and group velocity in human calcaneus, *Ultrasound Med Biol* **26**:641–646, 2000.
10. Droin P, Berger G, and Laugier P, Velocity dispersion of acoustic waves in cancellous bone, *IEEE Trans Ultrason Ferroelectr Freq Control* **45**:581–592, 1998.
11. Beyer RT, *Non-Linear Acoustics*, Naval Systems Command, Department of the Navy, 1974.
12. Gong XF, Zhu ZM, Shi T, and Huang JH, Determination of the acoustic nonlinearity parameter in biological media using FAIS and ITD methods, *J Acoust Soc Am* **86**(1):1–5, 1989.
13. Zhang J and Dunn F, In vivo *B*/*A* determination in a mammalian organ, *J Acoust Soc Am* **81**(5):1635–1637, 1987.
14. Fatemi M and Greenleaf JF, Real-time assessment of the parameter of nonlinearity in tissue using "nonlinear shadowing," *Ultrasound Med Biol* **22**(9):1215–1228, 1996.
15. Kim D, Greenleaf JF, and Sehgal CM, Ultrasonic imaging of the nonlinear parameter *B*/*A*: Simulation studies to evaluate phase and frequency modulation methods, *Ultrasound Med Biol* **16**(2):175–181,1990.
16. Cain CA and Houshmand H, Ultrasonic reflection mode imaging of the nonlinear parameter *B*/*A*. II: Signal processing, *J Acoust Soc Am* **86**(1):28–34, 1989.
17. Zhang D, Gong X, and Ye S, Acoustic nonlinearity parameter tomography for biological specimens via measurements of the second harmonics, *J Acoust Soc Am* **99**(4, Pt 1):2397–2402, 1996.

CHAPTER 6

REFLECTION AND TRANSMISSION

Synopsis: Thus far we have analyzed the propagation of waves in a homogeneous medium. In this chapter we shall describe the effect of a discontinuity in the medium. We shall learn about the acoustic impedance and its significance. We shall calculate the reflection and transmission coefficients from boundaries that separate two materials with different acoustic properties. We shall describe the angular dependency of these coefficients. We shall discuss the optimal conditions for transmission of waves through a "matching" layer, and we shall present mode conversion in solid materials.

6.1 THE ACOUSTIC IMPEDANCE

6.1.1 The Relation Between Particle Velocity and Pressure

In the field of fluid mechanics (e.g., reference 1), a relation is defined between laminar flow (i.e., with no turbulence) and a function defined as the "velocity potential" function φ. This function fulfills the following condition:

$$\overline{\nabla}\varphi \triangleq \overline{U} \tag{6.1}$$

where φ is a scalar function whose gradient equals the velocity vector $\overline{U}$. For the one-dimensional case where η is the displacement along the y coordinate, this function will fulfill the following relation:

Basics of Biomedical Ultrasound for Engineers, by Haim Azhari

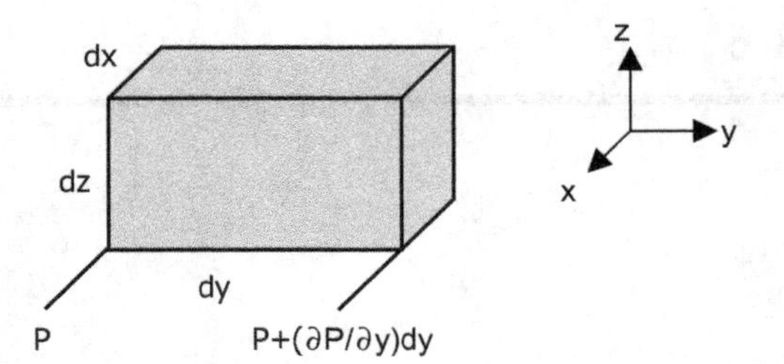

Figure 6.1. A wave traveling through a material element will induce a gradient in the pressure applied at its two opposite faces along the wave propagation direction.

$$\frac{\partial \varphi}{\partial y} = \dot{\eta} \tag{6.2}$$

We shall make use of this concept in the context of ultrasonic waves, to describe the particles velocity. Consider a material element, such as the one depicted in Fig. 6.1, through which a wave is passing. As was shown in Chapter 3, the dynamic state equation was given by

$$-\frac{\partial P}{\partial y} dxdydz = \rho_0 dxdydz\ddot{\eta} \tag{6.3}$$

where P is the pressure and ρ_0 is the density at the state of equilibrium (i.e., with no waves). Or, after dividing by its volume as in Eq. (3.10), we obtain

$$-\frac{\partial P}{\partial y} = \rho_0 \ddot{\eta} \tag{6.4}$$

which can be rewritten using a potential flow function as

$$\Rightarrow -\frac{\partial P}{\partial y} = \rho_0 \cdot \frac{\partial \dot{\eta}}{\partial t} = \rho_0 \cdot \frac{\partial^2 \varphi}{\partial t \partial y} \tag{6.5}$$

Integrating both sides of this equation by y yields

$$-P = \rho_0 \frac{\partial \varphi}{\partial t} + g(t) + \text{const} \tag{6.6}$$

where $g(t)$ is a function that depends only on time. This equation should be valid even when there is no wave in the medium—that is, when $\partial\varphi/\partial t = 0$. This implies that $-P_0 = g(t) + \text{const}$ when the rest pressure is P_0. However, by definition, when there are no temporal changes in the pressure, there are no waves in the medium. Hence, it follows that $g(t) = 0$ or that $-P_0 = \text{const}$. Thus, if we define the pressure rise over the medium element as the "excessive" pressure $\mathbb{P}$, we obtain

$$\boxed{(P-P_0)\equiv \mathbb{P}=-\rho_0\cdot\frac{\partial\varphi}{\partial t}} \tag{6.7}$$

6.1.2 An Exemplary Function φ

Consider a harmonic plan wave, for which the particle velocity is given by

$$\dot{\eta}=\dot{\eta}_0\cos(\omega t-ky) \tag{6.8}$$

we seek a velocity potential function φ that fulfills the condition $\partial\varphi/\partial y=\dot{\eta}$. Therefore, in order to find this equation, we have to integrate Eq. (6.8):

$$\varphi \triangleq \int\dot{\eta}dy=\int\dot{\eta}_0\cos(\omega t-ky)dy=-\left(\frac{\dot{\eta}_0}{k}\right)\sin(\omega t-ky) \tag{6.9}$$

This is the required function. Indeed, there is also an integration constant that should have been added to the right-hand side. However, this constant is not important since we are interested only in the derivatives of this function. Also, it is easy to see that φ satisfies the wave equation (see Chapter 1).

6.1.3 Definition of the Acoustic Impedance

When the wave passes through the material element, a pressure gradient is formed as shown above [see Eq. (6.7)]. This pressure gradient will induce strain and motion in the medium. In an analogy to Ohm's law for electric circuits, we may consider the excessive pressure $\mathbb{P}$ as an equivalent to voltage and the particles velocity as analogous to current (see Fig. 6.2). Using this analogy, the acoustic impedance is defined as the ratio between the excessive pressure and the particle's velocity. Using the definition for the potential flow function given in Eq. (6.9) and its relation to the pressure which is given in Eq. (6.7), we obtain by substitution the following relation:

$$\mathbb{P}=-\rho_0\frac{\partial\varphi}{\partial t}=+\rho_0\underbrace{\left(\frac{\omega}{k}\right)}_{=c}\dot{\eta}_0\cos(\omega t-ky) \tag{6.10}$$

As recalled, the ratio between the angular frequency and the wave number is equal to the speed of the wave; hence we can write

$$\mathbb{P}=\rho_0\cdot C\cdot[\dot{\eta}_0\cos(\omega t-ky)] \tag{6.11}$$

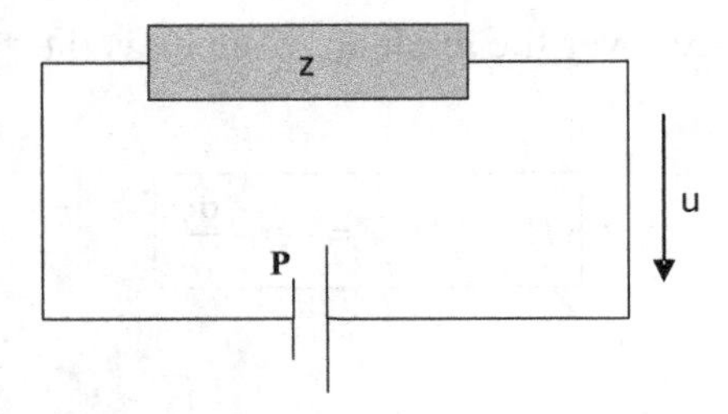

Figure 6.2. The acoustic impedance Z is defined, in an analogy to the electric impedance using Ohm's law, as the ratio between the excessive pressure $\mathbb{P}$ and the particles velocity u.

Based on the definition given in Eq. (6.8), the term in the square brackets is simply the particles velocity, which was marked as $u \equiv \dot{\eta}$. Thus, the excessive pressure is given by

$$\mathbb{P} = \rho_0 \cdot C \cdot u \tag{6.12}$$

Consequently, the acoustic impedance is therefore given by

$$\boxed{Z \triangleq \frac{\mathbb{P}}{u} = \rho_0 \cdot C} \tag{6.13}$$

It is important to note that this is true only for a *planar wave*!

In order to calculate the acoustic impedance for a *spherical wave*, we need to make use of the definition given in Chapter 1. Again we shall assume without any loss of generality that the wave is harmonic, that is,

$$\varphi = -\frac{\varphi_0}{r} \cdot e^{j(\omega t - kr)} \tag{6.14}$$

Taking the derivative of this function along the wave propagation direction will yield by definition the particles velocity:

$$u = \frac{\partial \varphi}{\partial r} = \varphi_0 \left(\frac{1}{r^2} + \frac{jk}{r} \right) e^{j(\omega t - kr)} \tag{6.15}$$

And since from Eq. (6.7) the relation

$$\mathbb{P} = -\rho_0 \frac{\partial \varphi}{\partial t} \tag{6.16}$$

is given, the excessive pressure over the element is given by

$$\mathbb{P} = \frac{j\omega\rho_0\varphi_0}{r} e^{j(\omega t - kr)} \tag{6.17}$$

Dividing this term by the term given above for the particles velocity yields the acoustic impedance for a spherical wave:

$$Z = \frac{\mathbb{P}}{u} = \frac{j\omega r\rho_0}{1 + jkr} \tag{6.18}$$

Or after a slight rearrangement, we obtain

$$\boxed{\Rightarrow Z = \frac{j\omega r\rho_0(1 - jkr)}{1 + k^2r^2}} \tag{6.19}$$

As can be noted, the acoustic impedance for a spherical wave is a complex number. This implies that the pressure and the particles velocity are not in the same phase (as is the case for the electric impedance in inductance or capacitance elements). However, if the condition $kr >> 1$ holds, then the acoustic impedance is given by

$$\boxed{Z \xrightarrow[kr \to \infty]{} \frac{k\omega r^2\rho_0}{k^2r^2} = \rho_0 \cdot c} \tag{6.20}$$

where we have used the following relation:

$$\left(\frac{\omega}{k} = C\right) \tag{6.21}$$

This implies that under this condition the acoustic impedance for a spherical wave is similar to that of a planar wave!

For most practical applications in biomedicine, there is no need for using the complex term since $kr >> 1$. This can be demonstrated by taking typical values used in ultrasonic imaging. For example, consider an ultrasonic transducer that transmits waves with a central frequency of 5 MHz and an object located at a distance of 10 mm away. And since the speed of sound is about 1500 m/sec, thewavelengthisabout $\lambda = 0.3$ mmand $k \cdot r = 2\pi/0.3 \cdot 10 = 209.4 >> 1$.

6.1.4 The Relation Between the Impedance and the Wave Intensity

The acoustic impedance plays a major role in determining the reflection and transmission coefficients (as will be shown in the following sections). However, at this point it is also worth noting its relation to the acoustic field intensity.

Consider a harmonic wave for which the particles displacement is given by $\eta_0 e^{j(\omega t - kx)}$. As was shown in Chapter 3 [Eq. (3.30)], the intensity is given by

$$I = \frac{1}{2}\rho_0 \cdot C\left(\eta_0^2 \omega^2\right) = \frac{1}{2}\rho_0 \cdot C \cdot u^2 \tag{6.22}$$

However, as was shown above for a planar wave and for a spherical wave under the condition of $kr >> 1$, the acoustic impedance is given by Eq. (6.13). Therefore the excessive pressure is given by

$$\mathbb{P} = \rho_0 C \cdot u = Z \cdot u \tag{6.23}$$

Substituting for u, we obtain

$$\Rightarrow I = \frac{1}{2} Z \cdot \frac{\mathbb{P}^2}{Z^2} = \frac{\mathbb{P}^2}{2Z} \tag{6.24}$$

One has to note that if a complex notation is used, then the real part of this term should be taken, that is,

$$\boxed{I = \mathrm{Re}\left\{\frac{\mathbb{P}^2}{2Z}\right\}} \tag{6.25}$$

6.2 SNELL'S LAW

When a wave traveling in one medium encounters a boundary of another medium for which the acoustic properties are different, part of its energy will be reflected and part of it will be transmitted. The reflected wave (an echo) will travel back into the first medium, and the transmitted wave will travel into the second medium (as shown schematically in Fig. 6.3). However, experimental observations show that while the first wave has a reflection angle that is equal to the incident angle, the transmitted wave will be deflected. This deflection is easily demonstrated in optics (e.g., reference 2)—for example, by observing the picture of the pencil immersed in a glass of water shown in Fig. 6.4. The relation between the angles of the incident and reflected waves obeys Snell's law, which is defined as [2]

$$\boxed{\frac{C_1}{\sin\theta_i} = \frac{C_2}{\sin\theta_T}} \tag{6.26}$$

where C_1 and C_2 are the waves speed in the two mediums and θ_i and θ_T are the angles of incidence and transmission, respectively. Importantly, it should be noted that the angles are measured relative to the normal to the boundary.

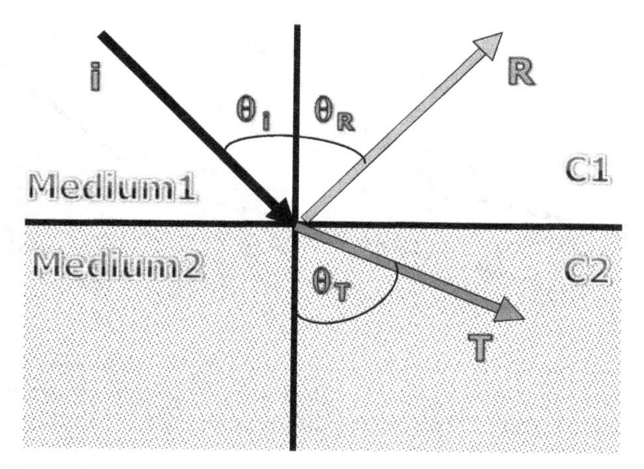

Figure 6.3. Schematic representation of the energy distribution resulting from the encounter of an incident wave (vector **i**) with a boundary. The reflected wave is denoted as vector **R**, and the through-transmitted wave is denoted as vector **T**. Note that the angle of the transmitted wave is different from the angle of the incident wave.

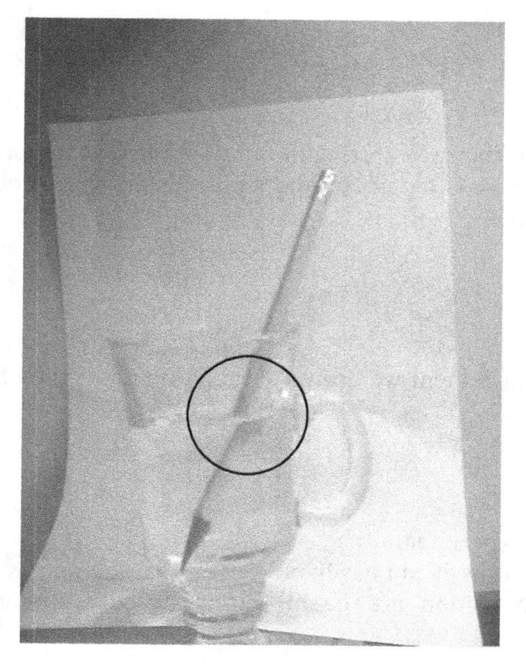

Figure 6.4. A simple experiment to demonstrate the wave deflection effect. The pencil seems broken due to the wave deflections resulting from the changes in the speed of light.

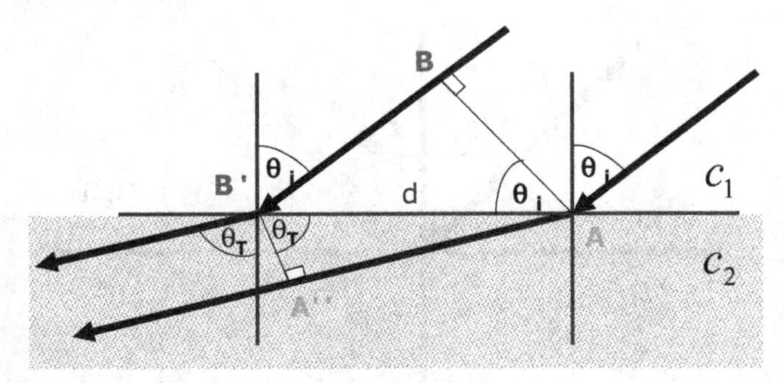

Figure 6.5. A plan wave whose front is represented by two points A and B encounters a boundary separating two mediums. The front of the through-transmitted wave (represented by points A′ and B′) will have a different angular orientation.

This law can be proven analytically. Using Fig. 6.5 for clarity, let us represent the front of a plan wave by the line connecting points A and B.

The angle of incidence is θ_i as shown. Upon penetration, the ray corresponding to point A is deflected and travels at speed C_2, while the ray corresponding to point B continues to travel at speed C_1 along the original direction. The time required for this ray to enter the second medium (point B') is given by

$$\Delta t = \frac{BB'}{C_1} = \frac{d \sin \theta_i}{C_1} \tag{6.27}$$

In the meantime the ray entering at point A has reached point A'. It can be easily shown that $AA' = d \sin \theta_T$. And since this distance should also equal to $c_2 \Delta t$, it follows that

$$d \sin \theta_T = C_2 \frac{d \sin \theta_i}{C_1} \tag{6.28}$$

And after rearrangement we derive the relation defined by Snell's Law,

$$\Rightarrow \frac{C_1}{\sin \theta_i} = \frac{C_2}{\sin \theta_T} \tag{6.29}$$

as was derived experimentally.

Using Snell's Law, it can easily be proven that the angles of incidence and the angle of reflection are equal simply by substituting $C_1 = C_2$, thus obtaining

$$\frac{C_1}{\sin \theta_i} = \frac{C_1}{\sin \theta_R}$$
$$\Rightarrow \theta_i = \theta_R \tag{6.30}$$

6.3 REFLECTION AND TRANSMISSION FROM BOUNDARIES SEPARATING TWO FLUIDS (OR SOLIDS WITH NO SHEAR WAVES)

6.3.1 Critical Angles

When a longitudinal wave propagating in a nonviscous fluid encounters another fluid that has different acoustic properties (also assumed nonviscous), part of its energy will be reflected in the form of a new wave (echo) as shown above. The angle of incidence will equal the angle of reflection: $\theta_i = \theta_r$. The rest will penetrate the second medium in the form of another wave. The angle of the transmitted wave θ_T will obey Snell's law [Eq. (6.26)]. If $C_2 > C_1$, then $\sin\theta_T > \sin\theta_i$ and naturally $\theta_T > \theta_i$. Now, suppose we start to increase incrementally the angle of incidence θ_i. As a result, the angle θ_T will increase by larger increments. Consequently, a point will be reached where $\theta_T = \pi/2$ (and of course $\sin\theta_T = 1$). At this point the penetrating wave will be able to travel only along the boundary layer. If we further increase θ_i, we shall obtain from Snell's law $\sin\theta_T > 1$. This implies that no wave will penetrate the second medium and that all the energy will be reflected. The corresponding angle is referred to as the *critical angle*. (This phenomenon can also be easily demonstrated in optics. For example, when looking at a transparent glass from a low angle, one can note that it reflects light like a mirror.)

6.3.2 Reflection and Transmission Coefficients

After defining the angular relations between the incident, reflected, and transmitted waves, let us now evaluate the magnitude of reflection and transmission in terms of amplitudes and intensities.

Referring to Fig. 6.6, we know that the pressure at the upper medium (from where the wave has arrived) is comprised of the pressures generated by the incident P_i and reflected waves P_R, while at the bottom the pressure is generated only by the transmitted wave P_T. At the boundary between the two mediums the pressure on both sides should be equal. Thus, we can write

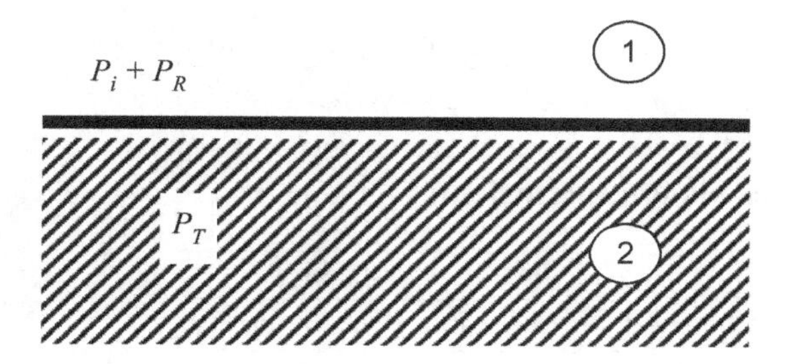

Figure 6.6. The pressure at upper medium (from which the wave has arrived) is comprised of the pressures generated by the incident and reflected waves. At the boundary between two mediums the pressure on both sides should be equal.

$$P_i + P_R = P_T \tag{6.31}$$

Also we know that as long as there is no detachment between the two layers, the two layers should move up and down together. (Note that detachment can generate cavitation bubbles—see Chapter 12.) Thus we can equate their displacements, that is,

$$\eta_i \cos\theta_i - \eta_r \cos\theta_r = \eta_T \cos\theta_T \tag{6.32}$$

where η_i, η_R, and η_T are the displacements of the incident reflected and transmitted waves, respectively. And since their displacements are the same, so should be their velocities (u). Thus, we can equate the temporal derivatives from both sides. And since $\dot{\eta} = u$, we can write

$$u_i \cos\theta_i - u_r \cos\theta_r = u_T \cos\theta_T \tag{6.33}$$

It should be pointed out that two layers can potentially slide relative to each other along the horizontal direction since (as we have assumed) the mediums cannot sustain shear stress.

Using the definition of the acoustic impedance (6.13), the pressure is related to the velocity by

$$P = (\rho C) \cdot u = Z \cdot u \tag{6.34}$$

Thus, we can write

$$\left(\frac{P_i}{Z_1}\right)\cos\theta_i - \left(\frac{P_r}{Z_1}\right)\cos\theta_i = \left(\frac{P_T}{Z_2}\right)\cos\theta_T \tag{6.35}$$

At this point we have two equations (6.31) and (6.35) relating the three pressure values at the boundary. Assuming that the pressure of the incident wave is known, we can solve these equations to obtain

Amplitude Reflection Coefficient:

$$\boxed{R \triangleq \frac{P_r}{P_i} = \frac{Z_2 \cos\theta_i - Z_1 \cos\theta_T}{Z_2 \cos\theta_i + Z_1 \cos\theta_T}} \tag{6.36}$$

Amplitude Transmission Coefficient:

$$\boxed{T \triangleq \frac{P_T}{P_i} = \frac{2Z_2 \cos\theta_i}{Z_2 \cos\theta_i + Z_1 \cos\theta_T}} \tag{6.37}$$

For the special case where $\theta_i = \theta_T = 0$—that is, when the wave is *perpendicular* to the boundary—we shall obtain

$$\boxed{R \triangleq \frac{Z_2 - Z_1}{Z_2 + Z_1}} \tag{6.38}$$

$$\boxed{T \triangleq \frac{2Z_2}{Z_2 + Z_1}} \tag{6.39}$$

Recalling that the pressure is related to the acoustic impedance through Eq. (6.34) and that the relation between the intensity and the pressure given by Eq. (6.24), we can use these parameters to calculate the intensity by

$$I = \frac{\text{Energy}}{\text{Area} \cdot dt} = \frac{P^2}{2Z} \tag{6.40}$$

Substituting Eq. (6.40) into Eq. (6.36) and Eq. (6.37), the energy/intensity reflection and transmission coefficients can be calculated by

Energy/Intensity Reflection Coefficient:

$$\boxed{\frac{\text{Reflected energy}}{\text{Incident energy}} = \frac{I_R}{I_i} = \frac{P_R^2}{2Z_1} \cdot \frac{2Z_1}{P_i^2} = \frac{P_R^2}{P_i^2} = R^2} \tag{6.41}$$

Energy/Intensity Transmission Coefficient:

$$\boxed{\frac{\text{Transmitted energy}}{\text{Incident energy}} = \frac{I_T}{I_i} = \frac{P_T^2}{2Z_2} \cdot \frac{2Z_1}{P_i^2} = \frac{Z_1}{Z_2} \cdot \frac{P_T^2}{P_i^2} = \frac{Z_1}{Z_2} \cdot T^2} \tag{6.42}$$

In this context it is worth noting the following special cases:

- If $Z_1 < Z_2$, then $R > 0$, which implies that the amplitude of the reflected wave is in phase with the incident wave.
- If $Z_1 > Z_2$, then $R < 0$, which implies that the amplitude of the reflected wave has an opposite phase (180°) relative to the incident wave.
- If $Z_1 << Z_2$, then $R \to 1$, which implies that the amplitude of the reflected wave is almost the same as that of the incident wave.
- If $Z_1 >> Z_2$ (as, for example, is the case for a fluid–air boundary), then $R \to -1$, which implies that the amplitude of the reflected wave is also almost the same as that of the incident wave. But, the reflected wave has an opposite phase (180°) relative to the incident wave.

The above relations are important in the context of ultrasonic imaging (see Chapter 9). Because gas has negligible acoustic impedance relative to tissues, very strong echoes will be reflected from organs containing gas. Thus, the chest for example has very limited acoustic windows that can be used for imaging (using standard equipment), in view of the fact that the lungs which contain air do not enable waves to pass through. (This fortunately is not the case for imaging embryos since their lungs are filled with a fluid.) Similarly, hairy surfaces can cause a problem since air can be trapped between the transducer and the skin. In such case, the poor contact will reduce the efficiency of wave transmission into and back from the body. This is the reason why jell is commonly spread over the skin prior to the application of an ultrasonic procedure.

On the other hand, gas reflectivity can be beneficial as a contrast agent. There are several commercially available echo-enhancing agents that are currently approved for clinical use. These agents are made of micro-bubbles containing gas. The bubbles are coated with albumin or a polymeric shell. This shell allows them to remain in the body for several minutes or longer in order to allow the completion of the diagnostic procedure. The solution containing the contrast material is injected into the vein and the blood circulates it in the body. Every organ into which the bubbles are washed will have an increased reflectivity. As a result, strong echoes will be viewed from this region. This fact allows the physician to discriminate between regions with normal perfusion and ischemic regions (i.e., with reduced or no blood supply).

Similarly, bones can present an obstacle in imaging. Their acoustic impedance is about 4–5 times higher than the average impedance of soft tissues. Thus, most of the energy of waves traveling in soft tissues will be reflected from bones. Furthermore, even the relatively small amount of energy that will penetrate into the bone will be rapidly absorbed due to its high attenuation coefficient. Gallstones, kidney stones, and metal implants are also strong reflectors due to their high acoustic impedance.

6.3.3 The Matching Layer

An important issue regarding ultrasonic transducers is the need for impedance matching. Since most of the ultrasonic transducers are made of piezoelectric materials (see Chapter 8), which typically have an acoustic impedance which is substantially different than that of soft tissues, the energy transmission from the transducer into the body will be inefficient. As a result, the transducer may absorb this energy and its temperature may rise (particularly in therapeutic applications), its sensitivity to detecting echoes will be reduced leading to poor SNR (in imaging applications). As will be shown in the following, the matching layer provides a good solution for this problem.

Consider the three-layer structure depicted in Fig. 6.7. Acoustic waves traveling from the first layer (e.g., the ultrasonic transducer), which has an acoustic impedance of Z_1, impinges with an intensity of I_1 upon the boundary separating it from a second layer. For simplicity, we shall assume that the boundary is planar and that the incident wave is perpendicular to it. As explained above, part of the energy will commonly be reflected and part of it will be transmitted. The transmitted wave will pass through the intermediate layer, which has

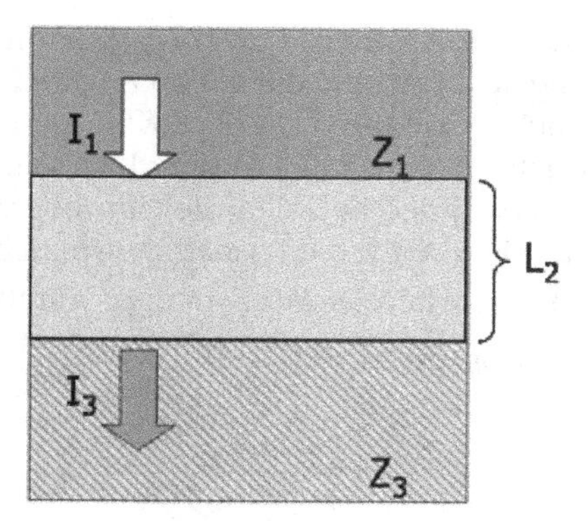

Figure 6.7. Schematic depiction of a matching layer positioned between the transducer **(top layer)** and the body **(bottom layer)**.

an acoustic impedance Z_2. The wave traveling through the second layer will then reach the second boundary corresponding to the third layer, which has an acoustic impedance of Z_3. Part of this energy will continue to travel into the third layer with an intensity of I_3. We would like to calculate the value of the transmission coefficient $\Upsilon \triangleq I_3 / I_3$, find its relation to the acoustic impedances, and find the terms for optimizing it. The solution for this problem was derived by Kinsler and Frey [3] and is given by

$$\Upsilon = \frac{I_3}{I_1} = \frac{4Z_1Z_3}{(Z_3+Z_1)^2 \cdot \cos^2(K_2L_2) + \left(Z_2 + \frac{Z_1Z_3}{Z_2}\right)^2 \cdot \sin^2(K_2L_2)} \tag{6.43}$$

where $K_2 = 2\pi/\lambda_2$ is the wave number for the intermediate layer.

Let us now consider three special cases. The first two cases are as follows:

(a) When the thickness of the intermediate layer is very small relative to the wavelength in this layer (i.e., $L_2 << \lambda_2/4$), the SIN term in the denominator will be negligible and the COS term will be approximately 1.

(b) When the thickness of the intermediate layer fulfills the condition $L_2 = n\lambda_2/2$, then again the SIN term in the denominator will be negligible and the COS term will be approximately 1.

For these two special cases the transmission coefficient will be *independent* of the acoustic impedance of the intermediate layer, that is,

$$\Upsilon = \frac{I_3}{I_1} = \frac{4Z_1Z_3}{(Z_3+Z_1)^2} \tag{6.44}$$

Nevertheless, it should be pointed out that this equation is valid for case (a) above, provided that the acoustic impedances *do not hold* the following condition: $Z_2 << Z_1 \cdot Z_3$. In such a case the term $Z_1 \cdot Z_3 / Z_2 \to \infty$ may cancel the effect resulting from the reduction in the value of the SIN term. Such case can occur when a small gas layer is trapped between the transducer and the tissue.

The third special case is when the intermediate (matching) layer thickness equals $\boxed{L_2 = (2n - 1)\lambda_2/4}$, where n is an integer number. In this case the transmission coefficient is given by

$$\Upsilon = \frac{I_3}{I_1} = \frac{4Z_1Z_3}{\left(Z_2 + \frac{Z_1Z_3}{Z_2}\right)^2} \tag{6.45}$$

However, if we choose an intermediate layer for which the acoustic impedance is given by $\boxed{Z_2 = \sqrt{Z_1Z_3}}$, the following transmission coefficient will be obtained:

$$\boxed{\Upsilon = \frac{I_3}{I_1} = \frac{4Z_1Z_3}{\left(\sqrt{Z_1Z_3} + \frac{Z_1Z_3}{\sqrt{Z_1Z_3}}\right)^2} = \frac{4Z_1Z_3}{\left(2\sqrt{Z_1Z_3}\right)^2} = 1} \tag{6.46}$$

This implies that all the energy leaving the transducer will be delivered into the body!!! This fact is used when designing an ultrasonic transducer. By attaching a thin matching layer to the ultrasonic transducer, its efficiency is increased substantially.

6.4 REFLECTION FROM A FREE SURFACE IN SOLIDS (MODE CONVERSION)

As discussed in Chapter 4, both longitudinal and shear (transverse) waves can coexist within a solid medium [4]. This fact may, under certain circumstances of reflection or transmission between different materials, lead to a conversion phenomenon [5]. In such case a longitudinal wave can be transformed into a shear wave and vice versa. Or two different waves can be generated from a single incident wave. Let us start with the simplest case depicted in Fig. 6.8.

Consider a longitudinal wave (depicted by the double headed arrow ↔ along the left vector) propagating in an isotropic solid medium with amplitude A_1. This wave encounters a free surface—that is, a planar boundary of vacuum or a medium with negligible acoustic impedance such as air. The angle of incidence is θ. From the point of incidence a longitudinal wave will be reflected (as expected) with an angle $\alpha = \theta$. The amplitude of the reflected longitudinal wave is B_1. However, in addition to that, a new wave will also be generated at the surface! This new wave is a shear wave (which is marked as ⇕) for which the amplitude is B_2. The angle of reflection of this new wave is β and is determined from Snell's Law.

The cause for the generation of this new shear wave stems from the fact the wave is actually consisted of a series of particle displacements. As was explained in Chapter 4, this displacement is associated with stresses within the solid material. Consequently, at an oblique angle of incidence, shear stresses can (under certain conditions) be generated at the edge. These shear stresses are the source for this shear wave. Importantly, it should be noted that shear waves (as light waves) have polarization that should be accounted for. In the case depicted in Fig. 6.8, the shear wave is *polarized in the X–Y plane*, since there is no motion along the Z direction.

Using Fig. 6.9 as a reference, let us now write the vector components for the wave numbers $\overline{K}$ for each of these three waves. Starting with the incident wave, we can write

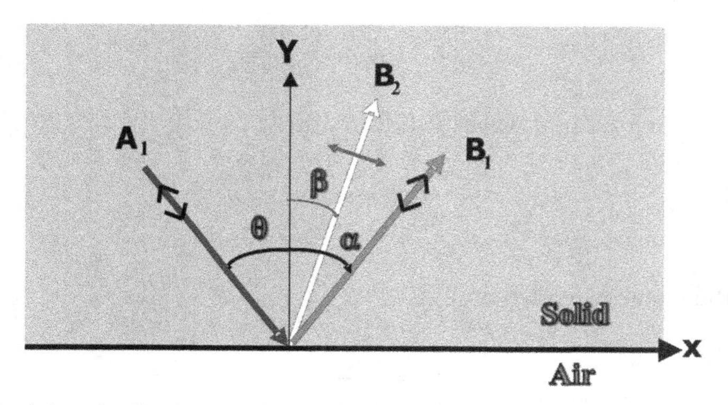

Figure 6.8. A longitudinal wave (depicted by the double-headed arrow ↔ along the left vector) propagating in a solid material with an amplitude A_1 encounters a free surface. As a result, two waves are reflected: a longitudinal wave (↔) with amplitude B_1 and a shear wave (↕) with amplitude B_2.

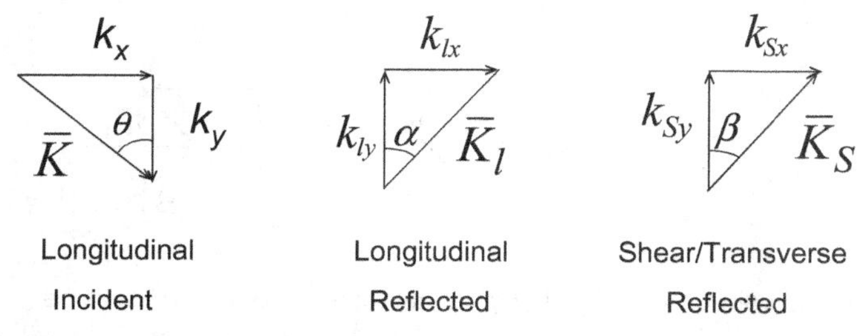

Figure 6.9. The corresponding components of the wave number vectors for each of the three waves.

$$k_x = k\sin\theta, \qquad (6.47)$$
$$k_y = k\cos\theta$$

For the reflected longitudinal wave the components are

$$k_{lx} = k_l \sin\alpha \qquad (6.48)$$
$$k_{ly} = k_l \cos\alpha$$

For the reflected shear wave the components are

$$k_{sx} = k_s \sin\beta \qquad (6.49)$$
$$k_{sy} = k_s \cos\beta$$

Also from Snell's law we can write

$$\frac{c_{\text{long}}}{\sin\theta} = \frac{c_{\text{long}}}{\sin\alpha} = \frac{c_{\text{shear}}}{\sin\beta} \qquad (6.50)$$

Thus for the longitudinal waves it follows that

$$\sin\theta = \sin\alpha$$
$$\Rightarrow \theta = \alpha \qquad (6.51)$$

And for the shear wave we obtain

$$\sin\beta = \frac{c_{\text{shear}}}{c_{\text{long}}}\sin\theta \qquad (6.52)$$

Now, using Fig. 6.10 as a reference, let us write down the displacement components for each of the three waves. Using the notations explained in Chapter 4, the directions X and Y will be designated by the index numbers 1 and 2, respectively.

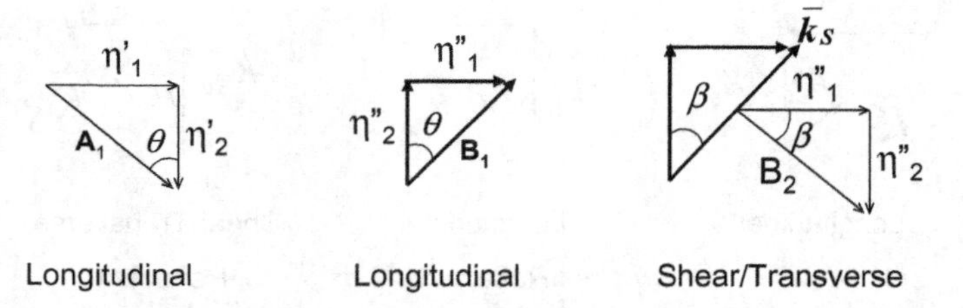

Figure 6.10. The displacement components along the x and y directions (marked as 1 and 2) for the incident and reflected longitudinal waves and for the reflected shear wave.

Hence, for the incident longitudinal wave (marked as η^{i}) the components are

$$\begin{aligned} \eta_1^i &= A_1 \sin\theta \\ \eta_2^i &= -A_1 \cos\theta \end{aligned} \tag{6.53}$$

For the reflected longitudinal wave (marked as η^{L}) the components are

$$\begin{aligned} \eta_1^L &= B_1 \sin\theta \\ \eta_2^L &= B_1 \cos\theta \end{aligned} \tag{6.54}$$

And for the reflected shear wave (marked as η^{S}) the components are (note that the displacement is orthogonal to the wave propagation direction)

$$\begin{aligned} \eta_1^S &= B_2 \cos\beta \\ \eta_2^S &= -B_2 \sin\beta \end{aligned} \tag{6.55}$$

Without loss of generality, we shall assume that the waves are harmonic; thus we can describe their displacements as

$$\eta_i = \eta_{0i} e^{j[\omega t - (k_x x + k_y y + k_z z)]} \tag{6.56}$$

where η_i is the displacement along the i direction, and η_{0i} is its corresponding amplitude. Substituting for the incident wave gives the following for the horizontal direction:

$$\eta_1^i = A_1 \sin\theta \cdot e^{j[\omega t - [k \cdot \sin\theta \cdot x + k \cdot \cos\theta \cdot y]]} \tag{6.57}$$

And for the vertical direction, we obtain

$$\eta_2^i = -A_1 \cos\theta e^{j[\omega t - [k \cdot \sin\theta \cdot x + k \cdot \cos\theta \cdot y]]} \tag{6.58}$$

Substituting for the reflected longitudinal wave gives

$$\begin{aligned} \eta_1^L &= B_1 \sin\theta \cdot e^{\{j[\omega t - (k_e \cdot \sin\theta \cdot x + k_e \cdot \cos\theta \cdot y)]\}} \\ \eta_2^L &= B_1 \cos\theta \cdot e^{\{j[\omega t - (k_e \cdot \sin\theta \cdot x + k_e \cdot \cos\theta \cdot y)]\}} \end{aligned} \tag{6.59}$$

Substituting for the reflected shear wave gives

$$\begin{aligned} \eta_1^S &= B_2 \cos\beta \cdot e^{\{j[\omega t - (k_s \cdot \sin\beta \cdot x + k_s \cdot \cos\beta \cdot y)]\}} \\ \eta_1^S &= -B_2 \sin\beta \cdot e^{\{j[\omega t - (k_s \cdot \sin\beta \cdot x + k_s \cdot \cos\beta \cdot y)]\}} \end{aligned} \tag{6.60}$$

Referring again to Chapter 4, it was shown there that the speed of sound for an isotropic solid material can be expressed in terms of its Lamé coefficients: λ', μ. Hence we can express the longitudinal wave numbers by

$$k = k_l = \frac{\omega}{c_{long}} = \omega\sqrt{\frac{\rho_0}{\lambda' + 2\mu}} \tag{6.61}$$

And we can express the shear wave number by

$$k_s = \frac{\omega}{C_{\text{shear}}} = \omega\sqrt{\frac{\rho_0}{\mu}} \tag{6.62}$$

For a free surface at $y = 0$, the stresses along both directions should equal zero; that is for the normal stress $\tau_{22} = 0$ and for the shear stress $\tau_{12} = 0$. Using the derivations given in Chapter 4, we obtain the following for the shear stress:

$$\tau_{12} = \mu\varepsilon_{12} = \mu\left(\frac{\partial\eta_1}{\partial y} + \frac{\partial\eta_2}{\partial x}\right) = 0 \tag{6.63}$$

And for the normal stress, we obtain

$$\tau_{22} = \lambda'\varepsilon_{11} + (\lambda' + 2\mu)\varepsilon_{22} = \lambda'\left(\frac{\partial\eta_1}{\partial x}\right) + (\lambda' + 2\mu)\left(\frac{\partial\eta_2}{\partial y}\right) = 0 \tag{6.64}$$

In addition, the displacements for the incident and reflected waves should be equal at the surface along both directions:

$$\begin{aligned} \eta_1^i &= \eta_1^L + \eta_1^S \\ \eta_2^i &= \eta_2^L + \eta_2^S \end{aligned} \tag{6.65}$$

Also, substituting Lamé coefficients into Eq. (6.52) yields

$$\sin\beta = \frac{c_{\text{shear}}}{c_{\text{long}}}\sin\theta = \sqrt{\frac{\mu}{\lambda' + 2\mu}}\sin\theta \tag{6.66}$$

As was shown by Arenberg [6], when substituting all these terms into Eqs. (6.63) and (6.64), we obtain after simplification the following pair of equations:

$$\begin{aligned} &(A_1 - B_1)\sqrt{\frac{\mu}{\lambda' + 2\mu}}\sin 2\theta + B_2\left(\frac{2\mu}{\lambda' + 2\mu}\sin^2\theta - 1\right) = 0 \\ &(A_1 + B_1)(\lambda' + 2\mu\cos^2\theta) - B_2 2\mu\sin\theta\sqrt{\left(1 - \frac{\mu}{\lambda' + 2\mu}\sin^2\theta\right)} = 0 \end{aligned} \tag{6.67}$$

From this set of equations we can derive the ratios B_1/A_1 and B_2/A_1 designating the reflection coefficients for the longitudinal and shear waves, respectively. In order to simplify the derivation, let us define four coefficients D_j that replace the constants in Eq. (6.67). Thus, these equations can be written as

$$\begin{aligned}(A_1 - B_1)\cdot D_1 + B_2 \cdot D_2 = 0\\(A_1 + B_1)\cdot D_3 - B_2 \cdot D_4 = 0\end{aligned} \tag{6.68}$$

And their solutions are

$$\boxed{\begin{aligned}\frac{B_1}{A_1} &= \frac{\left(D_1 + \dfrac{D_2 \cdot D_3}{D1}\right)}{\left(D_1 - \dfrac{D_2 \cdot D_3}{D1}\right)},\\ \frac{B_2}{A_1} &= \frac{(2D_1)}{\left(\dfrac{D_1 \cdot D_4}{D_3} - D_2\right)}\end{aligned}} \tag{6.69}$$

It is important to note that D_j depends on the angle of incidence θ and that $D_2 < 0$. Hence, the reflected wave's amplitudes (absolute value) will always be smaller than that of the incident wave.

6.5 REFLECTION AND TRANSMISSION FROM A LIQUID–SOLID BOUNDARY

When ultrasonic waves encounter a liquid–solid boundary, several possibilities may take place according to the direction of propagation, speed of sound velocity ratio, and the angle of incidence. This stems from the fact that three waves can be generated from such an encounter and that each wave has its own critical angle. It will be assumed in the following that the waves are planar and that only longitudinal waves can exist within the fluid medium and that the boundary is planar too.

6.5.1 Case #1: From a Fluid to a Solid

Let us consider the first case as shown in Fig. 6.11. A longitudinal wave propagates in the fluid medium and encounters a solid plane. As a result, a longitudinal wave will be reflected back into the fluid and two waves are expected to be generated and transmitted into the solid medium: a longitudinal wave and a shear wave.

The following notations are used here:

I_L is the amplitude of the incident longitudinal wave,
C_{L1} is the speed of the incident longitudinal wave,

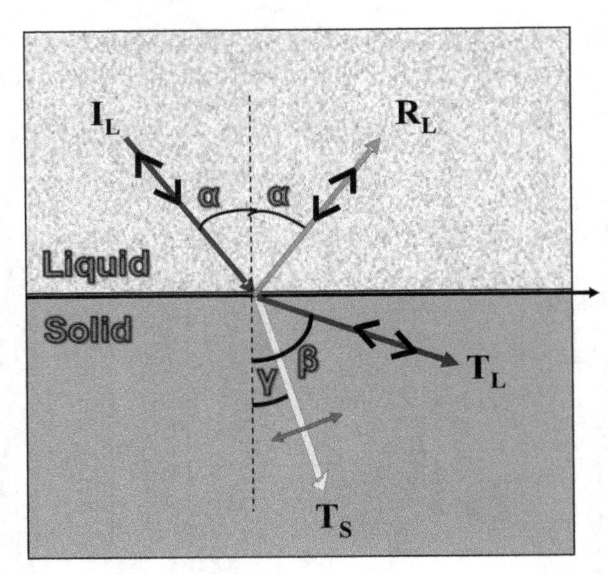

Figure 6.11. A longitudinal wave I_L propagates in the fluid and encounters a solid plane. A longitudinal wave will be reflected R_L and two waves are expected to be transmitted into the solid medium: a longitudinal wave T_L and a shear wave T_S.

α is the angle of incidence for the incident longitudinal wave,
R_L is the amplitude of the reflected longitudinal wave,
T_L is the amplitude of the transmitted longitudinal wave,
C_{L2} is the speed of the transmitted longitudinal wave,
T_s is the amplitude of the transmitted shear wave,
C_{S2} is the speed of the transmitted shear wave,
β is the angle for the transmitted longitudinal wave,
γ is the angle for the transmitted shear wave,
ρ_1 is the fluid density,
ρ_2 is the solid density,

The solution for this case, which was given by Mayer [7], is

$$\boxed{\begin{aligned} \left(\frac{R_L}{I_L}\right)^2 &= \left[\frac{1-G(1-2A)}{1+G(1-2A)}\right]^2 \\ \left(\frac{T_L}{I_L}\right)^2 &= \frac{4BG}{[1+G(1-2A)]^2} \\ \left(\frac{T_S}{I_L}\right)^2 &= \frac{4\left(\frac{\rho_2}{\rho_1}\right)\cdot D}{[1+G(1-2A)]^2} \end{aligned}} \tag{6.70}$$

where the constants are defined as

$$A = \sin\gamma \cdot \sin 2\gamma \cdot \left[\cos\gamma - \frac{C_{s2}}{C_{L2}} \cos\beta \right]$$
$$B = (\cos 2\gamma)^2$$
$$G = \frac{\rho_2 \cdot C_{L2} \cdot \cos\alpha}{\rho_1 \cdot C_{L1} \cdot \cos\beta} \tag{6.71}$$
$$D = \left(\frac{C_{s2}}{C_{L1}} \right)^2 \cdot \sin 2\alpha \cdot \sin 2\gamma$$

and the angles can be derived from Snell's law.

It is important to note that there is a corresponding critical angle for each of the waves in the solid. At these angles, total reflection may occur. Also, it is important to note that the reflection coefficient can be treated as a complex number. Hence, beyond the first critical angle an imaginary component may appear. For example, in Fig. 6.12 the reflection coefficient is plotted as a function of the angle of incidence, for a water/aluminum boundary [8]. For clarity, the real, imaginary, and absolute components of the reflection coefficient are plotted separately.

As can be noted, there is a total reflection at the first critical angle α_1 (corresponding to the longitudinal wave in the solid). However, beyond this angle

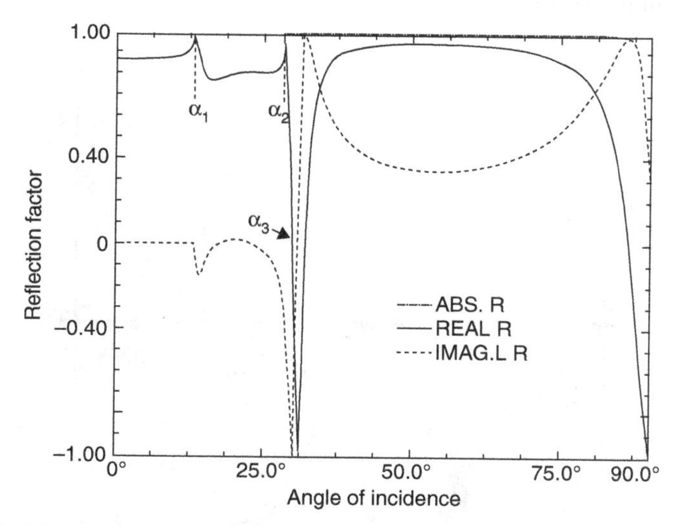

Figure 6.12. The theoretical reflection coefficient (real imaginary and absolute values) for a water/aluminum surface. The angles α_1, α_2, and α_3 correspond to the first second and third critical angles, respectively. Experimental measurements match the pattern of the real component. From Azhari [8].

an imaginary component appears and the value of the reflection coefficient drops. There is another total reflection at the second critical angle α_2 (corresponding to the shear wave in the solid). However, beyond this angle the value of the real component drops steeply to become zero at the third critical angle α_3 (corresponding to the appearance of surface waves in the solid) and then rises again to total reflection. Experimental observations have shown that the actual pattern measured by conventional ultrasonic transducers corresponds to the real component of the reflection coefficient.

6.5.2 Case #2: From a Solid to a Fluid

A longitudinal wave propagating in the solid medium encounters a solid plane (Fig. 6.13). As a result (provided that no critical angles have been reached), a longitudinal wave will be reflected back into the solid together with a shear wave that will be generated at the point of encounter. In addition, a longitudinal wave will be transmitted into the fluid medium.

The following notations are used here:

I_L is the amplitude of the incident longitudinal wave,
C_{L1} is the speed of the incident longitudinal wave,
α is the angle of incidence for the incident longitudinal wave,
R_L is the amplitude of the reflected longitudinal wave,
R_S is the amplitude of the reflected shear wave,
θ is the angle for the reflected shear wave,

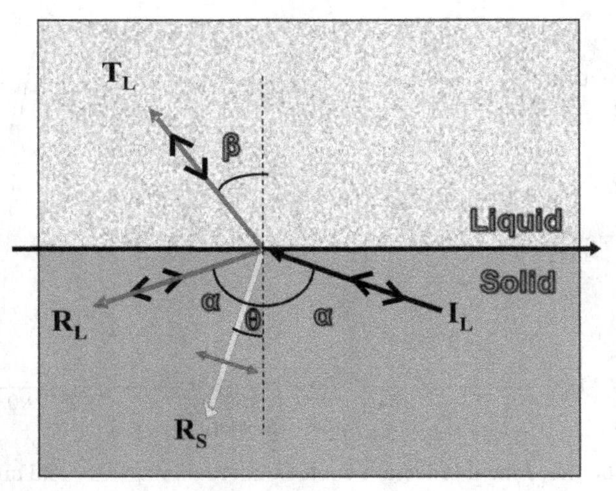

Figure 6.13. A longitudinal wave I_L propagates in the solid and encounters a fluid medium. Two waves are expected to be reflected: a longitudinal wave R_L and a shear wave R_S. Only one longitudinal wave T_L will be transmitted.

C_{S1} is the speed of the transmitted shear wave,
T_L is the amplitude of the transmitted longitudinal wave,
C_{L2} is the speed of the transmitted longitudinal wave,
β is the angle for the transmitted longitudinal wave,
ρ_1 is the solid density,
ρ_2 is the fluid density.

The solution for this case was also derived by Mayer [7] and is given by

$$\boxed{\begin{aligned} \left(\frac{R_L}{I_L}\right)^2 &= \left[\frac{E-F-G}{E+F+G}\right]^2 \\ \left(\frac{R_S}{I_L}\right)^2 &= \frac{4EF}{[E+F+G]^2} \\ \left(\frac{T_L}{I_L}\right)^2 &= \frac{4GE}{[E+F+G]^2} \end{aligned}} \tag{6.72}$$

where the following constants are used:

$$\begin{aligned} G &= \frac{\rho_2 \cdot C_{L2} \cdot \cos\alpha}{\rho_1 \cdot C_{L1} \cdot \cos\beta} \\ E &= (\cos 2\theta)^2 \\ F &= \left(\frac{C_{s1}}{C_{L1}}\right)^2 \cdot \sin 2\alpha \cdot \sin 2\theta \end{aligned} \tag{6.73}$$

The solution for a solid–solid interface is also given by Mayer [7].

6.5.3 An Exemplary Application

In order to demonstrate the possible applications of these equations, a noninvasive method for estimating bone density is introduced herein. The reduction in bone density is a common phenomenon that occurs in elderly population. However, due to a disease called osteoporosis, which affects mainly women, the rate by which bone density reduces is high. Consequently, the bone may become porous and brittle and can break easily. This may lead to complications, pain, and discomfort to the patient. Thus, a noninvasive tool for assessing bone density may detect the disease at early stages and may be used for monitoring the treatment.

A possible solution to the problem, based on a system for scanning flat bone surfaces, was suggested in Azhari [8]. The system comprised two transducers moving on a circular track. Both transducers were focused on the circle's central point, which was set to coincide with the bone surface. The first

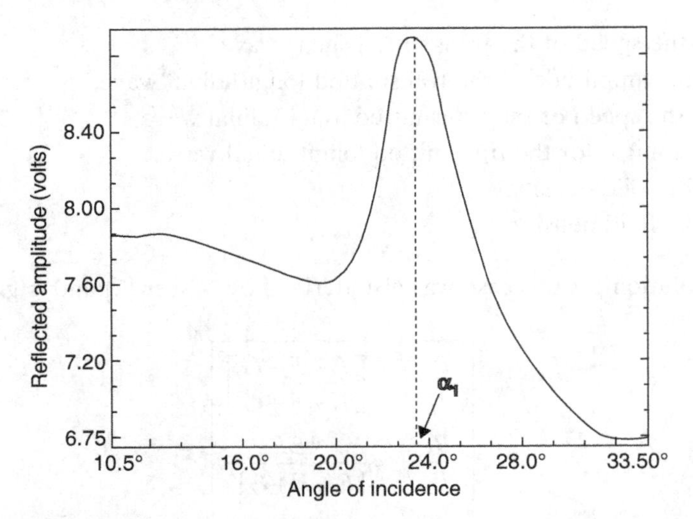

Figure 6.14. An exemplary scan of a bovine tibia bone taken from Azhari [8]. The amplitude of a reflected 5-MHz wave is depicted as a function of the angle of incidence. The first critical angle α_1 is clearly visible. The speed of sound for the longitudinal wave can be calculated from this angle using Snell's law.

ultrasonic transducer was used for transmitting waves toward the bone surface, and the other was used for reception of the reflected waves. The device was set so that both transducers were symmetrically positioned relative to the normal to the bone surface. Consequently, the angle of wave incidence and the angle of the reflected wave matched the lines connecting the transducers with the central point. The transducers scanned a range of angles. The output of the system was a plot depicting the reflected wave amplitude as a function of the angle of incidence.

The system was used for scanning human and bovine tibia bones immersed in water. An exemplary scan of a bovine bone is depicted in Fig. 6.14. As can be noted, the first critical angle manifested by a peak in reflection amplitude is clearly visible. From this angle the speed of sound (longitudinal waves) for the bone can be derived using Snell's law. Bone density can potentially be estimated by measuring the additional critical angles and by applying parameter estimation to Eq. (6.70).

REFERENCES

1. Sabersky RH, Acosta AJ, and Hauptmann EG, *Fluid Flow*, 2nd edition, Macmillan, New York, 1971.
2. Lipson SG, Lipson H, and Tannhauser DS, *Optical Physics*, 3rd edition, Cambridge University Press, New York, 1995.

3. Kinsler LE and Frey P, *Fundamentals of Acoustics*, John Wiley & Sons, New York, 1962.
4. Auld BA, *Acoustic Fields and Waves in Solids*, R. E. Krieger, Malabar, FL, 1990.
5. Beyer RT and Letcher SV, *Physical Ultrasonic*, Academic Press, New York, 1969.
6. Arenberg DL, Ultrasonic solid delay lines, *J Acoust Soc Am* **20**:1–26, 1948.
7. Mayer WG, Energy partition of ultrasonic waves at flat boundaries, *Ultrasonics* **3**(2):62–68, 1965.
8. Azhari H, An ultrasonic method for determination of bone density, M. Sc. Dissertation, Tel Aviv University, July 1984.

CHAPTER 7

ACOUSTIC LENSES AND MIRRORS

Synopsis: In this chapter the principles of acoustic lenses and mirrors will be presented. We shall start by reviewing the focusing principles used in optics and point out to the differences between optical and acoustic lenses. Then we shall derive the conditions needed for building ellipsoidal and spherical acoustic lenses. The inaccuracies in the focal region of spherical lenses will be discussed. And then the design principles of zone lenses will be introduced. Finally the principles of acoustic mirrors will be presented.

7.1 OPTICS

Let us start by understanding why do convex lenses focus light. Consider the prism shown in Fig. 7.1. A ray of light impinging upon a triangular prism with an incident angle of θ_1 will be deflected when entering the prism to an angle θ_2 according to Snell's law. The ray will continue its flight to the opposite wall where it will have an incident angle of θ_2'. When leaving the prism, it will be deflected again to an angle of θ_1'. The angular difference between the incident and exiting rays is δ. This angle δ is called the deflection angle. Let us now calculate its magnitude.

From basic geometry we know that δ is an external angle to the triangle formed by continuing the trajectories of the incident and exiting rays. Therefore, its value is equal to the two triangle angles that are not proximal to it:

Basics of Biomedical Ultrasound for Engineers, by Haim Azhari

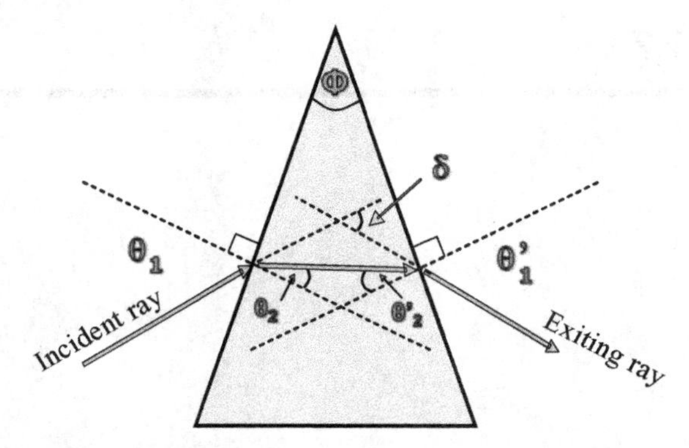

Figure 7.1. A light ray impinging upon a triangular prism with an incident angle of θ_1 will be deflected and leave the prism with an angle θ_1'.

$$\delta = (\theta_1 - \theta_2) + (\theta_1' - \theta_2') = \theta_1 + \theta_1' - (\theta_2 + \theta_2') \tag{7.1}$$

Also, we know that the sum of all four angles of the diamond shape formed by continuing the normal lines to the prism faces should equal 360°, thus,

$$\begin{aligned} &[180° - (\theta_2 + \theta_2')] + 90° + 90° + \phi = 360° \\ \Rightarrow \quad &\theta_2 + \theta_2' = \phi \end{aligned} \tag{7.2}$$

It follows then the deflection angle is given by

$$\Rightarrow \quad \delta = (\theta_1 + \theta_1') - \phi \tag{7.3}$$

For small incident angles ($\theta < 20°$) we can write the approximation $\sin\theta \approx \theta$. Thus, Snell's law can be written as

$$\frac{\sin\theta_1}{\sin\theta_2} \approx \frac{\theta_1}{\theta_2} = n \tag{7.4}$$

where n is the refractive index, which is defined as

$$n \equiv \frac{C_{\text{vacuum}}}{C_{\text{lens}}} \tag{7.5}$$

and C_{vacuum} is the speed of light in vacuum and C_{lens} is the speed of light in the lens.

If a horizontal beam of light impinges upon the left prism face and ϕ is relatively small, then θ_1 will be small and θ_2 will be even smaller. The same thing could be said regarding angles θ_1' and θ_2'. Hence the following approximation can be used:

$$\frac{\theta_1}{\theta_2} = n \quad \Rightarrow \quad \theta_1 = \theta_2 \cdot n \tag{7.6}$$

and

$$\frac{\theta_1'}{\theta_2'} = n \quad \Rightarrow \quad \theta_1' = \theta_2' \cdot n \tag{7.7}$$

Therefore

$$(\theta_1 + \theta_1') = n \cdot (\theta_2 + \theta_2') = n \cdot \phi \tag{7.8}$$

And the deflection angle δ is given by

$$\begin{aligned} &\delta = (\theta_1 + \theta_1') - \phi = n \cdot \phi - \phi \\ &\Rightarrow \delta = (n-1)\phi \end{aligned} \tag{7.9}$$

Studying this equation, we realize that the deflection angle is practically independent of the incident angle θ_1 (provided that $\theta_1 \leq 20°$). Thus, the deflection angle δ is linearly related to the prism head angle ϕ. So by increasing the head angle the deflection angle increases.

An optical biconvex lens, for example, can be approximated by virtually stacking many small slices taken from different prisms with a varying prism head angle as depicted in Fig. 7.2. Thus, for the central slice ϕ equals 0 and is getting larger and larger toward the edges of the lens (as depicted). Consequently, the deflection angle δ is negligible at the center and is increasing toward the lens edges. If the geometry is planed properly, parallel rays will be deflected from different parts of the lens so as to reach a single point located at distance F from the lens as depicted in Fig. 7.3.

Optical lenses made of polished glass which have two spherical surfaces (convexed or concaved) have been manufactured for centuries. The relation between the focal distance F and the radius of curvature for these surfaces has also been known for many years. This quantitative relation is known as the "Lens-maker's equation" and is given (for thin lenses) by [1]

$$\boxed{\frac{1}{F} \approx (n-1)\left(\frac{1}{R_1} + \frac{1}{R_2}\right)} \tag{7.10}$$

where F is the focal distance (note that $1/F$ is called "diopter" when F is measured in meters), n is the refractive index, and R_1 and R_2 are the two radiuses of curvature for the lens two surfaces as shown in Fig. 7.4.

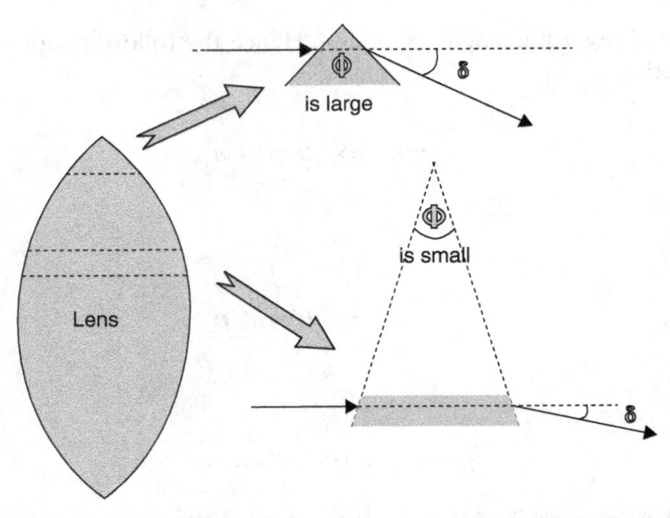

Figure 7.2. An optical biconvex lens can be approximated by virtually stacking many small slices taken from different prisms with a varying prism head angle ϕ. Thus, $\phi \to 0$ at the center and is largest at the edges as depicted.

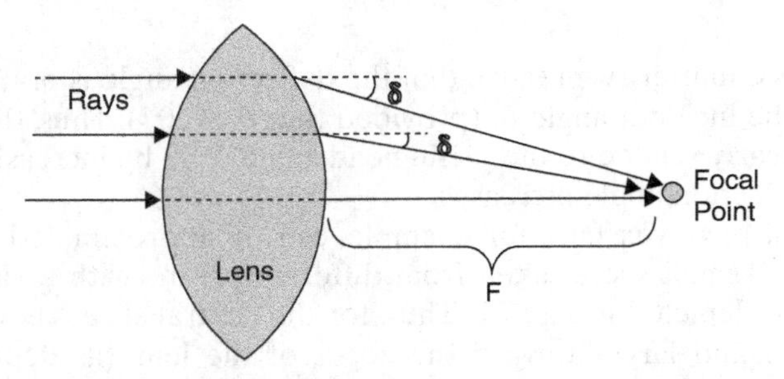

Figure 7.3. An optical lens focuses a parallel light beam to a single focal point by deflecting differently the individual light rays as depicted.

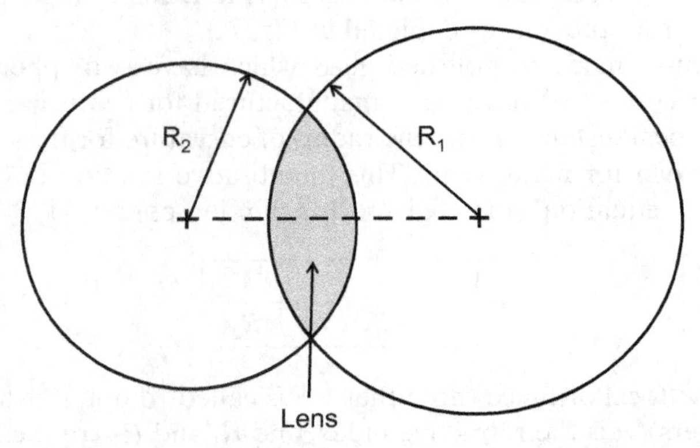

Figure 7.4. The common volume enclosed by two spherical surfaces forms the lens geometry, whose focal point is given by Eq. (7.10).

The use of the "Lens-maker's equation," Eq. (7.10), is based on the following signs convention of the lens radii R_1 and R_2: A *convex* surface is assigned a *positive* sign, and a *concaved* surface is assigned a *negative* sign. Also, a real focal point is marked as positive and an imaginary (virtual) focal point is marked as negative. For clarity, let us examine the following three examples assuming that $R_1 < R_2$ for all cases:

1. *A Biconcave Lens*

$$\frac{1}{f} = (n-1)\left(-\frac{1}{R_1} - \frac{1}{R_2}\right) < 0$$

Since $f < 0$

$\Rightarrow$ A scattering lens (imaginary focus)

2. *A Convex–Concave Lens Type I*

$$\frac{1}{f} = (n-1)\left(\frac{1}{R_1} - \frac{1}{R_2}\right) > 0$$

Since $f > 0$

$\Rightarrow$ A focusing lens(real focus)

3. *A Convex–Concave Lens Type II*

$$\frac{1}{f} = (n-1)\left(-\frac{1}{R_1} + \frac{1}{R_2}\right) < 0$$

Since $f < 0$

$\Rightarrow$ A scattering lens(imaginary focus)

To complete the picture, we need to refer to lenses such as those shown in Fig. 7.5, for which one of the surfaces is flat. For these cases we consider the planar surface as a curved surface with an infinitely large radius of curvature. Thus

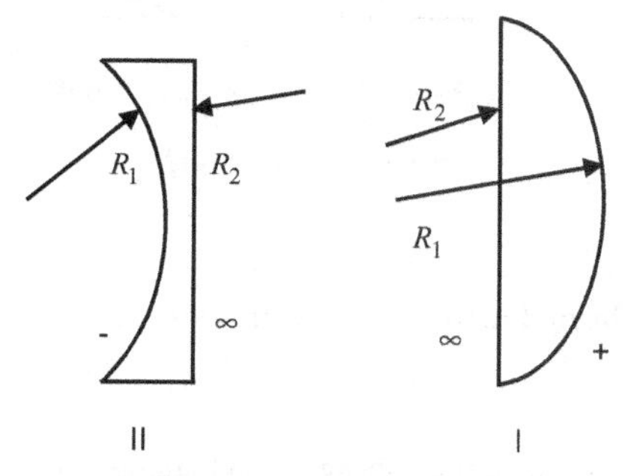

Figure 7.5. (Left) A concave–planar lens will scatter light. **(Right)** A planar–convex lens will focus light.

for the cases shown in Fig. 7.5 we can substitute into Eq. (7.10) the following value $1/R_2 = 0$.

Consequently, the focal distance is given by

$$\frac{1}{f} = (n-1)\frac{1}{R_1} \quad (7.11)$$

7.2 OPTICS AND ACOUSTICS

Commonly, in the context of biomedical applications, soft tissues can be closely approximated by water. Thus, when designing a lens or a mirror one may assume a typical speed of sound of $c_{\text{medium}} \simeq 1500\,\text{m/sec}$. Acoustic lenses are usually made of plastic materials* (i.e., "Perspex," for which $c_{\text{lense}} \simeq 2670\,\text{m/sec}$).

Consequently, the relative refractive index (i.e., the ratio: $n = c_{\text{medium}}/c_{\max}$) will be smaller than 1. As a result, we shall obtain from Snell's law $\theta_1 < \theta_2$ for the prism shown in Fig. 7.6.

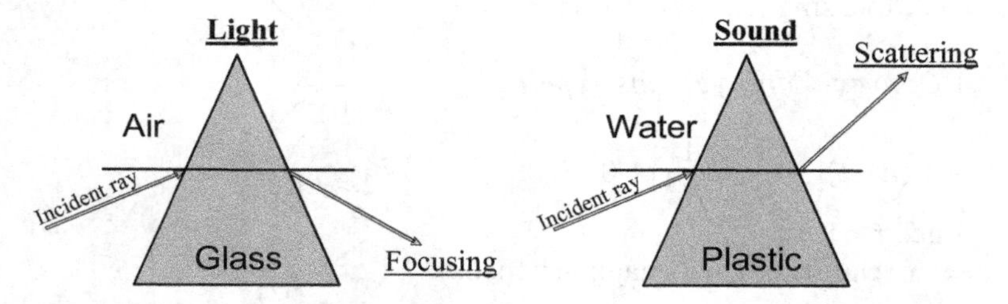

Figure 7.6. (Left) A glass prism will deflect a light ray in air in a manner that would enable focusing. **(Right)** A plastic prism will deflect a sound ray in water in a manner that would cause scatter.

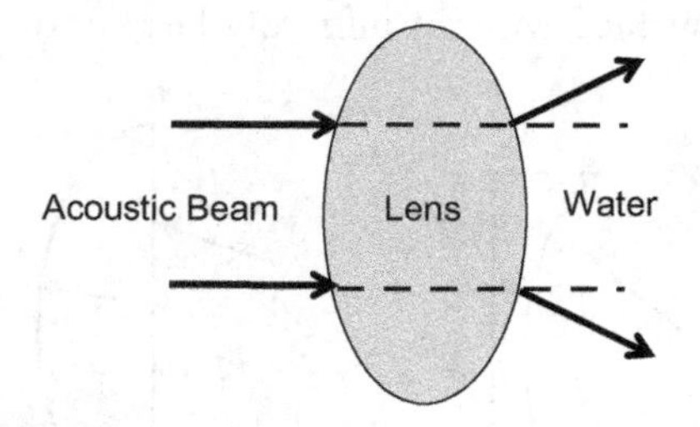

Figure 7.7. A plastic biconvex lens will scatter sound waves in water.

Note: Glass has relatively high acoustic impedance and attenuation. Hence, a glass lens may be less efficient.

It follows from the above that a biconvex plastic lens in water will scatter a sound beam as shown schematically in Fig. 7.7. Hence, for focusing sound waves in water or in soft tissues we must use *concave* lenses (assuming that the speed of sound in the lens is greater than that of the medium).

Another notable difference between sound and light optics is that there is no reference medium for sound waves. While for electromagnetic waves the propagation speed in vacuum serves as a reference (where its value is fixed and maximal), there is no reference speed for sound. Indeed, in medical applications it is common to use water as a reference. However, the speed of sound in water can change due to various parameters such as temperature, solvents concentrations, and so on. Thus, the refractive index is defined in acoustics as the reciprocal value of the medium's speed of sound:

$$\boxed{n = \frac{1}{C}} \tag{7.12}$$

Now that we have understood the basic differences between sound and light optics (in our context of course), let us investigate the conditions required to focus acoustic waves. We would like to know what the optimal contour is or what are the geometrical constraints for the lens–medium surface in order to achieve focusing.

Referring now to Fig. 7.8, consider a beam of sound rays (a spherical wave) that originates at point P on the left face of a planar–concave lens and travels toward the concaved surface. At point A on the boundary, it is deflected and reaches the focal point B in the medium of interest.

Clearly, in order to achieve focusing at point B, the time required for the ray to travel directly through the center along the distance $S_1 + S_2$, should be equal to the time needed for the other ray to travel through the path P–A–B, whose length is $d_1 + d_2$. Thus, if n_2 is the refractive index for the lens and n_1 is the refractive index for medium of interest, then the following equation should apply:

$$n_2 S_1 + n_1 S_2 = n_2 d_1 + n_1 d_2 \tag{7.13}$$

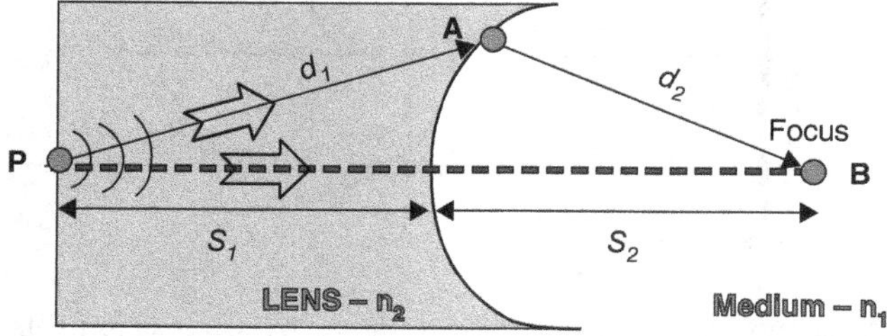

Figure 7.8. A spherical wave that originates at point P on the left face of a planar–concave lens is focused at point B in the medium.

The path $d_1 + d_2$ naturally varies according to the ray angle and the geometry of the lens. But on the other the path $S_1 + S_2$ is constant. Thus, the left-hand side of Eq. (7.13) must be constant. Consequently, the condition for focusing can be simply written as

$$\boxed{n_2 d_1 + n_1 d_2 = \text{const}} \tag{7.14}$$

Similarly, by observing Fig. 7.9, it can easily be concluded that a planar wave impinging upon the left-hand side of the lens must also fulfill the condition given by Eq. (7.14).

To find a suitable lens geometry for focusing planar waves, it was suggested to use a family of geometrical shapes called "Conics." This family of geometries is generated by intersecting a cone and a plane at different angles. An interesting property of this family is its fixed relation to a guiding line called "Directrix." Referring to Fig. 7.10, this ratio is given by

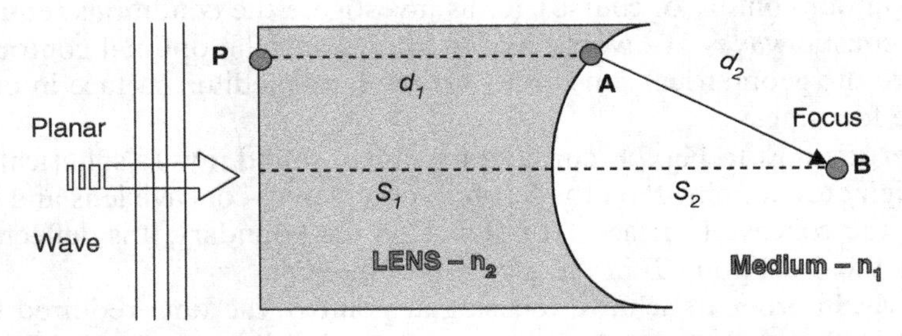

Figure 7.9. A planar wave reaches the left face of a planar–concave lens. All its acoustic "rays" must reach point B at the same time in order to achieve focusing.

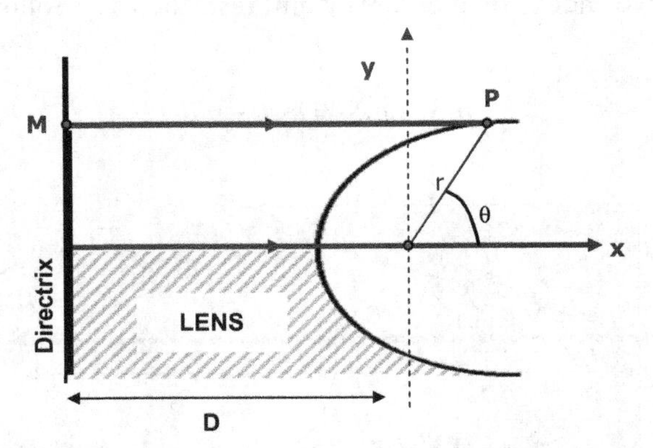

Figure 7.10. A planar–concave lens can be designed by using its left (planar) face as a Directrix line. The planar wave is assumed to propagate from left to right and its front is parallel to the lens plan.

$$\frac{r}{\overline{MP}} = \text{const} \triangleq \varepsilon \tag{7.15}$$

This property can be utilized for designing a planar–concave lens. If the coordinate system's point of origin is selected properly (Fig. 7.10), it can be shown (see, for example, reference 2) that the radius of the curve on the right-hand side is given by

$$\boxed{r(\theta) = \frac{\varepsilon \cdot D}{1 - \varepsilon \cdot \cos(\theta)}} \tag{7.16}$$

where D is the distance from the point of origin to the Directrix line, and ε is called the "eccentricity" of the curve.

We may obtain three different geometries according to the value of ε:

- If $\varepsilon < 0$, the shape is an ellipse.
- If $\varepsilon = 1$, the shape is a parabola.
- If $\varepsilon > 1$, the shape is a hyperbola.

Naturally, by rotating this curve around its axis of symmetry a corresponding three-dimensional shape is obtained. In the following we shall see how this family of curves relates to focusing devices.

7.3 AN ELLIPSOIDAL LENS

As stated above, in order to focus sound waves in water or soft tissues using a plastic lens, it is necessary that its geometry be concave. Hence, let us consider a plastic block from which an ellipsoidal shape has been removed. Consequently, a planar–concave shape is obtained as depicted in Fig. 7.11. As recalled, an ellipsoidal shape has two focal points F_1 and F_2.

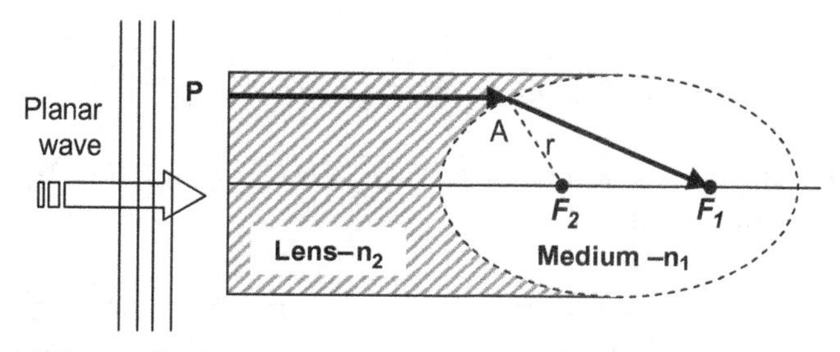

Figure 7.11. A planar–ellipsoidal lens will focus a planar wave at the ellipsoid's focal point F_1.

Consider the moment when the wave front reaches the lens's left face. An acoustic ray that penetrates the lens at point P will travel to point A, where it will be deflected and travel through the target medium. We would like to find out the required conditions for this arbitrary ray to reach point F_1. Taking the left face of the lens as the Directrix line and positioning the center of our coordinate system at F_2, it follows that the following condition must be fulfilled:

$$n_2 \cdot \overline{AP} + n_1 \cdot \overline{F_1A} = \text{const} \tag{7.17}$$

And since for an ellipsoidal shape the condition,

$$\overline{F_2A} = \varepsilon \cdot \overline{AP} \tag{7.18}$$

applies [Eq. (7.15)], we can substitute Eq. (7.18) into Eq. (7.17) and obtain

$$n_2\left(\frac{\overline{F_2A}}{\varepsilon}\right) + n_1\left(\overline{F_1A}\right) = \text{const} \tag{7.19}$$

Taking n_1 as a common factor yields

$$\Rightarrow \text{Left-hand side} = n_1\left[\overline{F_1A} + \frac{\overline{F_2A}}{\varepsilon}\frac{n_2}{n_1}\right] \tag{7.20}$$

As noted above, the speed of sound in the lens is higher than that of the medium; thus,

$$C_2 > C_1 \Rightarrow n_2 < n_1 \tag{7.21}$$

From this relation it follows that $n_2/n_1 < 1$. As recalled, for an ellipsoidal shape the eccentricity needs to be $\varepsilon < 1$. Thus, if we choose the eccentricity to be exactly $\varepsilon = n_2/n_1$, we shall obtain

$$\text{Left-hand side} = n_1\left[\overline{F_1A} + \overline{F_2A}\right] \tag{7.22}$$

But for an ellipse the relation $\overline{F_1A} + \overline{F_2A} = \text{Const}$ applies by definition! Therefore the conclusion is

> ∴ An ellipsoidal shape for which $\varepsilon = \dfrac{n_2}{n_1}$ will focus acoustic waves at F_1.

7.4 SPHERICAL LENSES

Although ellipsoidal lenses provide good focusing, they are more complicated to manufacture. Spherical lenses, on the other hand, can be easily produced with conventional workshop tools. This fact makes spherical lenses more attractive. However, their focusing capabilities are generally inferior to those of ellipsoidal lenses. In this section we would like to define the geometrical constrains for building a spherical lens and evaluating the quality of its focal zone. As stated above, we continue to assume that the speed of sound in the lens C_2 is higher than that of the medium C_1, and hence

$$\begin{aligned} & C_2 > C_1 \\ & n_1 = \frac{1}{C_1}, \qquad n_2 = \frac{1}{C_2} \\ & \Rightarrow n_2 < n_1 \end{aligned} \tag{7.23}$$

Consider a lens with a concaved spherical surface such as the one shown in Fig. 7.12. A source point P located within a concaved–spherical lens transmits spherical waves. The lens's spherical surface radius is R and the focal point is B. The center point for the spherical surface is at O. A ray traveling along an arbitrary path $\overline{PAB}$ forms an angle φ between the horizontal axis and line $\overline{AO}$. The time t required for the ray to reach the focal point P is given by

$$t = n_2 \cdot d_1 + n_1 \cdot d_2 \tag{7.24}$$

Applying the trigonometric law of cosines yields

$$\begin{aligned} t = & \left[(R+S_1)^2 + R^2 - 2R(R+S_1)\cos\varphi\right]^{1/2} n_2 + \\ & \qquad (\Delta AOP) \\ & \left[(S_2 - R)^2 + R^2 - 2R(S_2 - R)\cos(\pi - \varphi)\right]^{1/2} n_1 \\ & \qquad (\Delta BOA) \end{aligned} \tag{7.25}$$

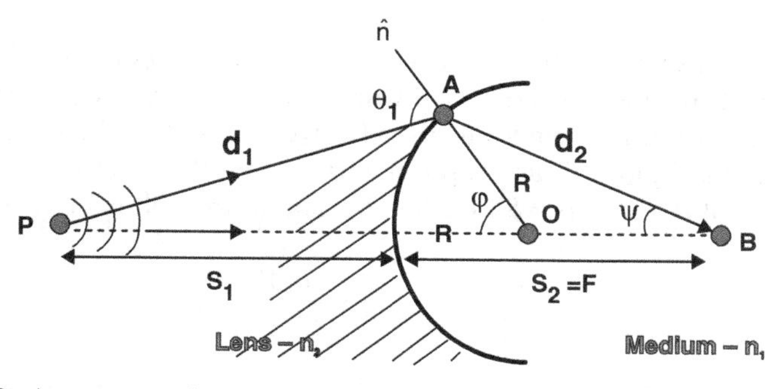

Figure 7.12. A source point P located within a concaved–spherical lens transmits spherical waves. The lens's spherical surface radius is R and the focus point is at point B.

In order to find the geometrical conditions for focusing the rays at point B, we shall apply Fermat's principle [3] (named after the French mathematician Pierre de Fermat), which states that

> "A ray of light (or sound) traveling between two points will travel through the path that can be traversed in the least time."

This law implies that the ray will reach the focal point B if the conditions (geometry and n_1, n_2) along the path which it travels through comply with the minimal time principle. And since the geometry is (supposedly) known, the only "free" variable is φ. Thus, in order to find the point of minima, we need to equate to zero the derivative of the travel time t [Eq. (7.25)] by φ:

$$\frac{dt}{d\varphi}=0 \tag{7.26}$$

which is explicitly given by

$$\frac{dt}{d\varphi}=\frac{n_2}{d_1}\frac{1}{2}[-2R(R+S_1)(-\sin\varphi)]+\frac{n_1}{d_2}\frac{1}{2}[-2R(S_2-R)(\sin\varphi)]=0 \tag{7.27}$$

or, after reordering,

$$\Rightarrow \quad \frac{n_2}{d_1}(R+S_1)-\frac{n_1}{d_2}(S_2-R)=0 \tag{7.28}$$

which can be written as

$$\Rightarrow \quad \frac{n_2}{d_1}+\frac{n_1}{d_2}=\frac{1}{R}\left(\frac{n_1 S_2}{d_2}-\frac{n_2 S_1}{d_1}\right) \tag{7.29}$$

At first glance, it seems as though this term is independent of φ, which implies that Fermat's principle is valid for all values of φ. However, one should recall that d_1 and d_2 are in fact functions of φ (as well as of R and S). At this point we shall make a significant assumption and assume that φ is so small that we can assume that $d_1 \approx S_1$; $d_2 \approx S_2$. Hence, we obtain the following condition:

$$\boxed{\frac{n_2}{S_1}+\frac{n_1}{S_2}\approx\frac{(n_1-n_2)}{R}} \tag{7.30}$$

As can be noted, this condition relates solely (under the above assumptions) to the focal distance and the surface radius of curvature.

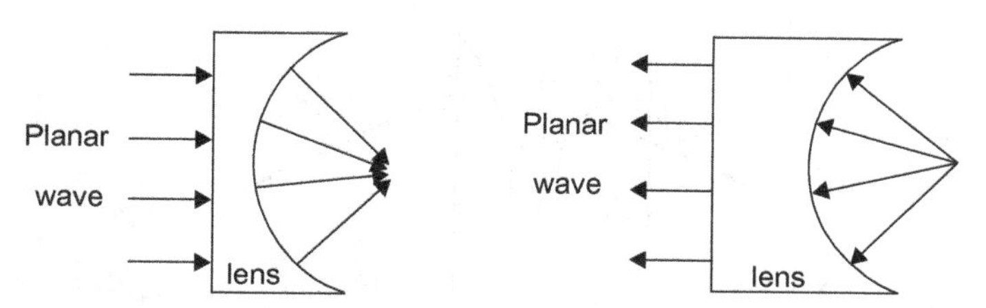

Figure 7.13. **(Left)** A spherical lens focuses a planar wave that impinges upon its flat surface. **(Right)** A spherical lens can transform a spherical wave that impinges upon its concaved surface into a planar wave.

Using Eq. (7.30), we can calculate the focal point for a planar–concaved–spherical lens such as the one shown in Fig. 7.13. As recalled this equation was developed for a point source (generating spherical waves). If we move the source to infinity (i.e., $S_1 \to \infty$), the radius of the transmitted spherical wave will be so large that the wave may be considered planar. Substituting this value into Eq. (7.30), the focal distance $S_2 \equiv F$ can be found from

$$\frac{n_1}{F} = \frac{n_1 - n_2}{R} \tag{7.31}$$

or, after rearrangement,

$$\boxed{F = \frac{R \cdot n_1}{n_1 - n_2}} \tag{7.32}$$

This equation may also be used by reversing the rays directions (as shown in Fig. 7.13). By so doing, we may transform a spherical wave into a planar one.

When using this equation for designing acoustic lenses, it is imperative to recall that its derivation was based on the assumption that $d \approx S$, which limits the value of the angle φ, and consequently ψ, as shown in Fig. 7.12. As the value of ψ increases, the rays arriving to the focal point will be slightly at off phase relative to the ray traveling along the center line. As a result, the quality of the focus will be degraded. The maximal value for ψ which limits the maximal off phase angle to $\pi/2$ is given in Wells [4]:

$$\lambda_1 \le \frac{F^2 \tan^4 \psi}{2Rn(n - 1 + n\tan^2 \psi)} \tag{7.33}$$

where $n = n_1/n_2$ and λ_1 is the wavelength in the medium.

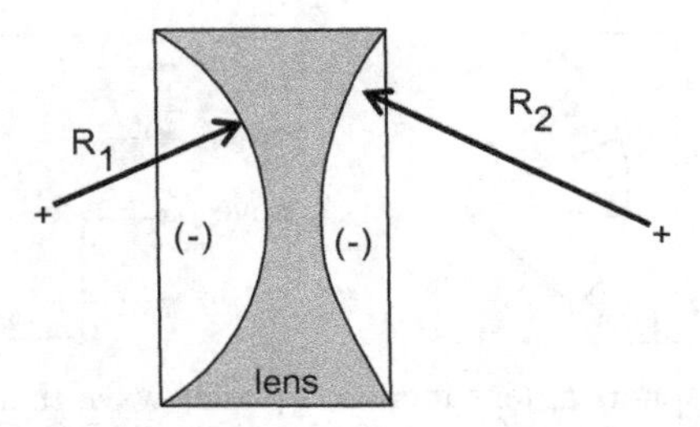

Figure 7.14. A bi-concave spherical lens can focus planar acoustic waves, provided that the speed of sound is higher in the lens than in the medium.

7.4.1 A Bi-Concave Lens

A bi-concave lens, such as the one depicted in Fig. 7.14, can be used for focusing planar acoustic waves in a similar manner by which a bi-convex lens focuses light.

As was shown by T. Tranóczy [5], a bi-concave lens that is constructed from two spherical surfaces can be used for that purpose. Again it is assumed that the speed of sound is higher in the lens than in the medium. The focal length F for such a lens is given by the following equation:

$$F = \frac{R_2}{1-(C_1/C_2)} \times \frac{R_1}{D[C_2/C_1-1]+R_1-R_2)} \tag{7.34}$$

where R_1 and R_2 are the radius of curvature for the two spherical surface as shown in Fig. 7.14. The diameter of the lens is D and the speed of sound is C_1 and C_2 for the medium and the lens, respectively.

When ignoring the attenuation, the pressure gain at the focal point was calculated by L. D. Rosenberg [5] and is given by the formula

$$\Rightarrow \quad G_p = \frac{\pi D^2}{2\lambda F} \times \frac{2Z_1 \cdot Z_2}{(Z_1+Z_2)^2} \tag{7.35}$$

where λ is the wavelength in the medium, and Z_1 and Z_2 are the acoustic impedances for the medium and the lens, respectively.

7.4.2 Focal Point Properties

As shown above, spherical lenses can focus sound waves only when certain conditions and approximations are made. Consequently, the quality of the focus is degraded (relative to an ellipsoidal shape). Instead of obtaining a focal point, these lenses create an elongated ellipsoid ("cigar")-shaped focal zone such as the one shown schematically in Fig. 7.15.

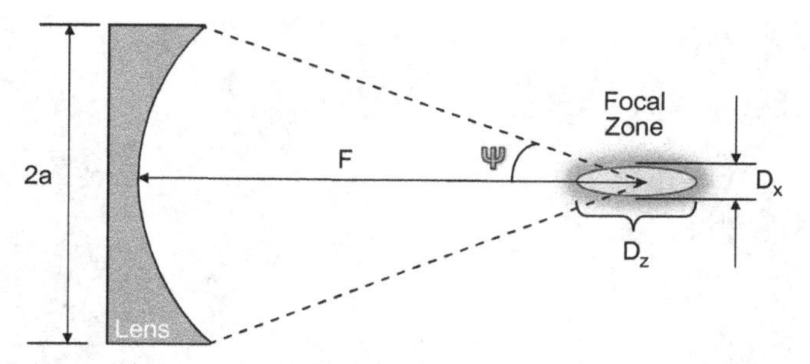

Figure 7.15. The focal zone for a spherical lens is an elongated ellipsoid ("cigar"-shaped).

As was shown by Fry and Dunn [6], the 3-db short diameter for the focal zone is given by

$$D_x = D_y \approx \frac{K_t \lambda F}{2a} = K_t \lambda \cdot \left[\frac{F}{2a}\right] \tag{7.36}$$

where F is the focal length, λ is the wave length in the medium, and $2a$ is the diameter of the lens. The term in the squared brackets is nondimensional and is called the "F-number." The factor K_t is also dimensionless and depends on the aperture angle ψ. For small angles (i.e., $\psi < 50°$), the value of this factor may be assumed to be $K_t = 1$. Under this condition the accuracy of Eq. (7.36) is about 20%.

The length of the focal zone also depends on the aperture angle. And for $\psi < 50°$ the following approximation can be used:

$$\begin{aligned} D_z &\approx K_a \cdot D_x \\ K_a &\approx 15(1 - 0.01\psi) \end{aligned} \tag{7.37}$$

where the angle ψ is given in degrees and the accuracy in that case is smaller than 15%.

The ratio between the transmitted intensity I_0 and the intensity at the focal point I_F is called the "gain factor" (G). If the losses caused by attenuation within the lens are ignored and for small aperture angles (i.e., $\psi < 15°$), the gain factor is given by

$$G = \frac{I_F}{I_0} = \frac{\pi}{4}\left(\frac{2a}{D_y}\right)^2 \tag{7.38}$$

Or by substituting the value for D_y from Eq. (7.36) the gain factor is given by

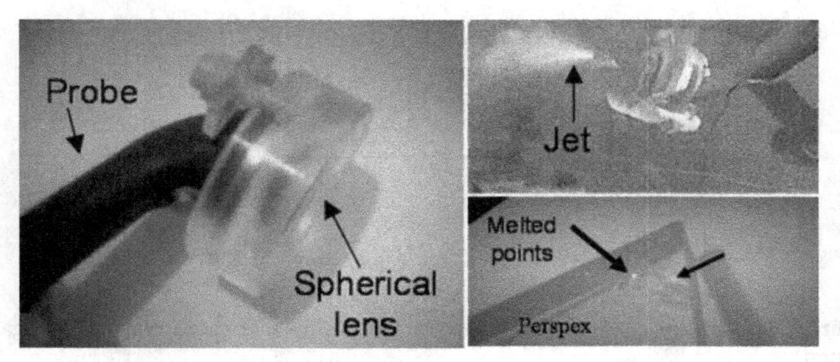

Figure 7.16. A spherical acoustic lens was attached to a physiotherapy probe **(left)**. The energy at the focal point was high enough to create a jet of water **(top right)** and melt a Perspex target when placed at its focal point **(bottom right)**. The melted points are marked by the arrows.

$$G = \frac{\pi}{4}\left(\frac{2a}{D_y}\right)^2 = \frac{\pi}{4}\left(\frac{4a^2}{\lambda \cdot F}\right)^2$$
$$\Rightarrow \boxed{G = 4\pi\left(\frac{a^2}{\lambda \cdot F}\right)^2} \tag{7.39}$$

An Example. To demonstrate the application of the formulas given in this chapter, a Perspex lens was built as shown in Fig. 7.16. The lens was attached to a physiotherapy probe that transmitted an intensity of $I_0 = 2.2\,\text{W/cm}^2$ (considered harmless). The lens has created a jet when placed in a water tank. When a Perspex target was placed at the focal point, it locally melted it as shown.

7.5 ZONE LENSES

In the previous sections we have demonstrated how ellipsoidal and spherical shapes can be utilized for building lenses. However, as can be noted from Figs. 7.11 and 7.13, these lenses are thin at the center and get thicker as the distance from the center increases. Consequently, waves passing closer to the lens's edge will be attenuated substantially. This may result in increased power loss of the beam and warming of the lens. Moreover, destructive interference may occur, further reducing the wave transfer efficiency. In his article, Tranóczy [5], has shown that the differences in power transmission between zones of different thicknesses may be as high as 3 db. This of course may degrade the quality of the focus. On the other hand, we know that constructive interference occurs when a wave passes through a medium for which the thickness equals to an integer multiplication (m) of half the wavelength, that is, $m \cdot \lambda/2$. As an alter-

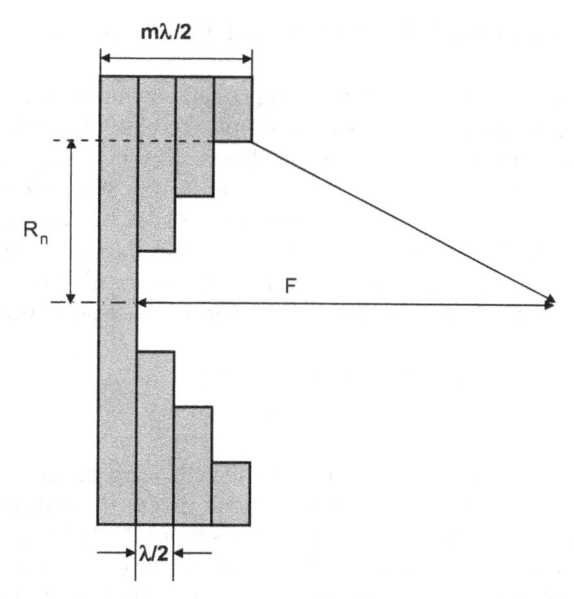

Figure 7.17. Suggested design for a zone lens. Each ring has an inner radius of R_n and a thickness of $\lambda/2$.

native geometry, Tranóczy [5] has presented a lens that is constructed from a set of stepped rings placed one atop the other as shown in Fig. 7.17. (Actually this lens could be manufactured from a single block.) The thickness of each ring is exactly one-half the wavelength. (This design resembles the Fresnel lens except for the surface curvature.)

In order to focus planar waves at a focal distance F, it is necessary that every wave leaving a ring for which the radius is R_n and the thickness is $m\lambda/2$ will reach the focal point at the same time (or with the same phase). By the way, the first ring could have a thickness of 0 (i.e., simply a hole). This condition leads to the following relation:

$$\frac{F}{C_2} = \frac{m\lambda}{2C_1} + \frac{1}{C_2} \cdot \sqrt{R_n^{\,2} + \left(F - \frac{m\lambda}{2}\right)^2} \tag{7.40}$$

where C_2 is the speed of sound in the lens, which is higher than the speed of sound for the medium C_1. And since the focal distance and the speeds of sound are known, the only unknown is the ring's radius R_n, which can be determined from

$$\boxed{R_n = \sqrt{\left(F - \frac{m\lambda C_2}{2C_1}\right)^2 - \left(F - \frac{m\lambda}{2}\right)^2}} \tag{7.41}$$

7.6 ACOUSTIC MIRRORS (FOCUSING REFLECTORS)

Sound waves can also be focused by using special acoustic "mirrors." A mirror is defined as a reflecting surface with a significantly high reflective index. For that purpose we can use a material that has a very high acoustic impedance or a very low one relative to the medium under consideration. The acoustic impedance for water and soft tissues is on the order of 1.5×10^6 (kg/sec·m^2). Thus, stainless steel, which has an acoustic impedance of about 45×10^6 (kg/sec·m^2), can be considered as a suitable material. The corresponding amplitude (pressure) reflection coefficient for a water–steel boundary is

$$R = \frac{(Z_2 - Z_1)}{(Z_2 + Z_1)} = \frac{(45 - 1.5)}{(45 + 1.5)} = 0.935 \tag{7.42}$$

In order to focus waves at a single point, the reflecting surface geometry must be selected so that all reflected rays will reach this point with the same phase. A suitable candidate is the family of conics shapes. Referring to Fig. 7.18, we choose a concaved surface for which the Directrix (guide line) is located at distance D from the focal point, as shown. A planar wave approaches the reflecting surface from the right side. Without any loss of generality, let us consider the moment at which the wave front is also located at a distance D from the focal point.

Consider an acoustic ray in the propagating planar wave which reaches point B on the reflecting surface. If the surface geometry is properly chosen,

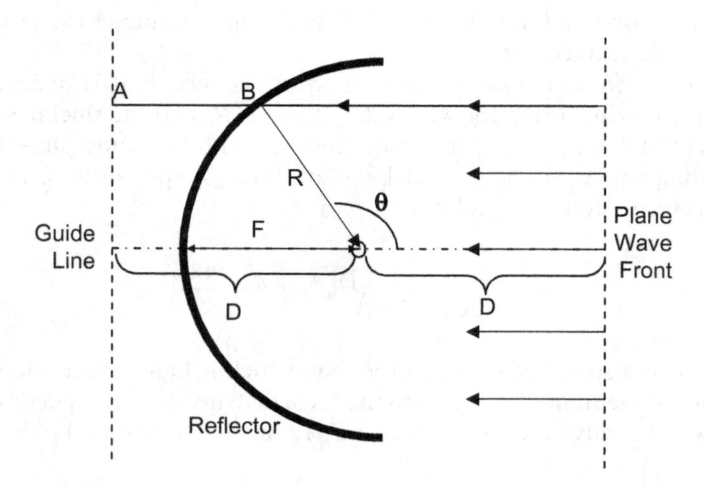

Figure 7.18. An acoustic mirror is constructed from a conics concave shaped reflecting surface. The Directrix line is located at distance D from the focal point O. For simplicity it is assumed that the front of the impinging planar wave is also located at distance D from the focal point as shown.

it will be reflected and will reach point O. The distance $\overline{AB}$ designates the distance to the Directrix. As recalled from Eq. (7.15), the following relation applies for a Conics-type surface:

$$\frac{R(\theta)}{\overline{AB}} = \text{const} \triangleq \varepsilon \tag{7.43}$$

where ε is the eccentricity index for the specific geometry. The radial distance $R(\theta)$ is given by Eq. (7.16). The time required for that ray to travel from the wave front to point B and to focal point O is given by

$$\begin{aligned} t &= \frac{1}{C}[(2D - AB) + R(\theta)] = \frac{1}{C}\left[2D - \frac{R(\theta)}{\varepsilon} + R(\theta)\right] \\ &= \frac{1}{C}\left[2D + R(\theta)\cdot\left(1 - \frac{1}{\varepsilon}\right)\right] \end{aligned} \tag{7.44}$$

where C is the speed of sound in the medium. In order to achieve focusing this time should be identical for all the rays constituting the wave. This implies that the geometry should be chosen so as to cancel the dependency on the angle θ. As explained at Section 7.2 above the eccentricity index for a parabolic surface is $\varepsilon = 1$. Thus, if we choose this geometry for the reflector under consideration, we obtain

$$t = \frac{1}{C}\left[2D + R(\theta)\cdot\left(1 - \frac{1}{1}\right)\right] = \frac{2D}{C} = \text{Const.} \tag{7.45}$$

This implies that the traveling time is constant for all the rays in the impinging beam. Hence, it can be stated that:

> A reflecting parabolic surface can serve as a focusing element!

The distance F to the focal point can be determined from Eq. (7.16) by substituting the values $\theta = \pi$ and $\varepsilon = 1$. By so doing, we obtain

$$\boxed{F = R(\theta = \pi) = \frac{D}{1 - \cos(\pi)} = \frac{D}{2}} \tag{7.46}$$

The pressure gain at the focal point was calculated by L. D. Rosenberg [5] and is approximately given by

$$G_p \approx \frac{\pi}{4}\left(\frac{a^2}{\lambda \cdot F}\right) \tag{7.47}$$

where a is the mirror's aperture and the complementary angle should be relatively small, that is, $180° - \theta \leq 30°$. As was shown by Tranóczy [5], the maximal intensity gain is obtained for $180° - \theta = 114°$ and is given by

$$G \approx \frac{1.5 \cdot \pi^2 \cdot F^2}{\lambda^2} \tag{7.48}$$

In some applications there is a need for creating a "silent zone" near the focusing surface—for example, if we wish to construct a system that combines a high-intensity therapeutic probe with an imaging probe or if we wish to bypass a sensitive organ (e.g., a major blood vessel or a bone). In such cases an assembly of reflecting surfaces can be devised—for example, by combining a hyperboling reflecting surface for scattering the waves and a complementary ellipsoidal surface for focusing them, or by combining a conical reflector and a parabolic one as depicted in Fig. 7.19.

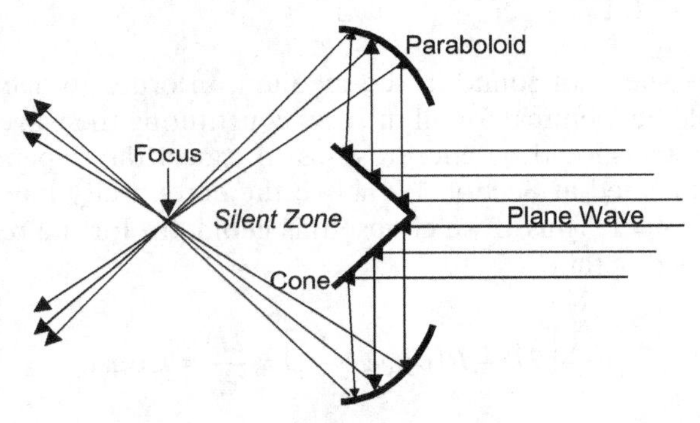

Figure 7.19. A reflecting assembly comprising a cone and a parabolic surface can be used for bypassing a sensitive region. The planar wave is deflected by the cone and focused by the surrounding parabolic surface. A "silent" zone is thus generated behind the reflecting cone.

REFERENCES

1. Haber-Schaim U, Dodge JH, Gardner R, Shore EA, *PSSC Physics*, 7th Ed., Kendall/Hunt Publishing Company, Dubuque, IA, 1991.
2. Spiegel MR, *Mathematical Handbook of Formulas and Tables*, Schaum's Outline Series in Mathematics, McGraw-Hill, New York 1968.
3. Lipson SG, Lipson H, and Tannhauser DS, *Optical Physics*, 3rd edition, Cambridge University Press, New York, 1995.
4. Wells PNT, *Biomedical Ultrasonics*, Academic Press, New York, 1977.
5. Tranóczy T, Sound focusing lenses and waveguides, *Ultrasonics* **vol. 3**, 115–127, 1965.
6. Fry WJ and Dunn F, Ultrasound: Analysis and experimental methods in biological research, in *Physical Techniques in Biological Research*, Vol. IV, Nastuk WL, editor, Academic Press, New York, 1962, pp. 261–394.

CHAPTER 8

TRANSDUCERS AND ACOUSTIC FIELDS

Synopsis: In this chapter the principles of ultrasonic transducers are explained. The corresponding acoustic fields of such transducers are analyzed in details. Basic terms such as "near field" and "far field" are introduced. The relationships between the transducer's size and frequency and directivity of the acoustic beam are presented. The principles of phased array transducers and their acoustic fields are explained. The principles of electronic beam steering, beam forming, and beam focusing are presented and the limitations of these procedures are discussed.

8.1 PIEZOELECTRIC TRANSDUCERS

There are many ways by which acoustic waves can be generated. In fact, any change in the pressure field or the stresses within the medium will induce such waves. Commonly, it is desirable to find such a source that would enable us to control the transmitted signal on the one hand but would also be sensitive enough to detect waves or echoes on the other hand. Piezoelectric materials are most suitable for this task. They have been used in ultrasonic applications for many decades [1], and their variety and new uses are steadily growing [2, 3]. The ultrasonic transducers used today are almost always made of piezoelectric materials. These materials can be shaped in various geometries, starting with cylindrical disc-type shapes, through rectangular transducers or an array of such, and on to flexible or tailor-made geometries. Their size also

Basics of Biomedical Ultrasound for Engineers, by Haim Azhari

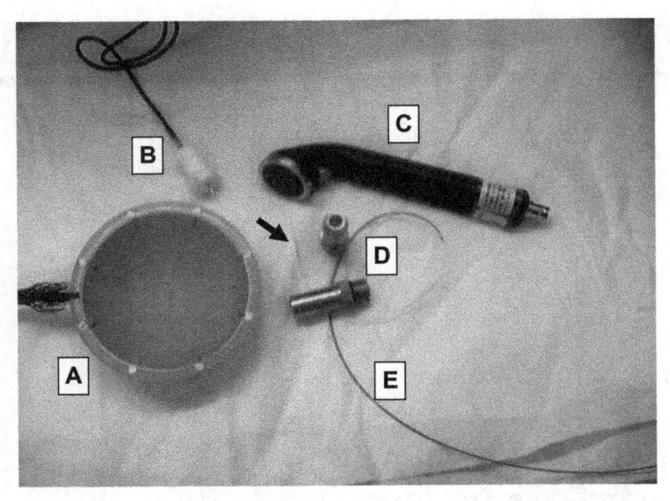

Figure 8.1. Exemplary collection of ultrasonic transducers: **(A)** High-intensity focused (HIFU) therapeutic transducer with a 100-mm diameter. **(B)** Needle-type hydrophone within a plastic housing. **(C)** Medium-sized physiotherapy transducer. **(D)** Two commercial laboratory-type transducers. **(E)** Intravascular ultrasound (IVUS) catheter. (The transducer is located near its tip and is marked by an arrow.)

varies within a very wide range. There are miniature transducers smaller than a needle head, which are assembled on a tip of a catheter or implanted within the body. And there are very large transducers that are designed to deliver high power for therapeutic applications (see Chapter 12). In short, the shape and size varies according to the specific application. An exemplary collection of transducers is depicted in Fig. 8.1. The basic principle upon which these piezoelectric transducers are based is the applied voltage-deformation duality, which is schematically depicted in Fig. 8.2.

If voltage is applied on two opposite faces of a piezoelectric material, a mechanical deformation (shortening, elongation or shear) is produced. On the other hand, if its opposite faces are mechanically deformed, a voltage is generated between the two faces of the transducer. Thus, if we take a piezoelectric material and solder two electrodes to its opposite faces as schematically shown in Fig. 8.3, we can cause it to deform and move its faces back and forth (or transversally in certain crystals). This motion will generate an acoustic wave that will be transmitted into the adjacent medium. On the other hand, when an acoustic wave impinges upon the transducer's surface, the corresponding change in pressure induces deformation and consequently generates voltage in the electrodes. Hence, the transducer can serve as both a detector and a receiver of acoustic waves. It is important to note that some transducers move transversally when voltage is applied to their surfaces. These transducers can be used for transmitting or receiving shear waves.

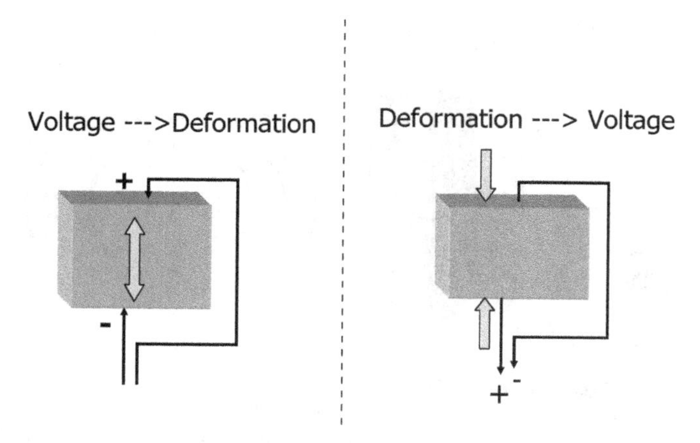

Figure 8.2. **(Left)** A piezoelectric transducer will deform mechanically and change its dimensions if voltage is applied on its opposite faces. **(Right)** The transducer will generate voltage if it's opposite faces are mechanically deformed.

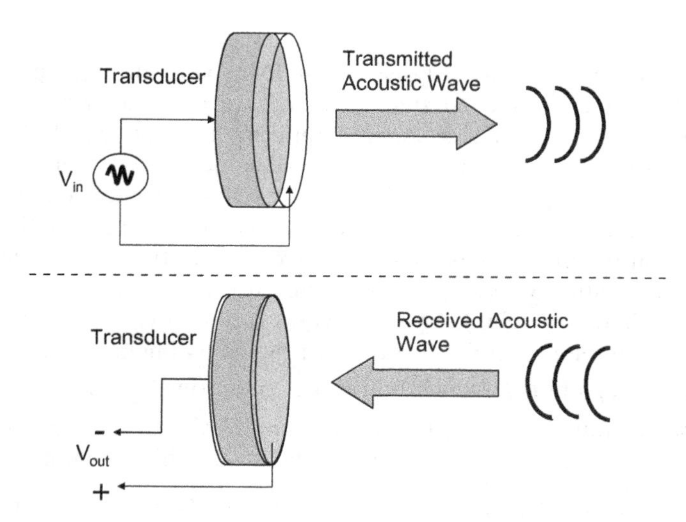

Figure 8.3. A piezoelectric material can operate as a transmitter **(top)** if voltage is applied to its opposite faces, or it can operate as a receiver by transforming the pressure applied on its faces into voltage **(bottom)**.

In order to achieve better efficiency, two additional layers are commonly bonded from both sides of the piezoelectric element, forming a sandwich structure such as the one shown in Fig. 8.4. At the front side facing the medium into which and/or from which acoustic waves are transmitted or received, an impedance matching layer is positioned. The purpose of this layer is to maxi-

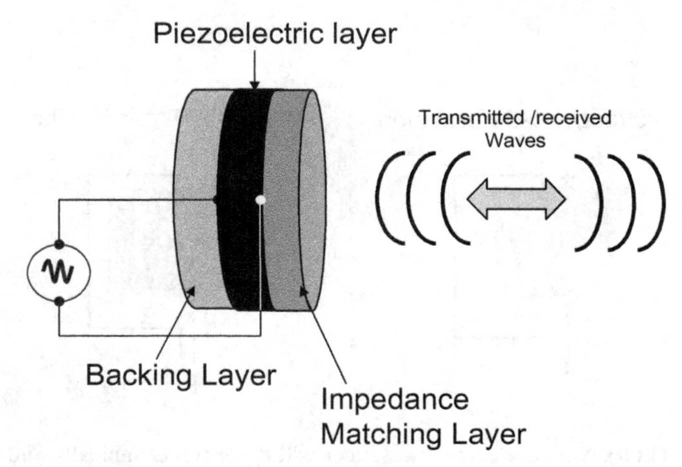

Figure 8.4. An ultrasonic transducer is commonly comprised of three layers: the piezoelectric crystal, which is located in the middle; an impedance matching layer facing the medium; and a backing layer that absorbs the energy transmitted to the back side of the transducer.

mize the efficiency of wave transmission from and into the piezoelectric element. This layer is needed since most of the piezoelectric elements are constructed of solid materials, such as ceramic crystals, for which the acoustic impedance is substantially higher than the acoustic impedance of soft tissues. Hence, if no matching layer is used, the energy transfer between the medium and the piezoelectric element will be very inefficient. As a result, the element will transform the energy into heat when working in transmit mode, and the signal-to-noise ratio (SNR) will be low when functioning in a receive mode.

The impedance matching layer provides a solution to the problem (as explained in Chapter 6). If the thickness of the matching layer attached to the piezoelectric element is L, which fulfills the condition $\boxed{L = (2n-1)\lambda/4}$, where n is an integer number and λ is the wavelength within this layer, and if the acoustic impedance of the matching layer is Z, which complies with the relation $\boxed{Z = \sqrt{Z_{\text{piezo}} \cdot Z_{\text{medium}}}}$, then, as shown in Chapter 6, the intensity transfer coefficient equals 1, which implies that all the acoustic energy will be transferred from the element to the medium and vice versa, despite the difference in their acoustic impedances.

The function of the backing layer is to absorb energy transmitted to the back (the "wrong" side). For that purpose a suitable material with proper acoustic impedance and attenuation coefficient is used to avoid reverberations of the piezoelectric element. This layer damps mechanically the oscillations; hence it may slightly reduce the transducer's sensitivity. Alternatively, an electric circuit is used to damp the oscillations electronically (see the following). In addition to that, a metal housing is commonly added to provide mechanical rigidity to the probe structure. The case is usually grounded, and coaxial cables are used to provide electric shielding and reduce the signal noise.

As stated above, a small electric circuit is usually connected in parallel to the piezoelectric element. This is commonly comprised of an RLC-type circuit with adjustable components. The purpose of this resonating circuit is to damp the piezoelectric element oscillations after transmission of an acoustic pulse. This is achieved by electrically loading the piezoelectric faces. This damping is particularly important in ultrasonic imaging, where a short pulse is desirable (see Chapter 9). Generally speaking, the shorter the transmitted pulse, the better the axial resolution.

As for the piezoelectric crystals from which the transducers are made, they may be obtained from natural crystals or manufactured compounds. In the past [1], natural crystals such as quartz and "Rochelle salt" (sodium potassium tartrate tetrahydrate) were used. Today, ceramic crystals containing lead zirconate titanate, which is called PZT for short, are the most popular source material. Also, piezoelectric polymers such as polyvinylidene difluride (PVDF) are currently available. Using modern materials, transducers can be manufactured in almost every shape and size including flexible devices and fibers. Interestingly, bones are also piezoelectric in nature.

The resonance frequency of a piezoelectric crystal is determined by its mechanical properties and density (which define the speed of sound in the crystal) and by its geometry. For example, consider a disc-shaped element that is designed for transmitting longitudinal waves. As was explained in Chapter 2, for a one-dimensional problem the ratio between the speed of sound and the thickness of the element determines the natural vibration frequencies [see Eq. (2.23)]. As explained there, standing waves are generated within the medium when the length of the medium (in this case its thickness) is an integer multiplication of half the wavelength. Similarly, if we take for example a piezoelectric crystal made of PZT4 (for which the speed of sound is $C = 4000\,\text{m/sec}$) and choose a thickness of 2 mm, a standing wave for which the wavelength is $\lambda = 4\,\text{mm}$ ($n = 1$) will be generated. Hence, the basic resonance frequency will be 1 MHz.

The crystal's resonance quality is determined by an index called "Q factor". The value of this Q factor is calculated (see Fig. 8.5) from

$$Q = \frac{f_0}{\Delta f} \tag{8.1}$$

where Δf is the spectral bandwidth at 50% the maximal value [full width at half-maximum (FWHM)]. It is worth noting that when using the spectral power, Δf corresponds to a bandwidth at $\sqrt{2}/2$, the maximal amplitude (recall that the power is proportional to the squared value of the amplitude). On the other hand, sometimes Δf is estimated by taking the band at 50% the maximal amplitude. Thus, when buying a transducer, one should pay

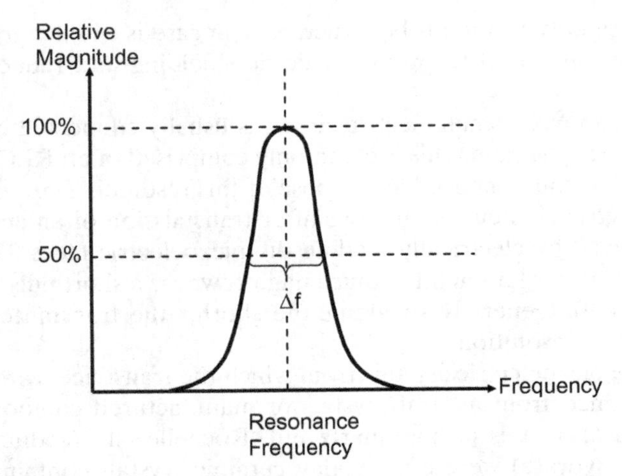

Figure 8.5. The "Q factor" for a given transducer is determined by the ratio of its resonance frequency and its spectral band (at its full width at half-maximum value).

attention to the exact definitions used by the manufacturer to characterize its product.

As can be noted from Eq. (8.1), the value of the Q factor is inversely proportional to the bandwidth. This implies that a narrow band transducer has a large Q factor and vice versa. Hence, for therapeutic applications it is preferred to use a transducer with a large Q factor. On the other hand, for imaging it is preferable to use a transducer that is suitable for transmitting short pulses. Thus, due to the time–frequency inverse duality a broadband and a small Q factor are desired.

8.2 THE ACOUSTIC FIELD

Thus far, we have discussed the propagation of acoustic waves without considering the geometry of their source. Most derivations were based on the assumption that the wave is planar and has an infinite wave front. In reality, however, wave sources have finite dimensions. Consequently, the acoustic field stemming from each source is also finite and varies in space. In order to calculate the acoustic field the properties of the transducer must be accounted for. A common approach for calculating the acoustic field of a general geometry source is to use the particles velocity potential function introduced in Chapter 6. For example (see Beyer and Letcher [4]), it can be assumed that the velocity potential function at the surface of a source (under certain conditions) is given by

$$\varphi_p(x, y, z) = -\frac{1}{2\pi} \cdot \iint_{\text{surface}} \frac{e^{-jkr}}{r} \cdot \frac{\partial \varphi}{\partial n} ds \tag{8.2}$$

where $\varphi_p(x, y, z)$ is the velocity potential function at an arbitrary point $P(x, y, z)$ within the field, and $\partial\varphi/\partial n$ is its derivative relative to the surface normal. For a harmonic wave of frequency ω assuming that the surface of the source moves with a uniform velocity $\dot{\xi}$, this derivative is given by

$$\begin{aligned} &\frac{\partial \varphi}{\partial n} = \dot{\xi} = \dot{\xi}_0 \cdot e^{j\omega t} \qquad \text{On the surface} \\ &\frac{\partial \varphi}{\partial n} = 0 \qquad \text{else} \end{aligned} \tag{8.3}$$

The corresponding pressure is therefore given (see Chapter 6 Eq. (6.7)) by

$$\mathbb{P} = -\rho_0 \cdot \frac{\partial \varphi}{\partial t} \tag{8.4}$$

In this book, however, we shall employ a different approach using Huygens' Principle, as will be explained in the following.

8.3 THE FIELD OF A POINT SOURCE

For clarity let us again rephrase Huygens' Principle [5], which states that: *Every source of waves for which a is much less than λ may be considered as a source for a spherical wave*! Where a is the size of the source and λ is the corresponding wavelength. If we employ this principle, we can describe any wave source by using a very large (in fact infinite) number of point sources, each of which is assumed to transmit a spherical wave.

As recalled, the intensity of an acoustic wave is defined as the power per unit area. For a point source transmitting a spherical wave of power W, the intensity at distance r is given by

$$I = \frac{W}{4\pi r^2} \tag{8.5}$$

To avoid the singularity at $r = 0$, let us define a reference distance r_0 at which the reference intensity is I_0 (see Fig. 8.6). Thus, the following relation holds:

$$\frac{I}{I_0} = \frac{r_0^2}{r^2} \tag{8.6}$$

Therefore Eq. (8.7) also holds:

$$\Rightarrow \frac{I}{I_0} = \frac{P^2}{P_0^2} \Rightarrow \frac{P}{P_0} = \frac{r_0}{r} \tag{8.7}$$

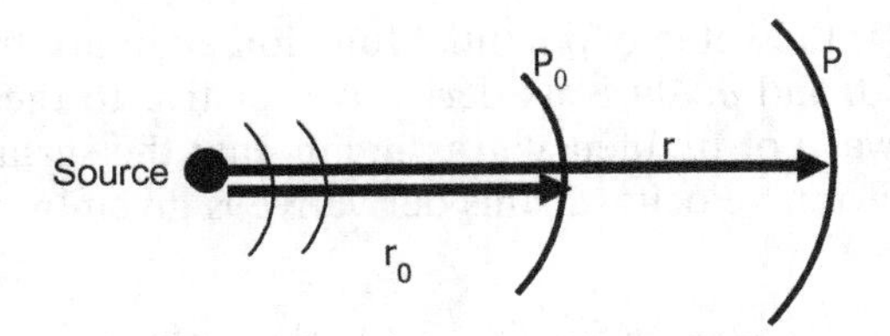

Figure 8.6. A point source transmits a spherical wave for which the reference pressure P_0 is known at a reference distance r_0.

As recalled, in Chapter 6 a velocity potential function φ was defined for a spherical harmonic wave of angular frequency ω. This function fulfills the relation: $\partial\varphi/\partial r = U$; and as was shown in Eq. (6.17), the corresponding pressure is given by

$$P = \frac{j\omega\rho_0\varphi_0}{r} e^{j(\omega t - kr)} \tag{8.8}$$

In order to extend this term from a *point* source to an infinitesimally small *surface element* of area ds, a reference amplitude A_0 is defined so as to fulfill the relation

$$P_s(r) = A_0 \cdot \frac{e^{j(\omega t - kr)}}{r} ds \tag{8.9}$$

where $P_s(r)$ is the corresponding field pressure at distance r. The constant A_0 defines the reference amplitude at the reference distance r_0. It is important to note that the physical units of A_0 are: pressure × distance/area.

8.4 THE FIELD OF A DISC SOURCE

Consider a disc source of radius a. It is required to determine the amplitude of the corresponding acoustic field at every point Q located within the medium. This point is located at a distance R from the center of the transducer and with an angle θ relative to the Z axis (the symmetry axis; see Fig. 8.7). The acoustic field can be calculated by employing Huygens' Principle. It can be assumed that the transducer's surface is comprised of numerous miniature surface elements each of which has an area of ds. Each transmitting element is located at radius r from the center and at an angle ψ from the X axis. The distance to point Q is d.

Each transmitting miniature element has an acoustic field that is given by Eq. (8.9). Because of the different distances that the waves transmitted from each element have to travel to reach point Q, each wave will have a different

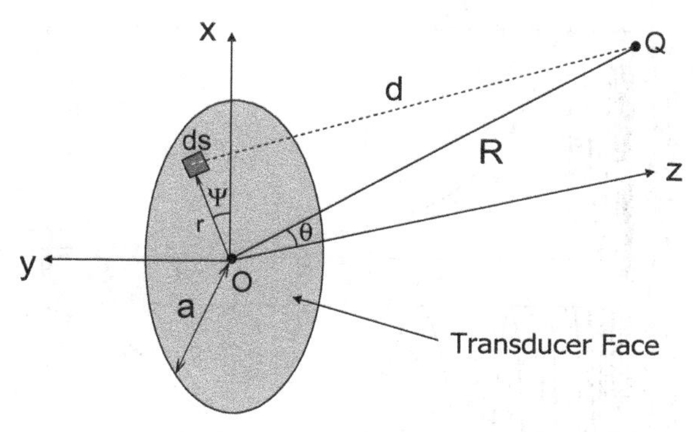

Figure 8.7. The corresponding reference axes used for calculating the acoustic field stemming from a disc transducer of radius *a*.

phase. Consequently, constructive and destructive interferences between all the waves reaching point *Q* will occur. Thus, in order to calculate the pressure at this point, we need to integrate all the contributions made by all elements of size *ds* located on the surface of the transmitting disc. For simplicity, it will be assumed that all the small elements on the surface are moving together with the same phase. This model is called the "plane-piston" model (for obvious reasons).

8.4.1 Near Field and Far Field

The general term for calculating the acoustic field is given by Eq. (8.15), which is introduced in the next section. For clarity, however, let us start with a simpler analysis by investigating the intensity of the field only along the axis of symmetry (the *z* axis). The solution in this case is given by (see, for example, Beyer and Letcher [4])

$$\boxed{\frac{I}{I_0} = \sin^2\left[\frac{\pi}{\lambda}\left\{(a^2+z^2)^{1/2} - z\right\}\right]} \tag{8.10}$$

where I and I_0 are the intensity at distance z and at the source respectively, and λ is the corresponding wavelength in the medium. When plotting the normalized intensity value I/I_0 versus the dimensionless term $\lambda z/a^2$, typical curves such as those depicted in Fig. 8.8 are obtained. As can be observed, the intensity changes rapidly with a nonuniform periodicity near the surface of the transducer. Points of alternating maxima and minima appear. The location of the points of maxima are given by

$$z_{\max} = \frac{4a^2 - \lambda^2(2m+1)^2}{4\lambda(2m+1)} \qquad m = 0, 1, 2, \ldots (\sin = \pm 1) \tag{8.11}$$

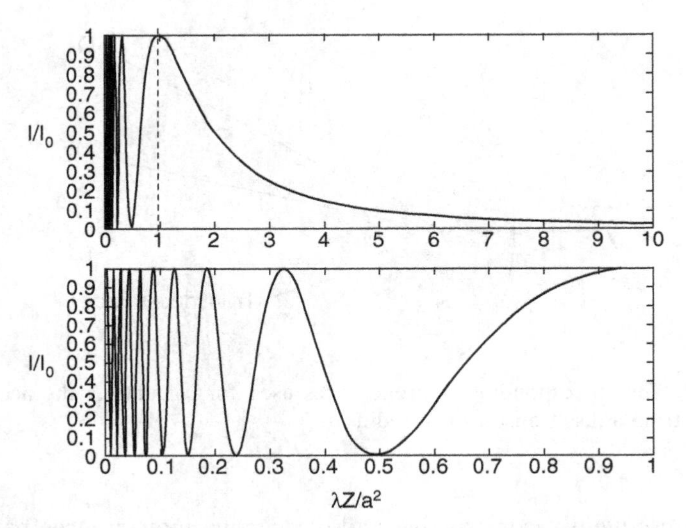

Figure 8.8. (Top) Normalized intensity (I/I_0) along the axis of symmetry for a transmitting disc source plotted versus the normalized distance $\lambda z/a^2$. **(Bottom)** "Zoomed-in" view of the range [0:1].

whereas location of the points of minima appear at

$$z_{\min} = \frac{a^2 - \lambda^2 n^2}{2n\lambda}, \qquad n = 1, 2, 3, \ldots (\sin = 0) \tag{8.12}$$

The last point at which the value of the normalized intensity reaches its maximal value (Z_{NF}) is given by

$$\boxed{Z_{NF} = \frac{4a^2 - \lambda^2}{4\lambda}} \tag{8.13}$$

which can be approximated for the case where $a^2 >> \lambda^2$ (i.e., the source is much bigger than the wavelength) by

$$\boxed{Z_{NF} = \frac{a^2}{\lambda}} \tag{8.14}$$

As can be noted (see Fig. 8.8), up to this point (where $\lambda z/a^2 = 1$) the field's intensity changes very rapidly. This region of the field is very inconvenient for performing measurements and imaging. It is difficult to determine in this region whether the changes in echo amplitudes, for instance, stem from discontinuities in the medium's properties or spatial distribution of reflectors in the medium or from the variations in the acoustic field. This region is called the "near field." After Z_{NF} the field becomes more predictable and decays monotonically. This region in general is more convenient for use and is called the "far field."

8.4.2 The Acoustic Far (Off Axis) Field

In order to calculate the acoustic field at an arbitrary point $Q(R, \theta)$ within the medium, we should, as stated above, integrate all the contributions made by all the small elements ds forming the surface of the transducer, according to Huygens principle. For simplicity the attenuation within the field is neglected here. As explained, each element ds is actually a source for a spherical wave. Accordingly, its contribution to the pressure field P_S is given by Eq. (8.9). The pressure at point $Q(R, \theta)$ in the field is therefore given by

$$P = \iint_{\text{surface}} P_s \, ds = A_0 \cdot \int_{r=0}^{a} \int_{0}^{2\pi} \frac{e^{j(wt-kd)}}{d} \cdot r d\psi \cdot dr \tag{8.15}$$

In order to calculate this integral, we shall use the drawing depicted in Fig. 8.9. Kindly note that although the calculation is made in the Y–Z plane, the solution is identical for any plane containing the Z axis due the symmetry of the problem.

Using the Pythagorean relation, the distance d from the transmitting dS element to point Q is given by

$$d^2 = h^2 + g^2 \tag{8.16}$$

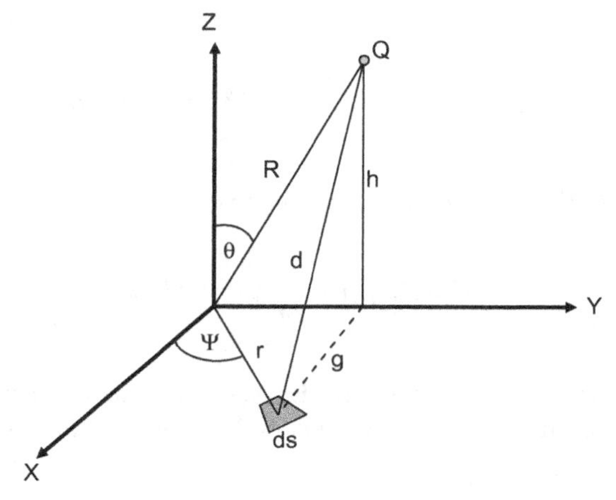

Figure 8.9. Schematic depiction of the geometrical parameters needed for calculating the acoustic pressure at point Q located within the Y–Z plane.

where

$$\begin{aligned} h &= R\cdot\cos\theta \\ g^2 &= (R\cdot\sin\theta)^2 + r^2 - 2r\cdot(R\cdot\sin\theta)\cdot\cos(90^\circ - \psi) \end{aligned} \tag{8.17}$$

Hence by substituting into Eq. (8.16), we obtain

$$\begin{aligned} d^2 &= (R\cdot\cos\theta)^2 + (R\cdot\cos\theta)^2 + r^2 - 2r\cdot(R\cdot\sin\theta)\cdot\cos(90^\circ - \psi) \\ &= R^2 + r^2 - 2r\cdot(R\cdot\sin\theta)\cdot\sin(\psi) \end{aligned} \tag{8.18}$$

If point Q is located at a sufficiently large distance from the transducer (i.e., $r << R$, the term r^2 can be neglected and the distance d can be approximated by

$$\begin{aligned} d^2 &\approx R^2 - 2R\cdot(r\cdot\sin\theta\cdot\sin\psi) + (r\cdot\sin\theta\cdot\sin\psi)^2 \\ &\approx [R - r\cdot\sin\theta\cdot\sin\psi]^2 \\ &\Rightarrow d \approx R - r\cdot\sin\theta\cdot\sin\psi \end{aligned} \tag{8.19}$$

Substituting this value into Eq. (8.15) yields

$$P = \int P_s\, ds = A_0 \cdot \int_{r=0}^{a}\int_{0}^{2\pi} \frac{e^{j(wt - k[R - r\cdot\sin\theta\cdot\sin\psi])}}{R - r\cdot\sin\theta\cdot\sin\psi} \cdot r d\psi \cdot dr \tag{8.20}$$

In order to further simplify this term and considering the fact that $r << R$, the second term in the denominator is neglected (naturally this term cannot be neglected from the exponential term). Thus after rearrangement we obtain

$$P = A_0 \cdot \frac{e^{j(wt - k\cdot R)}}{R} \int_{r=0}^{a}\int_{0}^{2\pi} e^{j(kr\cdot\sin\theta\cdot\sin\psi)} \cdot r d\psi \cdot dr \tag{8.21}$$

From reference 6, the following relation is given:

$$I_0(x) = J_0(jx) = \frac{1}{2\pi}\int_{0}^{2\pi} e^{(x\cdot\sin\psi)} d\psi \tag{8.22}$$

where I_0 and J_0 are the corresponding Bessel functions of order 0. Recalling that $J_0(-x) = J_0(x)$ and considering $(jkr\sin\theta)$ as the argument of the Bessel function yields

$$P = A_0 \cdot \frac{e^{j(wt - k\cdot R)}}{R} \int_{r=0}^{a} 2\pi \cdot J_0(kr\sin\theta)\cdot r\cdot dr \tag{8.23}$$

Next, let us multiply and divide the integral by $k\sin(\theta)$ and also define a new integration parameter $\beta \triangleq kr\sin(\theta)$ and obtain

$$P = 2\pi \cdot A_0 \cdot \frac{e^{j(wt-k\cdot R)}}{R} \cdot \frac{1}{(k\sin\theta)^2} \int_{\beta=0}^{\beta=kr\sin(\theta)} \beta \cdot J_0(\beta) \cdot d\beta \tag{8.24}$$

From the above-mentioned mathematical tables we know that $\int \beta \cdot J_0(\beta) \cdot d\beta = \beta \cdot J_1(\beta)$, where J_1 is the Bessel function of order 1. Therefore it follows that

$$P = 2\pi \cdot A_0 \cdot \frac{e^{j(wt-k\cdot R)}}{R} \cdot \frac{ka\sin\theta \cdot J_1(ka\sin\theta)}{(k\sin\theta)^2} \tag{8.25}$$

Finally by multiplying and dividing by the transducer's radius a and after rearrangement, we obtain

$$\boxed{P(R,\theta) = \left[\pi a^2 \cdot A_0\right] \cdot \left[\frac{e^{j(wt-k\cdot R)}}{R}\right] \cdot \left[\frac{2 \cdot J_1(ka\sin\theta)}{(ka\sin\theta)}\right]} \tag{8.26}$$

This equation expresses the acoustic far field of a disc-shaped transmitting transducer, assuming that the axis of symmetry is aligned with the Z axis and perpendicular to the X–Y plane, and in accordance with the assumptions made above.

Studying Eq. (8.26), it can be noted that the field is determined by a multiplication of three terms that are separated by squared brackets for clarity. The first term in this equation determines the amplitude of the pressure field and is dependent on the transducer's area and the set transmission intensity. The second term is the characteristic function of a spherical wave (see Chapter 1). The denominator of this equation causes the pressure in the field to decrease with the distance. (It should be pointed that thus far we have ignored the attenuation in the medium, but this effect has also to be accounted for.) The exponential expression is complex and modulates the phase as a function of the distance and the time. The third term in Eq. (8.26) determines the amplitude of the field as a function of the angular direction (azimuth). This term is called the "directivity function" of the field and is depicted graphically in Fig. 8.10. As can be noted from Fig. 8.10, the directivity function is symmetrical. It has a main lobe with a maximum at $x = 0$, where its value equals 1. It also has side lobes whose amplitude decays as one moves away from the center. This function has zeros at $ka\sin\theta = \pm 3.83, \pm 7.02, \pm 10.17$.

Using Eq. (8.26), the acoustic field can be calculated and mapped for various transducer diameters, frequencies, and mediums. For example, in Fig. 8.11, several acoustic maps are depicted. The maps were generated using computer simulations and the calculations were performed at the Z–Y plane. The calculations were made for the acoustic field in water. The maps depict the changes resulting from changes in the transducer's diameter and frequency.

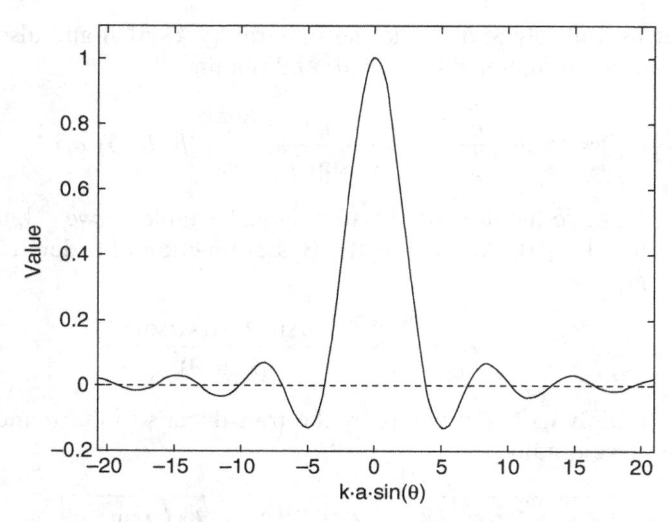

Figure 8.10. The profile of a directivity function for a disc-shaped transducer.

In the left-hand column of Fig. 8.11, the changes stemming from changes in the transducer's radius a for a 1-MHz frequency are displayed. For clarity, the amplitude was presented in a logarithmic scale (i.e., $\log(P)$). It should be pointed out that the reliability of the simulations decreases in regions that are proximal to the transducer due to the assumptions made above. As can be noted, for a small-diameter transducer the field is almost nondirectional (omnidirectional). As the diameter is increased, the main lobe of the field becomes narrower and the directivity of the field improves. Nonetheless, the side lobes become more noticeable.

On the right-hand side of Fig. 8.11, the influence of the frequency increase on the acoustic field is demonstrated for a transducer with a radius of 2.5 mm. As can be observed, the main lobe of the field becomes narrower and the directivity of the field improves. The side lobes also become narrower as can be noted.

When using ultrasound for imaging, our preference is usually for a more directional field. This implies that we would commonly prefer a main lobe that is as narrow as possible. We would also prefer a field with negligible side lobes since these lobes may invoke echoes from reflectors whose location is ambiguous (see Chapter 9).

The width of the field's main lobe can be determined from the directivity function given by the right-hand side term of Eq. (8.26) and depicted in Fig. 8.10. As recalled, the directivity function is comprised of a Bessel function divided by its argument. (It is worth noting that this function is called "jinc"—see, for example, reference 7—because its structure resembles

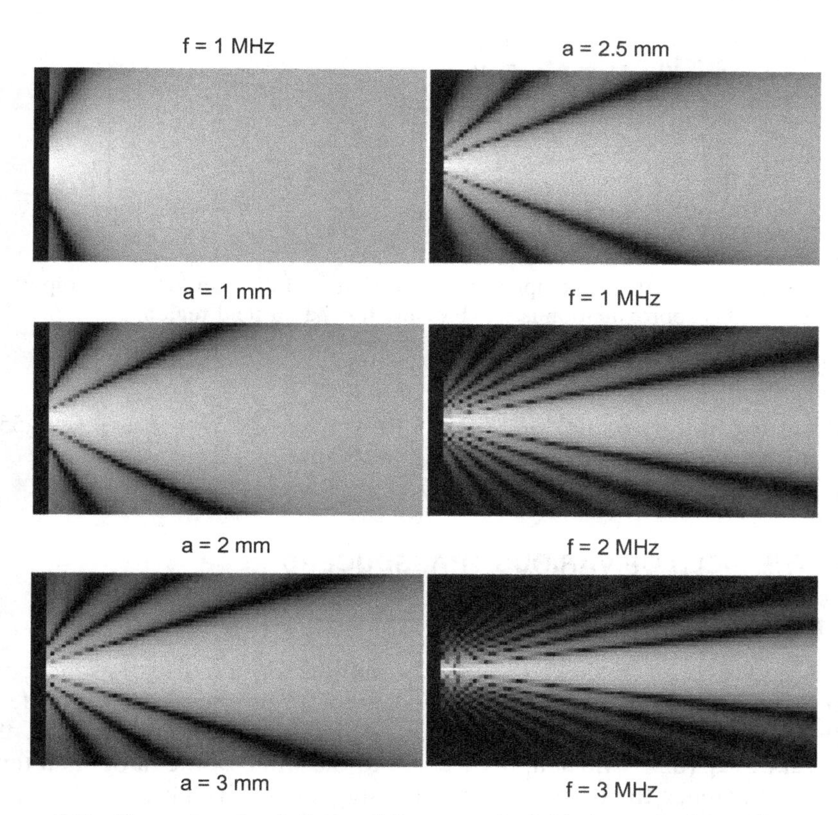

Figure 8.11. Computer simulations of the acoustic fields in water for a disc source. **(Left column)** The effect of increasing the transducer's radius on the field is demonstrated. **(Right column)** The effect of increasing the transducer's frequency on the field is demonstrated. (*Note*: The pressure is depicted using a logarithmic scale.)

the well-known "sinc" function.) This directivity function has a first zero value at

$$ka \sin\theta = 3.83 \tag{8.27}$$

Assuming that the main lobe of the field can be approximated by a cone, as shown schematically in Fig. 8.12, the cone head angle θ can be determined from

$$\boxed{\theta = \sin^{-1}\left(\frac{3.83}{ka}\right) = \sin^{-1}\left(\frac{0.61\lambda}{a}\right)} \tag{8.28}$$

As an example, the calculated cone head angle approximation using Eq. (8.28) for the main lobe of a disc transducer with a diameter of 10 mm (i.e., $a = 5$ mm)

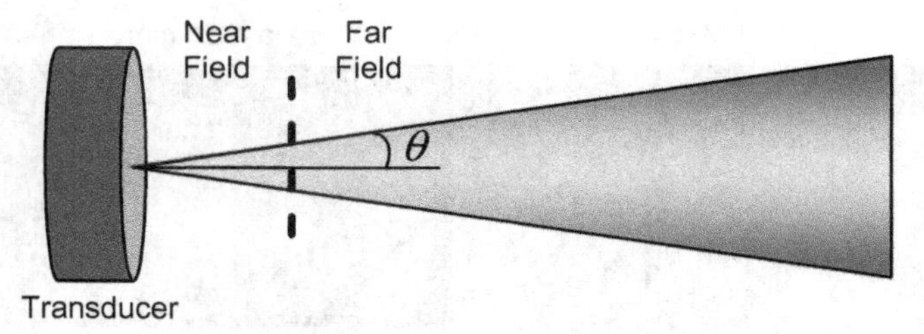

Figure 8.12. A conical-shaped approximation of the field's main lobe. It is important to note that this approximation is reliable only for the far-field region.

transmitting at a frequency of 1 MHz into water ($\lambda = 1.5$ mm) is $\theta = 10.55°$. The corresponding far field starts at $Z_{NF} = 16.3$ mm.

8.5 THE FIELD OF VARIOUS TRANSDUCERS

8.5.1 The Field of a Ring Source

The equations derived above [Eqs. (8.23) and (8.26)] for a disc transducer can also be used for calculating the acoustic field of a ring transducer. Consider a ring transducer of thickness T which is negligible in size. For this case the integral of Eq. (8.23) will simply be equal to the value of the integrated term at radius a. Hence the corresponding acoustic field is given by,

$$P_{ring} = \left[2\pi a T \cdot A_0 \cdot \frac{e^{j(wt - k\cdot R)}}{R} \right] \cdot J_0(ka \sin\theta) \tag{8.29}$$

From this equation we can deduce that the directivity function for a ring transducer is a Bessel function of order 0. The profile of this function is depicted in Fig. 8.13.

To further illustrate the differences between the two types of transducers, the calculated acoustic pressure fields for the two transducers are depicted in Fig. 8.14. (In this case the amplitude scale is linear.) As can be observed, the main lobe is narrower and the side lobes are enhanced. Again it should be pointed out that the calculations were made based on the assumptions stated above.

8.5.2 The Field of a Line Source

Although disc-shaped transducers are very popular, there are many applications in which other geometrical configurations are advantageous. The simplest

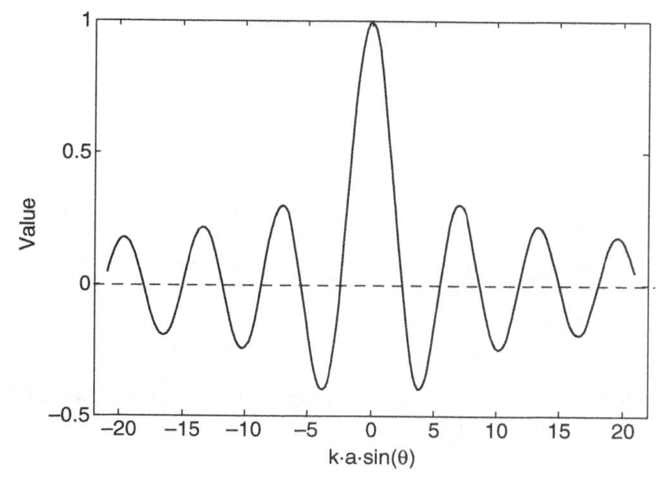

Figure 8.13. The profile of the directivity function for a ring transducer. As can be noted, the main lobe is slightly narrower than that of a disc-shaped transducer. However, the side lobes in this case are much stronger.

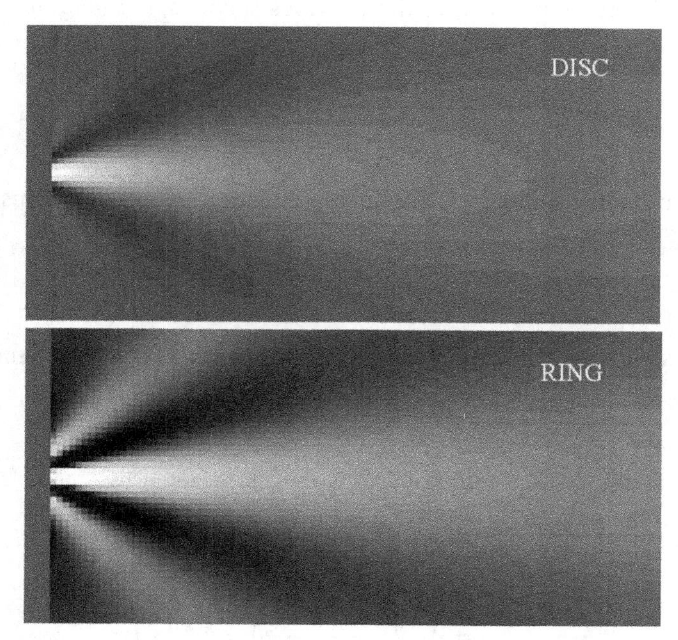

Figure 8.14. Simulated acoustic fields in water for a disc transducer **(top)** and a ring transducer **(bottom)**. The transducer's radius was $a = 2.5\,\text{mm}$ and the frequency was 1 MHz for both cases. Note that in this case the scale is *linear*.

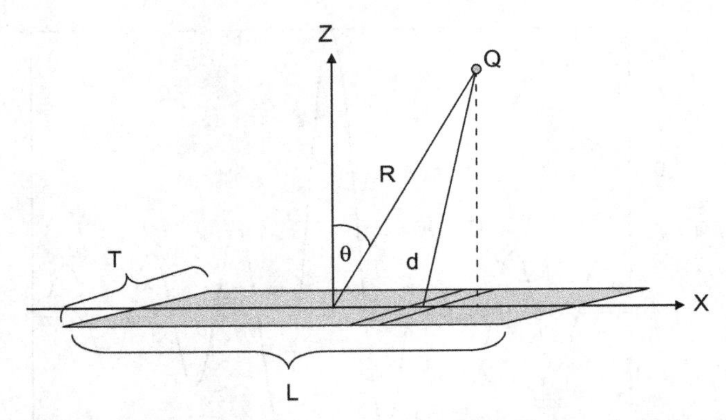

Figure 8.15. Illustration of the coordinate system utilized for calculating the acoustic field stemming from a line source.

geometry is naturally a straight line. Consider a source of length L and thickness T. It is assumed that the thickness is negligible so that the transducer can be assumed to be a line source. In order to calculate the corresponding acoustic field, we shall use the coordinate system depicted in Fig. 8.15. For simplicity the calculations are made only for the field containing the source. Using the same considerations outlined in the previous sections, the field at point Q is given by

$$P_Q(R,\theta) = A_0 \int_{-L/2}^{L/2} \frac{e^{j(\omega t - kd)}}{d} \cdot T\,dx \tag{8.30}$$

The distance from an infinitesimal surface element $T\,dx$ on the transmitter to pint Q is given by

$$d^2 = (R\cdot\sin\theta - x)^2 + (R\cdot\cos\theta)^2 \tag{8.31}$$

Using the same approximation applied in Eq. (8.19), it can be assumed that

$$d \approx R - x\cdot\sin\theta \tag{8.32}$$

The acoustic field pressure at point Q is therefore given by

$$\begin{aligned} P_Q(R,\theta) &= A_0\cdot T\cdot \int_{-L/2}^{L/2} \frac{e^{j[\omega t - k(R - x\sin\theta)]}}{R}\cdot dx \\ &= A_0\cdot T\cdot \frac{e^{j(\omega t - kR)}}{R}\cdot \int_{-L/2}^{L/2} e^{jkx\sin\theta}\cdot dx \end{aligned} \tag{8.33}$$

for which the solution is given by

$$P_Q(R,\theta)=\left[A_0\cdot T\cdot L\cdot\frac{e^{j(wt-k\cdot R)}}{R}\right]\cdot\frac{\sin\left(k\frac{L}{2}\sin\theta\right)}{\left(k\frac{L}{2}\sin\theta\right)}$$
$$=\left[A_0\cdot T\cdot L\cdot\frac{e^{j(wt-k\cdot R)}}{R}\right]\cdot\left[\operatorname{sinc}\left(k\frac{L}{2}\sin\theta\right)\right] \quad (8.34)$$

From this equation we conclude that the profile of the directivity function in the plane containing the transducer is described by the familiar sinc function.

8.5.3 The Field of a Rectangular Source

Another popular geometry is the rectangular shape. Such geometry is often used in phased array transducers, where each element is actually a small rectangular transmitter. This geometry is convenient for manufacturing line arrays. Indeed, in many applications the size of each element may be considered sufficiently small to allow a point source approximation. However, sometimes in order to achieve better calculation of the field, the actual size of the source element needs to be accounted for. So let us study now the acoustic far field of a rectangular transducer.

Consider a rectangular transducer such as the one depicted in Fig. 8.16. The length of the transducer is a and its width is b as shown. The distance from the center of the transducer to an arbitrary point Q in the medium is R. The angle between R and the Z axis is θ. The angle between its projection L and the X axis is ψ. The distance between a transmitting surface element $dS = dx \cdot dy$ is d. The height of point Q above the transducer is h.

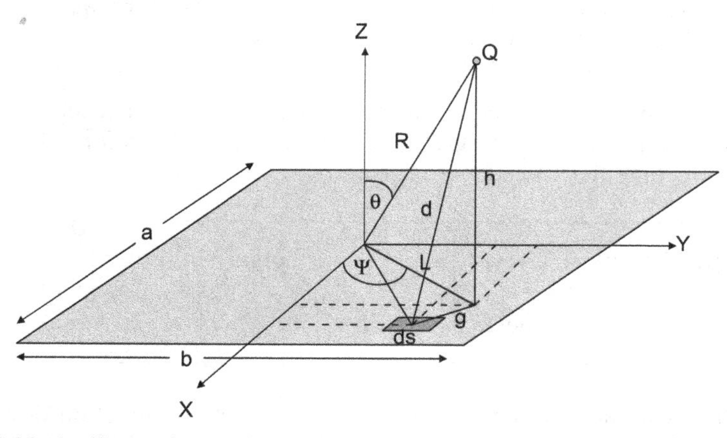

Figure 8.16. An illustration of the coordinate system utilized for calculating the acoustic field stemming from a rectangular source.

From simple trigonometry we have

$$d^2 = h^2 + g^2 \tag{8.35}$$

And also,

$$\begin{aligned} h &= R \cdot \cos\theta \\ L &= R \cdot \sin\theta. \end{aligned} \tag{8.36}$$

In addition, it is easy to see that

$$g^2 = (x - R \cdot \sin\theta \cdot \cos\psi)^2 + (y - R \cdot \sin\theta \cdot \sin\psi)^2 \tag{8.37}$$

By substituting these terms into Eq. (8.35) and after a slight simplification, we obtain

$$d^2 = R^2 - 2R \cdot (x \cdot \sin\theta \cdot \cos\psi + y \cdot \sin\theta \cdot \sin\psi)^2 + (x^2 + y^2) \tag{8.38}$$

Since we are studying the far field, we can assume that $R >> x$ and $R >> y$. Thus, the expression can be further simplified as was done in Eq. (8.19), to yield

$$d \approx R - (x \cdot \sin\theta \cdot \cos\psi + y \cdot \sin\theta \cdot \sin\psi) \tag{8.39}$$

Substituting this term into Eq. (8.15) yields

$$\begin{aligned} P(R,\theta,\psi) &= \iint P_s\, ds = A_0 \cdot \frac{e^{j(wt-k\cdot R)}}{R} \cdot \int_{x=-a/2}^{a/2} \int_{y=-b/2}^{b/2} e^{j(K[x\cdot\sin\theta\cdot\cos\psi + y\cdot\sin\theta\cdot\sin\psi])} \cdot dx \cdot dy \\ &= A_0 \cdot \frac{e^{j(wt-k\cdot R)}}{R} \cdot \int_{x=-a/2}^{a/2} e^{j(Kx\cdot\sin\theta\cdot\cos\psi)} \cdot dx \int_{y=-b/2}^{b/2} e^{j(Ky\cdot\sin\theta\cdot\sin\psi])} \cdot dy \end{aligned} \tag{8.40}$$

If we study carefully this equation, we can easily see that each of the two integrals is similar to the integral given in Eq. (8.33). Therefore, the acoustic field of a rectangular transducer is given by

$$\begin{aligned} P(R,\theta,\psi) = [A_0 \cdot a \cdot b] \cdot \left[\frac{e^{j(wt-k\cdot R)}}{R}\right] \cdot \left[\operatorname{sinc}\left(k\frac{a}{2}\sin\theta\cos\psi\right)\right] \\ \cdot \left[\operatorname{sinc}\left(k\frac{b}{2}\sin\theta\sin\psi\right)\right] \end{aligned} \tag{8.41}$$

From this expression we can deduce that the directivity function of a rectangular transducer is comprised of a multiplication of two sinc functions. Also, it can be observed that if the transducer is very thin, then for $\psi \approx 90^\circ$ the term

$\cos\psi = 0$ and $\sin\psi = 1$. Thus, by taking $a = T$ and $b = L$, we obtain an expression that is equal to Eq. (8.34).

8.6 PHASED-ARRAY TRANSDUCERS

8.6.1 The General Field from an Array Source

Array transducers are very common today and particularly in medical ultrasonic imaging. To calculate the field from an array source, we again make use of the Huygens principle and utilize Eq. (8.9) as the basis for computing the field in a manner that resembles (8.15). Thus, in the most general case, where each element has an arbitrary shape, the acoustic field stemming from one element in the array is given by

$$P(x, y, z) = \iint_{\text{surface}} P_s\, ds = \iint_{\text{surface}} A_0(\mu, \eta) \cdot \frac{e^{j(wt - kd(\mu,\eta,x,y,z) + \varphi(\mu,\eta))}}{d(\mu, \eta, x, y, z)} \cdot ds \tag{8.42}$$

where $d(x, y, z)$ is the distance between point (μ, η) on the transmitting surface and the point Q in the medium for which the field is calculated. $A_0(\mu, \eta)$ is the amplitude at point (μ, η), and $\varphi(\mu, \eta)$ is the corresponding phase. If the array is comprised of N elements, then the field will be given by their summation,

$$P(x, y, z) = \sum_{i}^{N} \iint_{\text{surface}-i} P_s\, ds_i = \sum_{i}^{N} \iint_{\text{surface}-i} A_{0i}(\mu, \eta) \cdot \frac{e^{j(wt - k \cdot d(\mu,\eta,x,y,z) + \varphi_i(\mu,\eta))}}{d(\mu, \eta, x, y, z)} \cdot ds_i \tag{8.43}$$

where i is an index designating a certain element in the array.

8.6.2 The Field of a Linear Phased Array

In many practical applications (e.g., medical imaging) it is common to use a linear phased array transducer. This type of an array is built from many small rectangular elements lined up with a very thin gap between them. (Sometimes they are manufactured by simply slicing a single long piezoelectric bar.) In order to calculate the acoustic field of such an array, let us assume that each element can be approximated by a point source as schematically depicted in Fig. 8.17.

Consider an array of length L which is comprised of N elements. The transmitting surface of each element is a. From Fig. 8.17 and this relation the distance between two adjacent elements is given by

$$\Delta x = \frac{L}{(N - 1)} \tag{8.44}$$

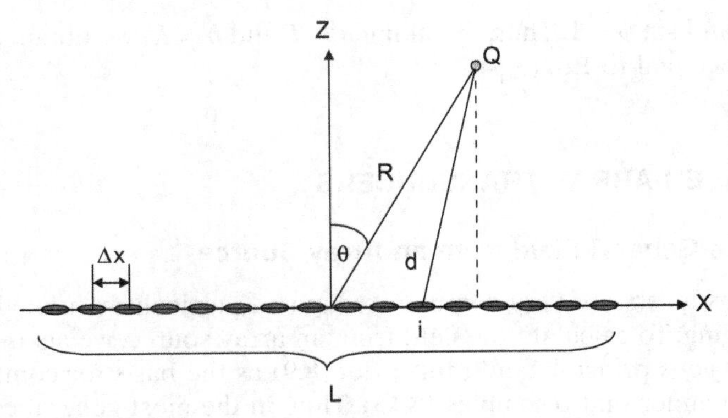

Figure 8.17. An illustration of the coordinate system utilized for calculating the acoustic field of a linear phased array.

Importantly, it should be noted that if the gap between the elements is negligible, then Δx represents the size of each element.

The coordinate of element i along the X axis is given by

$$x_i = (i-1)\cdot \Delta x - \frac{L}{2} \tag{8.45}$$

In order to calculate the acoustic field pressure at point Q in the X–Z plane, we need to calculate its distance d from the transmitting element number i. This distance is given from the cosine theorem by

$$\begin{aligned} d(R,\theta)^2 &= R^2 + x_i^2 - 2R\cdot x_i \cdot \cos(90° - \theta) \\ &= R^2 + x_i^2 - 2R\cdot x_i \cdot \sin(\theta) \end{aligned} \tag{8.46}$$

Substituting this term into Eq. (8.43) while avoiding the integration over the surface of the individual element (we have assumed that it is sufficiently small to be consider as a point source), we obtain

$$P(R,\theta) = \sum_{i}^{N} A_{0i} \cdot \frac{e^{j(wt - kd(R,\theta) + \varphi_i)}}{d(R,\theta)} \cdot a \tag{8.47}$$

It is important to note that in this case we have not used the far-field approximation. However, we have neglected the actual shape of the individual element. Thus, if we would like to achieve a more accurate solution, we need to integrate over the surface of each element. Alternatively, if the array is comprised of disc elements or rectangular elements, Eq. (8.26) and Eq. (8.41) can be used respectively while assuming the far-field approximation for each element.

8.6.3 Far-Field Approximation for a Linear Phased Array

Assuming that $R >> x_i$ and then using the same approximation applied in the previous sections, we can write

$$d(R,\theta) \approx R - x_i \cdot \sin(\theta) \tag{8.48}$$

Substituting this term into Eq. (8.43) while assuming that each element is a point source yields

$$P(R,\theta) = \sum_i^N A_{0i} \cdot \frac{e^{j(wt - kd(R,\theta)+\varphi_i)}}{d(R,\theta)} \cdot a = a \cdot \frac{e^{j(wt-kR)}}{R} \sum_i^N A_{0i} \cdot e^{j(k \cdot x_i \cdot \sin\theta + \varphi_i)} \tag{8.49}$$

where the value of x_i is calculated from Eq. (8.45).

If we further assume that the amplitude and the phase of all the transmitting elements is the same, then after some further derivation the following expression is obtained:

$$P(R,\theta) = N \cdot A_0 \cdot a \cdot \frac{e^{j(wt-kR)}}{R} \cdot \left\{ \frac{\sin\left[k \frac{N \cdot \Delta x}{2} \sin(\theta)\right]}{N \sin\left[k \frac{\Delta x}{2} \sin(\theta)\right]} \right\} \tag{8.50}$$

The advantage of this equation over Eq. (8.47) is that the summation is replaced by a continuous function from which the general behavior of the field can be analyzed. However, the accuracy in the field calculations is compromised. Thus, considering the fact that current computation times are rather short even for large arrays, I personally prefer to use Eq. (8.47) for field calculations as was indeed done for the acoustic field maps shown in the following sections.

8.6.4 Grating Lobes for a Linear Phased Array

When designing and building a linear phased array, we must be aware of the fact that with an improper design, strong disturbing grating lobes may appear. In order to understand the source of this phenomenon, we need to investigate the denominator of Eq. (8.50). As can be observed, this term contains a sine expression that can be equal to zero not only when $\theta = 0$. When this happens, the function reaches its maximal value. This implies that in addition to the main lobe (where $\theta = 0$) we may obtain strong undesired side lobes. In order to find out the condition for these lobes appearance, let us examine the condition for

$$\sin\left[k \frac{\Delta x}{2} \sin(\theta)\right] = 0 \tag{8.51}$$

Obviously the condition is given by

$$k\frac{\Delta x}{2}\sin(\theta) = m\cdot\pi, \quad m = 0, \pm 1, \pm 2, \ldots \tag{8.52}$$

Thus one may expect the first side lobe to appear when $m = 1$, which leads to

$$\begin{aligned} k\frac{\Delta x}{2}\sin(\theta) &= \frac{2\pi}{\lambda}\cdot\frac{\Delta x}{2}\cdot\sin(\theta) = \pi, \\ \Rightarrow \sin(\theta) &= \frac{\lambda}{\Delta x} \end{aligned} \tag{8.53}$$

Commonly, we would like to avoid these side lobes. Hence, we must require that the above condition be fulfilled at least when $\theta = \pi/2$, which leads to

$$\begin{aligned} \sin\left(\frac{\pi}{2}\right) &\le \frac{\lambda}{\Delta x} \\ \Rightarrow 1 &\le \frac{\lambda}{\Delta x} \end{aligned} \tag{8.54}$$

Or in other words,

$$\boxed{\Delta x \le \lambda} \tag{8.55}$$

This implies that the size of each element (including one gap size) should be smaller than the wavelength in the medium. To demonstrate this condition, let us examine the simulated acoustic fields depicted in Fig. 8.18. These two maps display the field for a linear array containing 64 elements and transmitting at a frequency of 5 MHz into water. The top image is for $\Delta x = 0.9\lambda$, while the bottom one is for $\Delta x = 1.1\lambda$. As can be noted, two strong side lobes occur in the second case.

8.6.5 Beam Steering with a Linear Phased Array

One of the main advantages of using a linear phased array is the ability to "shape" the acoustic field electronically. This ability allows us to move the beam and steer it; and as will be explained in the following section, focus is at arbitrarily chosen focal lengths.

Let us start with the simplest form of steering which is a lateral (i.e., parallel to the array) motion of the beam as illustrated in Fig. 8.19. This type of beam

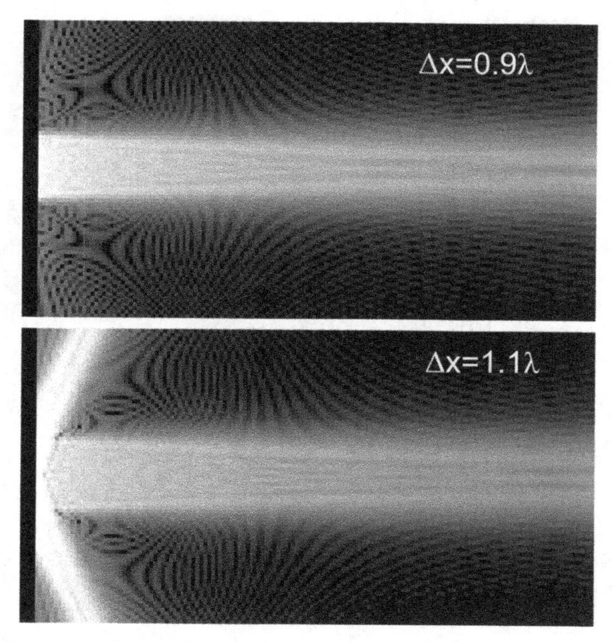

Figure 8.18. Computer simulations for the acoustic field of a 64-element linear array transducer transmitting at frequency of 5 MHz into water. The pressure values are depicted using a logarithmic scale. **(Top)** The field when $\Delta x = 0.9\lambda$. **(Bottom)** The field when $\Delta x = 1.1\lambda$. Note the appearance of two strong side lobes for the second case.

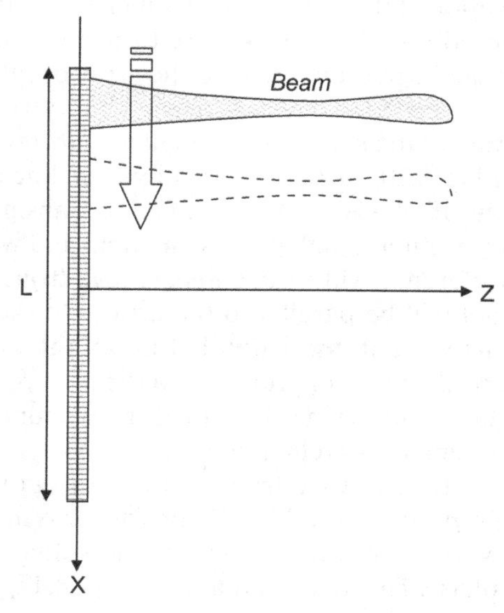

Figure 8.19. Lateral beam motion is obtained by turning on and off subgroups of elements within the linear phased array.

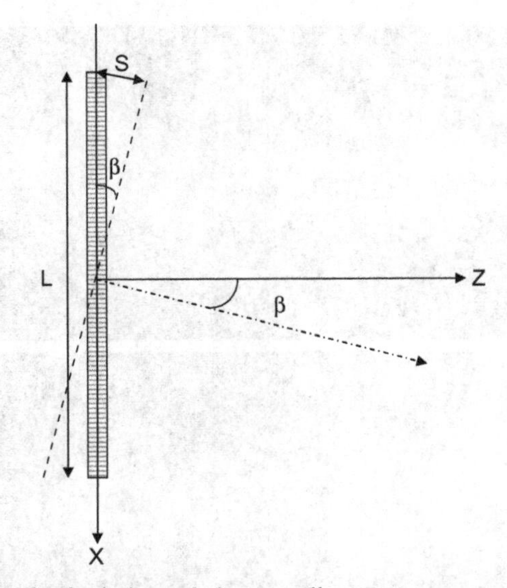

Figure 8.20. Schematic depiction of the coordinate system used for calculating the steering of an acoustic field.

motion can be achieved by turning on a subgroup of elements in the array—that is, elements numbered i to $i + m$. After transmission, the same or other subgroup (or even the entire array) can be used for receiving the resulting echoes. Upon completion of this step, another subgroup containing for example the elements numbered $i + 1$ to $i + m + 1$ are turned on and the procedure is repeated. The beam can be perpendicular to the array or tilted by any arbitrary angle.

In order to understand the principles of angular steering, let us consider the drawing depicted in Fig. 8.20. As described above, the linear array is of length L and has N elements. Each element is assumed to transmit a spherical wave. If all the elements transmit a signal at the same time and with the same phase, the constructive interference will form a linear wave front in the transducer's plane. This wave front will be parallel to the array. On the other hand, if the first element starts its wave transmission before all the other elements, then the front of its spherical wave will propagate a distance $S = c \cdot t$. The phase of the wave at this distance will be $\varphi = k \cdot s$. (This is of course in addition to the initial phase and the temporal cycling phase.)

In order to form a straight wave front that is propagating along the direction defined by angle β (as denoted by the broken line in Fig. 8.20), we must ensure that all the waves transmitted from the individual elements reach this line with the same phase. This implies that [see Eq. (8.47)]

$$\omega \cdot t - k \cdot S_i + \varphi_i = \omega \cdot t + \text{const} \tag{8.56}$$

where S_i is the distance that the spherical wave transmitted from element number i has traveled. For simplicity, we can set the constant on the right-hand side of Eq. (8.57) to be zero. Hence, we obtain

$$-k \cdot S_i + \varphi_i = 0$$
$$\Rightarrow \varphi_i = k \cdot S_i \quad (8.57)$$

Making use of Eq. (8.45), we can calculate the initial phase value that is needed to be added to element number i from

$$\boxed{\begin{aligned}&\varphi_i = k \cdot x_i \cdot \sin(\beta)\\ &\Rightarrow \varphi_i = \left[(i-1)\cdot \Delta x - \frac{L}{2}\right]\cdot k \cdot \sin(\beta)\end{aligned}} \quad (8.58)$$

To demonstrate the effect, we have combined Eq. (8.58) with Eq. (8.47); and using a computer simulation, the acoustic pressure maps depicted in Fig. 8.21 were obtained.

In the images depicted in Fig. 8.21, angular beam steering via Eq. (8.58) is clearly demonstrated. The fields were calculated for a linear phased array transducer with 64 elements, transmitting at a frequency of 3 MHz into water. The length of the array transducer is 20 mm. The pressure is displayed using a logarithmic scale. The beam was steered to the following directions: $\beta = -20°:0°:20°$, corresponding to the top, middle, and bottom images, respectively.

Equation (8.58) applies to a continuous wave transmission. However, in many applications (particularly in imaging), short acoustic pulses are transmitted. In such a case the phase has to be replaced by time delayed. The corresponding time delay for element number i (i.e., Δt_i) can be calculated from

$$\varphi_i = 2\pi f \cdot \Delta t_i = k \cdot S_i = \frac{2\pi}{\lambda}\cdot S_i$$
$$\Rightarrow \Delta t_i = \frac{2\pi}{2\pi \cdot f \cdot \lambda} = \frac{S_i}{C}$$
$$\Rightarrow \Delta t_i = \left[(i-1)\cdot \Delta x - \frac{L}{2}\right]\cdot \frac{\sin(\beta)}{C} \quad (8.59)$$

8.6.6 Maximal Steering Angle for a Linear Phased Array

In Eq. (8.55) we have shown the constrain for the size of the elements in a linear phased array relative to the wavelength, needed to avoid the grating lobes. This constraint did not include angular steering of the beam. By substituting Eq. (8.58) into Eq. (8.49), the corresponding far-field pressure approximation for a beam steered by angle β is obtained:

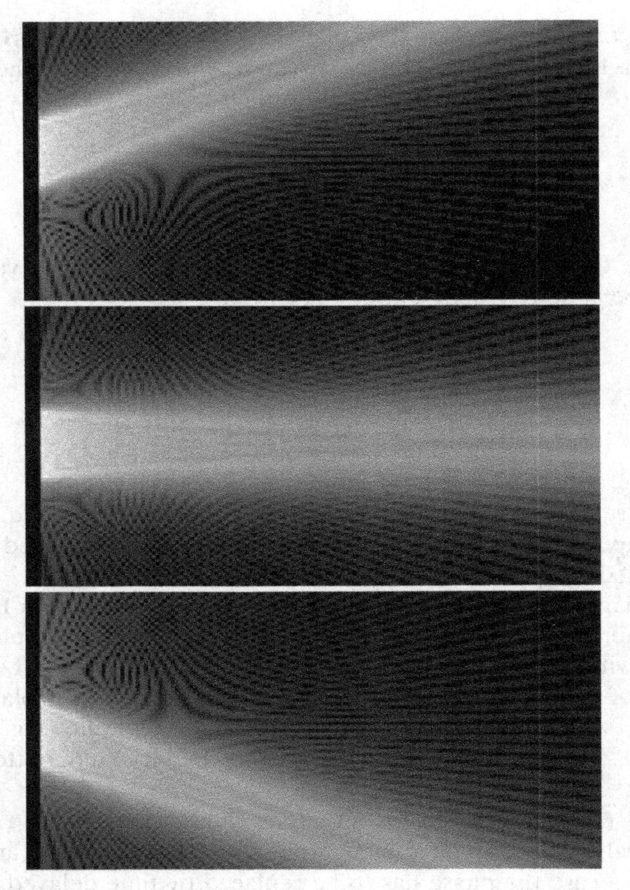

Figure 8.21. Calculated acoustic fields depicting angular beam steering using a linear phased array transducer. The array has 64 elements and its length is 20 mm. The transmission was at a frequency of 3 MHz into water. The pressures are displayed using a logarithmic scale. The beam was steered to angles $\beta = -20°:0°:20°$, corresponding to the top middle and bottom images, respectively.

$$P(R,\theta) = a \cdot \frac{e^{j(wt-kR)}}{R} \sum_{i}^{N} A_{0i} \cdot e^{j(k \cdot x_i \cdot \sin\theta + k \cdot x_i \cdot \sin\beta)} \tag{8.60}$$

This expression can also be described by (if all amplifudes are equal)

$$P(R,\theta) = N \cdot A_0 \cdot a \cdot \frac{e^{j(wt-kR)}}{R} \cdot \left\{ \frac{\sin\left[k \dfrac{N \cdot \Delta x}{2} \cdot (\sin\theta + \sin\beta)\right]}{N \sin\left[k \dfrac{\Delta x}{2} (\sin\theta + \sin\beta)\right]} \right\} \tag{8.61}$$

As can be noted, this function reaches its maximal value when the denominator equals zero. Hence, the corresponding condition for that is given by

$$k\frac{\Delta x}{2}(\sin\theta+\sin\beta)=m\cdot\pi, \qquad m=0,\pm1,\pm2,\ldots \tag{8.62}$$

This implies that there are many strong lobes (for every m value) that can appear within the acoustic field. Naturally, these lobes are undesirable. Thus, if we require that the first lobe for which $m = 1$ (in addition to the main one for which $m = 0$) will appear only at $\theta = \pi/2$ (at the bottom of the field), the constraint for the maximal steering angle β_{max} is given by

$$\frac{2\pi}{\lambda}\cdot\frac{\Delta x}{2}\left(\sin\left(\frac{\pi}{2}\right)+\sin(\beta_{max})\right)=\pi$$

$$\Rightarrow \sin(\beta_{max})=\left(\frac{\lambda}{\Delta x}-1\right) \tag{8.63}$$

Or more explicitly:

$$\boxed{\beta_{max}=\sin^{-1}\left(\frac{\lambda}{\Delta x}-1\right)} \tag{8.64}$$

It should be pointed out that this condition must be kept in addition to the condition given by Eq. (8.54), that is, $1 \le \lambda/\Delta x$. Otherwise, side lobes will occur anyway.

Just as a demonstrative example, consider a linear phased array for which $\Delta x = 1$ mm and we would like to steer it in water, where the speed of sound is assumed to be $C = 1500$[m/sec]. If the transmission frequency is 1 MHz, then the corresponding wavelength is $\lambda = 1.5$ mm. By substituting these values into Eq. (8.64), it follows that the maximal steering angle for this array is 30°.

8.6.7 Beam Forming with a Linear Phased Array

As can be noted, it was demonstrated above how an acoustic beam stemming from a linear phased array can be moved and steered. However, using similar principles, we can employ this type of an array also for shaping the acoustic beam. This procedure is called "beam forming." The most common application of beam forming is electronic focusing of the beam—that is, achieving focusing without a focusing hardware.

Consider the array transducer depicted in Fig. 8.20. It is required to focus the field with this array at a focal distance F along the Z axis. For the element transmitting from the axes origin, the corresponding phase (in addition to the cycling temporal phase and the initial phase) is simply $\varphi_0 = k \cdot F$. In order to obtain a focusing effect, it is desired that transmitted waves from all the other elements will have the same phase at the focal point. To calculate the phase of element number i, we need to multiply its distance to the focal point by the wave number and add its initial phase φ_i. This value should equal φ_0. Hence mathematically the relation between the two phases is given by

$$k \cdot \sqrt{x_i^2 + F^2} + \varphi_i = k \cdot F \tag{8.65}$$

Or after rearranging the terms, we obtain

$$\boxed{\varphi_i = k \cdot \left(F - \sqrt{x_i^2 + F^2}\right)} \tag{8.66}$$

This equation applies for a continuous wave transmission. In case pulse transmission is needed (as in imaging), the initial phase needs to be replaced by the corresponding time delay as was done in Eq. (8.59). This yields

$$\boxed{\Delta t_i = \frac{1}{C} \cdot \left(F - \sqrt{x_i^2 + F^2}\right)} \tag{8.67}$$

where negative time delays are simply set by shifting the reference time point (i.e., by starting the transmission with the most negative time delay).

In order to complete the beam forming procedure, we should combine the beam translation and angular steering with the focusing. This can be easily achieved by adding the phase obtained by Eq. (8.58) with the phase obtained from Eq. (8.66). The combined effect is depicted in Fig. 8.22.

8.7 ANNULAR PHASED ARRAYS

Another type of a phased array transducer that has gained popularity is the annular array. This array (as its name implies) is commonly comprised of a set of concentric rings. The inner ring (or actually a disc) has the smallest outer radius, and the following rings are arranged according to a predefined requisite (typically the equal-area condition). The gap between the rings is commonly considered negligible. The rings can be positioned atop a single flat surface or piled atop a curved surface to achieve an initial focal zone as explained in Chapter 7.

A schematic depiction of an annular array comprised of a set of concentric rings is depicted in Fig. 8.23. The array can be tiled upon a curved surface as shown on the right-hand side. Commonly, the design request is for an array for which all the elements have the same transmission area. (This is convenient for designing the back-end electronics since all the elements are activated with the same power.) Let us now use the drawing on the left side of Fig. 8.23 to determine the size of the requested outer radius for each ring, assuming a flat

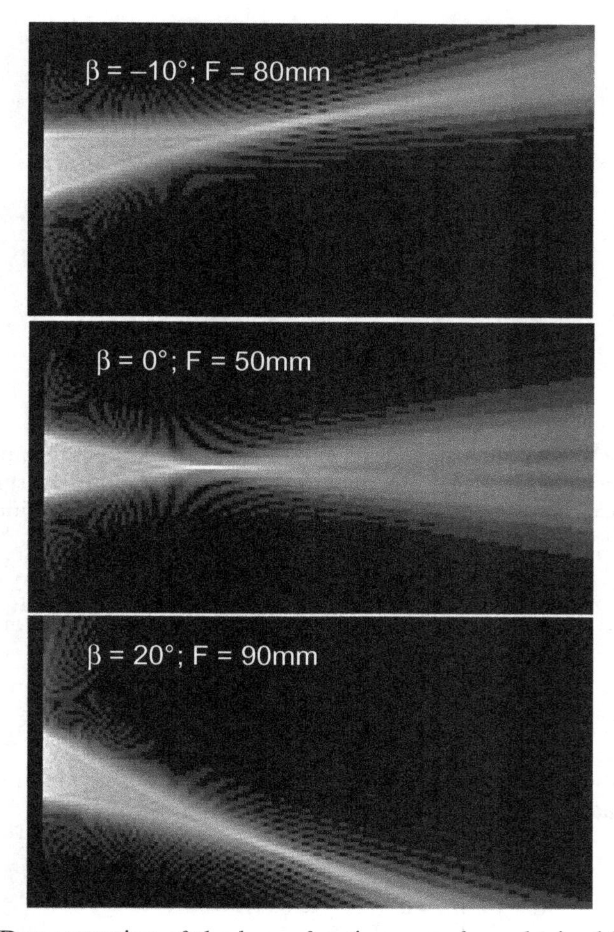

Figure 8.22. Demonstration of the beam-forming procedure obtained by combining angular steering and focusing. The depicted acoustic field was calculated for a 64-element array 20 mm in length transmitting at a frequency of 3 MHz. (The pressure scale is linear.) Note how the focal distance can be varied and the angular orientation controlled by the phased array.

transducer. Staring with the second ring, the request for equal areas leads to the equation

$$\pi\left(r_2^2 - r_1^2\right) = \pi r_1^2$$
$$\Rightarrow r_2^2 = 2r_1^2$$
$$\Rightarrow r_2 = \sqrt{2} \cdot r_1 \tag{8.68}$$

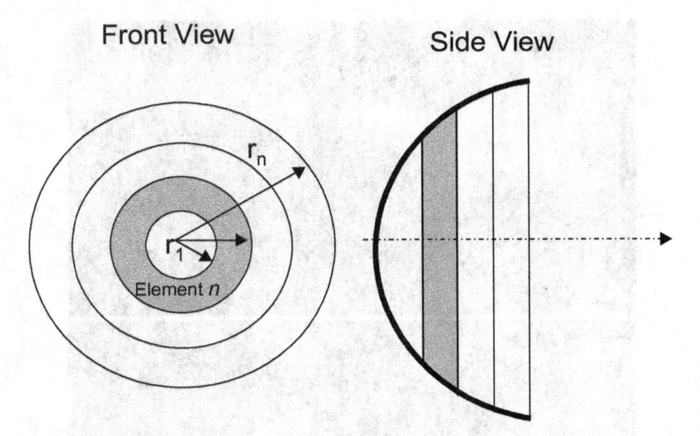

Figure 8.23. Schematic depiction of an annular array. The array is comprised of a set of concentric piezoelectric rings. The rings can be positioned atop a flat surface or upon a curved surface such as the one depicted on the right side, to achieve initial focusing.

Proceeding to the other rings, it can be easily shown that the general condition is that

$$r_n = \sqrt{n} \cdot r_1 \tag{8.69}$$

And if we have a constraint on the size of the outer radius, which should equal a, then the radius of every ring number n is given by

$$\boxed{r_n = a \cdot \sqrt{\left(\frac{n}{N}\right)}, \qquad n = 1:N} \tag{8.70}$$

Regardless of the fact that the rings have equal areas or not, the field pressure P_n transmitted from individual ring number n is given by [see Eq. (8.23)]

$$P_n = A_n \cdot \frac{e^{j(wt-k\cdot R)}}{R} \int_{r=r_{n-1}}^{r_n} 2\pi \cdot J_0(kr\sin\theta) \cdot r \cdot dr \tag{8.71}$$

where A_n is the corresponding amplitude for this particular ring (again it should be recalled that the units are [pressure × distance/area]). The overall field is of course provided by summation of all the contributions made by all the rings, that is,

$$P(R,\theta) = \sum_{n=1}^{N} P_n(R,\theta) \tag{8.72}$$

8.7.1 Steering the Focal Point of an Annular Array

Annular arrays can be useful in several applications. One option is to transmit with a subset of the rings in order to achieve a sharp main lobe and "beam form" differently from the reception to avoid echoes from the side lobes.

Another useful application is steering the focal point of a high-intensity focused ultrasound (HIFU) during a therapeutic procedure. HIFU offers an excellent tool for noninvasive and minimal invasive surgery procedures. Using HIFU, selective hyperthermia and thermal ablation can be obtained within the body in a totally noninvasive manner (see Chapter 12). Thermal ablation is achieved by focusing high acoustic energy into a small focal zone. Focusing can be achieved by using proper hardware as described in Chapter 7. However, with conventional transducers the focal distance cannot be changed. Annular rings, on the other hand, can be used to control the focal distance electronically, thus enabling much more flexibility, as explained in the following.

Consider an annular array of diameter $2a$, such as the one shown schematically in Fig. 8.24. In this case the array is shaped as a concaved surface. The radius of curvature for the surface is R_c. It is required to focus the transmitted acoustic beam at point F located on the transducer's axis of symmetry (the X axis) at a distance F from the surface. Next, consider a point G located on the surface of ring number n. The radius of that particular ring from the axis of symmetry is r_n and its distance from the Y axis is S_n, as described. The value of S_n is given by

$$S_n = R_c - \sqrt{R_c^2 - r_n^2} \tag{8.73}$$

The distance that a wave transmitted from point G has to travel in order to reach the focal point F is marked as d_n and its value is given by

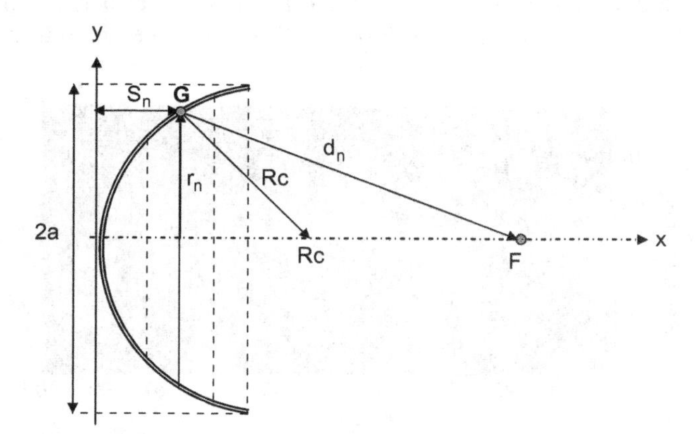

Figure 8.24. Schematic cross section through an annular array. The array is shaped as a concaved surface. The radius of curvature for the surface is R_c. It is required to focus the transmitted acoustic beam at point F located on the transducer's axis of symmetry (the X axis). Point G is located on the surface of ring number n.

$$d_n = \sqrt{(F - S_n)^2 + r_n^2} \tag{8.74}$$

Naturally, due to symmetry, this value is the same for all other points located on ring number n. In order to obtain focusing at point F, the phase of every wave transmitted from any point on that ring should equal the phase reaching this point from the transducer's central point (the axes origin). Practically, this cannot be obtained since the ring transmits from all its surface points with the same phase value. On the other hand, the distance to the focal point varies as one moves from points located on the inner side of the ring to the outer side of the ring. Hence, as a compromise we shall require that the *average phase* of waves reaching the focal point from the inner side of the ring and those arriving from the outer side of the ring will be equal to the phase of a wave reaching the focal point from the axes origin (the central point). This can be expressed mathematically as

$$\varphi = k \cdot F = k \cdot \left(\frac{d_n + d_{n-1}}{2} \right) + \varphi_n \tag{8.75}$$

Following rearrangement of this equation, the needed phase required for ring n in order to achieve focusing at point F is given by:

$$\boxed{\varphi_n = k \cdot \left[F - \left(\frac{d_n + d_{n-1}}{2} \right) \right]} \tag{8.76}$$

It should be noted that this is a generic expression that does not assume equal area rings. An example for applying Eq. (8.76) is depicted in Fig. 8.25. In this figure the calculated acoustic field obtained for an annular array comprised of 10 rings all having the same area is depicted. The array is shaped as a concaved

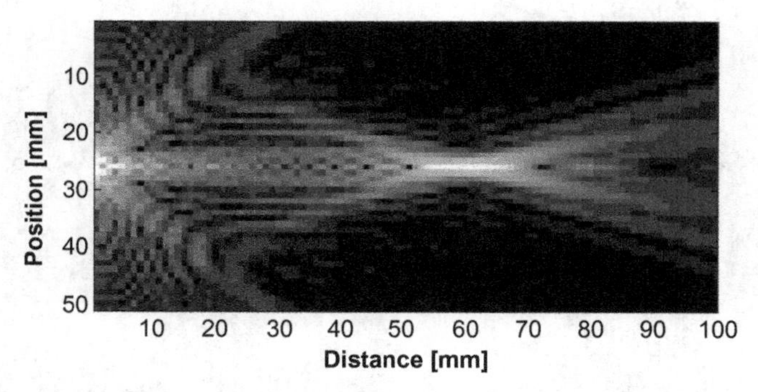

Figure 8.25. A computer simulation of the acoustic field in water obtained for an annular array shaped as a concaved surface of R_c = 50 mm. All rings have the same area. The focal distance is F = 60 mm and the frequency is 1.5 MHz. The pressure scale is linear.

surface with a radius of curvature: $R_c = 50$ mm. The diameter of this array is 50 mm, that is, $a = 25$ mm. The transmission frequency is 1.5 MHz, and the medium is assumed to have a speed of sound $C = 1500$ m/sec (water). The distance to the focal point is $F = 60$ mm. The array is located on the left side of the field and transmits toward the right side. The field is mapped at the plane containing its axis of symmetry. Each pixel is 1 mm × 1 mm in size. The absolute pressure is depicted using a linear scale.

An excellent and thorough analysis of acoustic fields stemming from various transducers is provided in a book by Bjøren A. J. Angelsen [8]. In that book, an approximation formula for an annular array with equal area rings is introduced. The approximation is obtained by utilizing Eq. (8.70) and replacing the square root by the first two terms of its series expansion. The obtained formula is

$$\boxed{\varphi_n = k \cdot S_N \cdot \left(\frac{2n-1}{2N}\right) \cdot \frac{(F - R_c)}{F}} \tag{8.77}$$

where S_N is the distance form the Y axis in Fig. 8.24 to the most external ring (ring number N out of the N rings).

Angelsen [8] also provides the formula for the preferred radius of curvature for an annular array with N equal area rings, when the desired focal steering range is predefined:

$$\boxed{R_c = \frac{2F_{\text{near}} \cdot F_{\text{far}}}{(F_{\text{near}} + F_{\text{far}})}} \tag{8.78}$$

where F_{near} and F_{far} are the required nearest and farthest focal distances, respectively.

Using the above equations, the focal zone can be steered along the array's axis of symmetry. However, it is important to note that the length of the focal zone increases with the distance, while its intensity is reduced. This is demonstrated in Fig. 8.26, where field profiles along the axis of symmetry are depicted. The fields were calculated for the array described in Fig. 8.25. The focal lengths were $F = 40$, 50, and 60 mm. The graphs depict the absolute pressure normalized to the focal pressure for a focal length of $F = 50$ mm. As can be noted, the width of the focal zone peak increases as the focal length is increased, while its amplitude decreases.

8.7.2 The Bessel Beam

Another quality of an annular array is its suitability to transmit a special beam called "the Bessel beam." Using conventional phased array transducers, a single defined focal zone can be obtained in each transmission (as demonstrated in Fig. 8.22). Moreover, the regions located outside the focal zone are

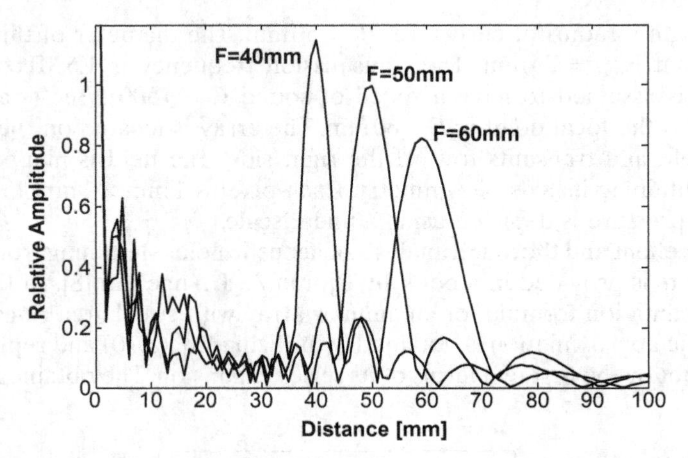

Figure 8.26. Three field pressure profiles obtained by computer simulation for an annular array with $N = 10$ rings transmitting into water. The array has a concaved surface with a radius of curvature $R_c = 50$ mm. The beam is focused at distances of $F = 40$, 50, and 60 mm, respectively. The frequency is 1.5 MHz.

degraded and yield poor spatial resolution. With the Bessel beam, on the other hand, the obtained main lobe is narrow for a very long range. For example, Lu and Greenleaf [9] have reported on an obtained 200-mm-long Bessel beam that has a diameter of only 1.27 mm. This was obtained using an annular array with 10 rings and an external diameter of 25 mm.

The Bessel beam is in fact a family of nondiffractive solutions to the wave equation. With an infinite aperture array, the acoustic pressure field for a Bessel beam of order 0 is given by [9]

$$U(x, y, z, t) = J_0(\alpha\rho) \cdot e^{j(\beta Z - \omega t)}, \qquad \text{where} \quad \alpha^2 + \beta^2 = k^2$$
$$\text{and} \quad \rho = \sqrt{x^2 + y^2} \tag{8.79}$$

where J_0 is a Bessel function of order 0, k is the wave number, and the parameter α fulfills the condition $\alpha \le k$ and has a positive and real value. The parameter β also has a real value. In the private case where $\alpha = 0$, the acoustic field would be that of a regular planar wave.

The Bessel beam has a main lobe whose width at 6 db the maximal value $B_{6\text{db}}$ is given by [9]

$$B_{6\,\text{db}} = \frac{3.04}{\alpha} \tag{8.80}$$

And its width between two zero pressures $B_{0\text{–}0}$ is given by

$$B_{0-0} = \frac{4.81}{\alpha} \tag{8.81}$$

The above equations are valid for an infinite aperture transducer. If the aperture is finite and has a radius of R, it was shown (see reference 9) that the maximal range of the beam $Z_{\max}$ is given by

$$Z_{\max} = R \cdot \sqrt{\left(\frac{k}{\alpha}\right)^2 - 1} \tag{8.82}$$

It should be further noted that above-presented equations assume an ideal continuous Bessel transducer. The annular array that has naturally only a finite number of rings merely approximates the ideal beam. For an annular array containing N rings with equal area, it was shown by Holm and Jamshidi [10] that for a prefocused array with focal distance F, the error in the phase at an infinite distance φ_∞ from the array is given by

$$\varphi_\infty = \frac{\pi}{N \cdot S} \qquad \text{where} \quad S = \frac{F \cdot \lambda}{R^2} \tag{8.83}$$

where S is the Fresnel number, which defines the ration between the focal distance to the length of the near field [see Eq. (8.14)]. The goal is to keep this phase error less than $\pi/2$. By setting the radius of the most inner ring to match the Bessel's function zero value, they have shown that the parameter α is given by [10]

$$\boxed{\alpha = \frac{2.405\sqrt{N}}{R}} \tag{8.84}$$

REFERENCES

1. Mason WP, *Piezoelectric Crystals and Their Application to Ultrasonics*, Van Nostrand, New York, 1950.
2. Setter N, editor, *Piezoelectric Materials in Devices: Extended Reviews on Current and Emerging Piezoelectric Materials, Technology, and Applications*, N. Setter Lausanne, Switzerland, 2002.
3. Kenji U, *Piezoelectric Actuators and Ultrasonic Motors*, Kluwer Academic Publishers, Boston, 1997.
4. Beyer RT and Letcher SV, *Physical Ultrasonic*, Academic Press, New York, 1969.
5. Ferwerda HA, *Huygens' Principle 1690–1990: Theory and Applications*, Proceedings of an International Symposium; Studies in Mathematical Physics, Blok HP, Kuiken HK, and Ferweda HA, editors, the Hague/Scheveningen, Holland, November 19–22, 1990.
6. *Schaum's Outline Series: Mathematical Handbook*, McGraw-Hill, New York, 1968.

7. Alizad A, Wold LE, Greenleaf JF, and Fatemi M, Imaging mass lesions by vibro-acoustography: Modeling and experiments. *IEEE Trans Med Imaging* **23**(9):1087–1093, 2004.
8. Angelsen BAJ, *Ultrasonic Imaging*, Emantec AS, Trondhein, Norway, 2000.
9. Lu JY and Greenleaf JF, *IEEE Trans Ultrason Ferroelectr Freq Control* **37**(5):438–447, 1990.
10. Holm S and Jamshidi GH, *IEEE Ultrasonic Symposium*, San Antonio, TX, November 1996.

CHAPTER 9

ULTRASONIC IMAGING USING THE PULSE-ECHO TECHNIQUE

Synopsis: In this chapter the basic principles of ultrasonic imaging are introduced. The basic definitions associated with reconstructed images are initially given. Then, models of wave scatter from soft tissues are discussed. The process of time gain compensation (TGC) is analyzed. The process of image reconstruction using the pulse-echo method is then thoroughly explained. Finally, special pulse-echo methods such as high harmonics, 3D, and invasive imaging techniques are presented.

9.1 BASIC DEFINITIONS IN IMAGING

9.1.1 Image and Data Acquisition

Using acoustic waves, images that depict certain qualities of the body may be obtained. The first step of the imaging process is *data acquisition*. In this stage of the process, information is collected by sampling the waves that pass through the body and/or scattered waves (echoes) generated by the interaction between the penetrating waves and the tissue. The strategy and the protocols of wave transmission and data collection should be defined a priori.

For example, in Fig. 9.1 a schematic depiction of various wave transmission and detection configurations are shown. As recalled from Chapter 8, the same piezoelectric transmitter can also serve as a receiver. Therefore, one can transmit and receive waves externally to the body (as represented by

Basics of Biomedical Ultrasound for Engineers, by Haim Azhari

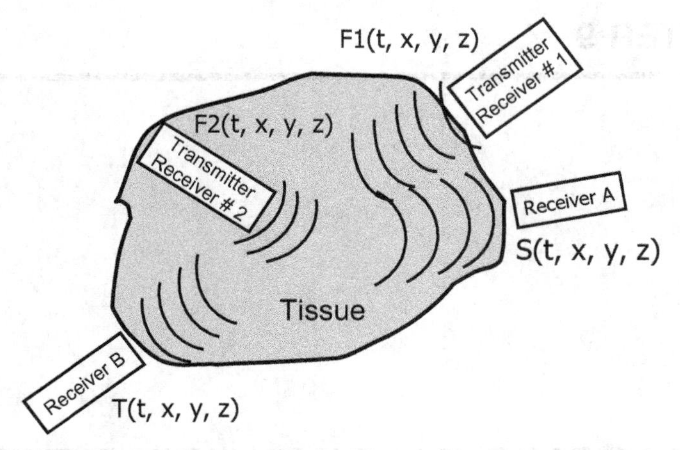

Figure 9.1. Schematic depiction of various configurations of wave transmission and detection that can be used externally or internally to the body.

transmitter/receiver #1), or even from within the body (as represented by transmitter/receiver #2), using the same probe. Each transmit/receive process has to be predefined by a known protocol $F(t, x, y, z)$ specifying the location of the transducer, the transmitted wave profiles and the temporal windows used for detection. Alternatively, the scattered waves $S(t, x, y, z)$ or the through-transmitted waves $T(t, x, y, z)$ can be detected by separate transducers (Receivers A and B, respectively). The transducers can be moved mechanically in space (manually or using a robotic system), and the beam can also be steered electronically using a phased array system as explained in Chapter 8.

After setting the protocol of the data acquisition process, we need to define what is the physical or physiological property of the tissue that will be imaged. This tissue property can be the reflectivity, attenuation, speed of sound, flow velocity, and so on. The obtained image will actually represent a *map of the specific tissue property* chosen. The mathematical relation between that chosen property and the measured data must be known in order to allow quantification of the property. Finally, we need a *reconstruction algorithm* that will transform the information obtained from the measured waves into a picture.

For example, an ultrasonic picture of a baby in her mother's womb is depicted in Fig. 9.2. This image was obtained by transmitting short acoustic pulses and receiving the echo train reflected from within the body. The tissue property depicted in this image represents (roughly) the local reflection coefficient. Strong reflectors are depicted by bright spots, and weak reflectors are depicted as dark regions. The scale used to match the brightness and the tissue property is defined through a transformation that yields a *gray level* lookup table. The user can modify this transformation arbitrarily, and this relation does not have to be linear.

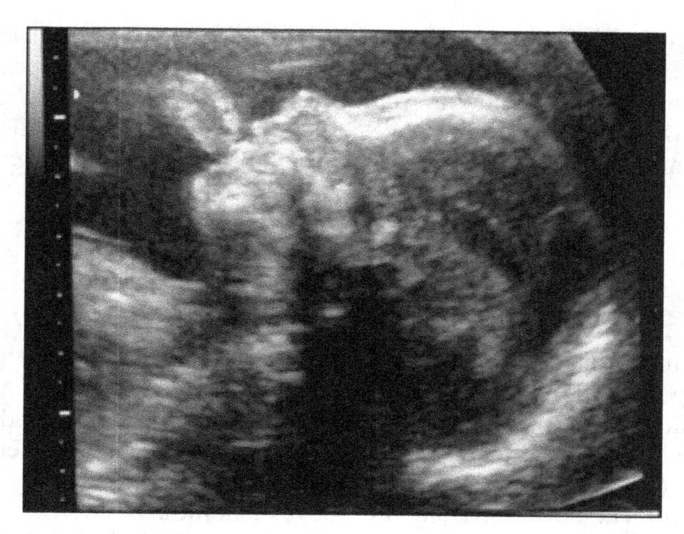

Figure 9.2. An ultrasonic image of a baby in her mother's womb. This image was obtained by registering the reflected echoes after transmitting short acoustic pulses into the body.

9.1.2 Image Contrast

As can be appreciated from the example shown in Fig. 9.2, the most edifying information is located in such zones where the gray level discrepancy between two neighboring regions is significant. For example, the profile of the baby's face is clearly visible due to the strong contrast between the weak reflectivity of the uterine fluid and the strong reflectivity of the baby's tissues. The discrepancy in gray levels between two neighboring regions is called *contrast*. The image contrast can be quantified using different definitions. One option is given by the equation

$$\text{Contrast} \triangleq E\left\{\left|\frac{I_{\text{ROI}} - I_{\text{background}}}{I_{\text{ROI}} + I_{\text{background}}}\right|\right\} \tag{9.1}$$

where I_{ROI} is the gray level value at the region of interest and $I_{\text{background}}$ is the gray level value at a reference region (i.e., background). The symbol $E\{\}$ represents the expected value operator. In some cases the contrast is calculated by substituting only I_{ROI} in the denominator. In other cases the square values of the gray levels are taken. Regardless of the definition used, one must keep the same standard when comparing two images or two scanners.

9.1.3 Signal-to-Noise Ratio

Noise in an image is defined as image features that originate from sources other then the imaged object. Such features may stem from random electronic signals that enter the system through the detectors, cables, or electric circuitry. They may also stem from physical events that the reconstruction process does not account for. A common metric for evaluating the quality of an obtained image is the signal-to-noise ratio (SNR). For a discrete image, this value, in terms of mean squares, can be assessed using the following formula [1]:

$$\mathrm{SNR} \triangleq E\left\{ \frac{\sum_i \sum_j I_0(i,j)^2}{\sum_i \sum_j [I_0(i,j) - I_1(i,j)]^2} \right\} \tag{9.2}$$

where summation is performed on the entire image, $I_1(i, j)$ is the gray level value at point $\{i, j\}$ in the reconstructed image (see definition of the term *pixel* in the following), and $I_0(i, j)$ is the gray level value at point $\{i, j\}$ in the "ideal" (i.e., noiseless) image. In many cases this value is quite high and therefore the "decibel" (db) scale is used (as explained in Chapter 5).

The problem resulting from the above definition is how to obtain the "ideal" image values $I_0(i, j)$. For that purpose a special object must be built. This special object is called a *phantom*. All the relevant physical properties of the phantom should be exactly known. The phantom can be made of solid or liquid elements or even from a biological substance. In order for the phantom to be useful, one should know with a sufficient accuracy what would be its corresponding $I_0(i, j)$ image. Sometimes instead of using a physical phantom, a virtual one can be generated using a computer emulation.

In Fig. 9.3a, for example, a virtual phantom generated by a computer is depicted. This numeric phantom is the well-known Shepp–Logan head phantom [2]. It is comprised of a set of embedded ellipses with predefined dimensions and gray levels. The same phantom is shown in Fig. 9.3b, but this time image "noise" was added. (The "noise" was generated using a special Matlab® dedicated function.) The image obtained by subtracting these two images is shown in Fig. 9.3c. The corresponding SNR for the image depicted

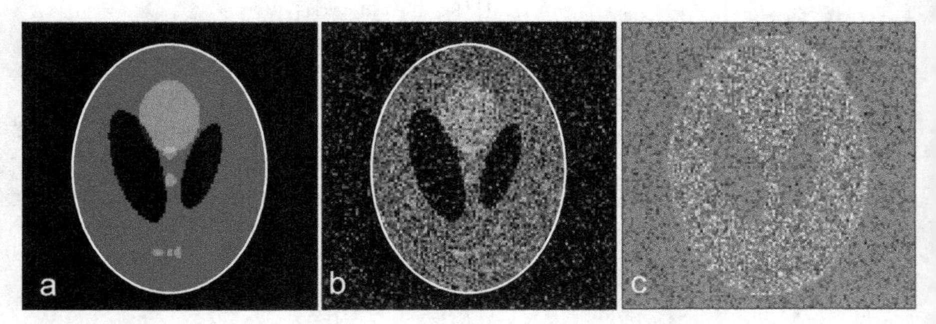

Figure 9.3. **(a)** Image of the Shepp–Logan phantom. **(b)** Same image with added noise. **(c)** Image obtained by subtracting two images.

in Fig. 9.3b can be calculated using Eq. (9.2) by substituting the values of the image shown in the numerator (Fig. 9.3c) and the values of the image shown in the denominator (Fig. 9.3c).

It is not always possible to estimate the SNR reliably. Building a good phantom is sometimes an art of its own. Therefore, an additional metric called the contrast-to-noise ratio (CNR) is sometimes used. Assuming that the image obtained has an important feature that we would like to present in a clear manner, it is imperative that the difference in gray levels between this specific feature and its background would be significantly greater than the random fluctuations in gray levels induced by the image noise. Naturally, it is required to have as large a CNR as possible. The corresponding value of the image CNR can be calculated by

$$\mathrm{CNR} \triangleq \frac{\left|\bar{I}_{\text{object}} - \bar{I}_{\text{background}}\right|}{\sqrt{\sigma^2_{\text{object}} + \sigma^2_{\text{background}}}} \tag{9.3}$$

where $\bar{I}_{\text{object}}$ and $\bar{I}_{\text{background}}$ are the mean gray levels in the object of interest and the background, respectively, and σ_{object} and $\sigma_{\text{background}}$ are the standard deviations within the object and the background, respectively.

9.1.4 Resolution

Another important parameter in the assessment of image quality is resolution. There are basically two important values:

I. *Spatial Resolution.* This is defined as the minimal distance separating two adjacent points having the same property in the object, which are displayed in the image as two separate points (or regions).
II. *Temporal Resolution.* This is defined as the minimal time required to complete the acquisition of the data needed to reconstruct one complete image of the object.

It is important to note that current imaging systems provide discrete representation of the images. Thus, every reconstructed image is actually stored in the computer as a matrix of numbers, i.e. $I(i, j)$, where column i represents the horizontal coordinate (x axis) and line j represents the vertical coordinate (y axis). Every cell in the matrix represents a small area in the object for which it is assumed that the corresponding imaged property may be considered uniform. This area is called *pixel*. The numerical value of the cell in the matrix is used for assigning the appropriate gray level to that pixel. When reconstructing an object in three dimensions (3D), the matrix is three-dimensional and each element in the matrix represents a small cube of the object. This cube is called *voxel*. The pixel or voxel size, therefore, sets the limit to the spatial resolution of the reconstructed image.

The property that sets the spatial resolution is the *impulse response* of the imaging system—or, more accurately, the characteristics of the image obtained when scanning an isolated point target. The obtained image in this case depicts a map of the combined effect of the physical processes involved and the reconstruction procedure. This map is known as the *point spread function* (PSF).

The profile of the PSF along any direction depicts the "smearing" of the information along that direction due to the overall reconstruction process. In the context of ultrasonic imaging, it is important to characterize the resolution along the beam direction, defined herein as the "axial resolution," and the resolution perpendicular to the beam, defined herein as the "lateral resolution."

In order to clarify these terms, a set of drawings is depicted in Fig. 9.4. An ultrasonic beam, which is schematically represented as an arrow, scans a point target (top row). The obtained image in this case is the corresponding PSF of the system, which is schematically represented as an ellipse. This ellipse-type image stems from the fact that the smearing along the lateral direction is typi-

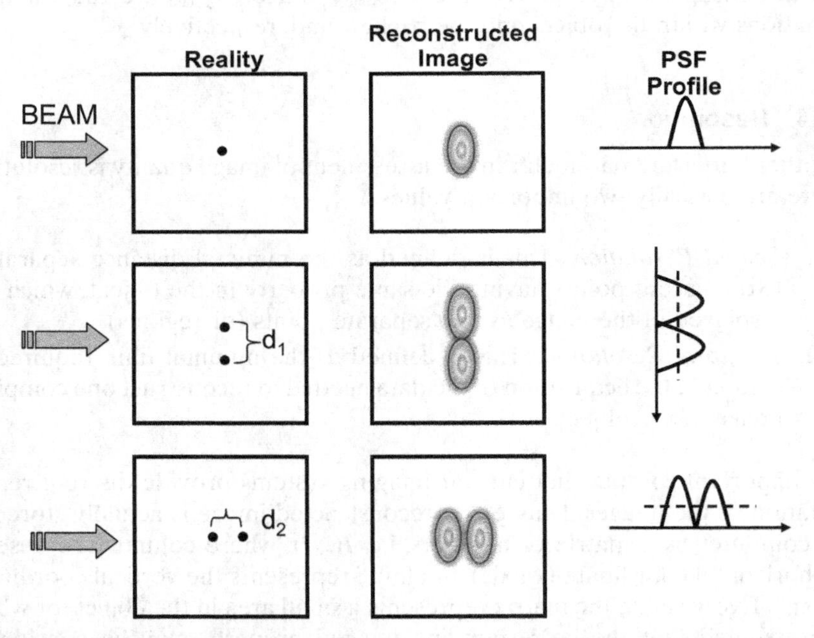

Figure 9.4. **(Top row)** Schematic representation of the PSF parameter. A point target is scanned by an acoustic beam (*left*). The corresponding image is represented by the dark ellipse (*center*). The profile along the axial direction is shown (*right*). **(Middle row)** The process of evaluating the lateral resolution d_1 is shown. This value can be derived from the two PSF lateral profiles of two adjacent point targets. **(Bottom)** A schematic depiction of the process used for estimating the axial resolution d_2, by scanning two point targets positioned one after the other.

cally much greater than along the axial direction. The profile of the PSF is schematically represented in the right-hand column.

In order to estimate the lateral resolution of a system, two point targets separated by distance d_1 along the lateral direction are scanned. If the resolution of the system is sufficient, the reconstructed image will depict two adjacent spots. By plotting the central profiles along the lateral direction of these spots, two adjacent (or even overlapping) profiles will be obtained. The two spots in the image may be considered separated if by cutting these profiles at 50% of their maximal amplitude the profiles do not overlap. Similarly, the resolution along the axial direction d_2 can be estimated by scanning two point targets positioned one after the other.

9.2 THE "A-LINE"

9.2.1 The Simple Model

The building block of pulse-echo based ultrasonic imaging is a signal that is called the "A-line." In order to reconstruct this A-line signal, let us make the following assumptions:

I. The transmitted wave is a very short pulse in time and may be approximated by a delta function.
II. The transmitted wave may be considered planar. Hence, the only sources of attenuation in the propagating wave amplitude are the transmission coefficients and the attenuation confinements of the medium.
III. The targets that the propagating wave encounters may be approximated by a set of parallel plates.
IV. The targets are positioned normal to the wave propagation direction; hence only axial reflections and transmissions occur.

In order to derive the mathematical model for the A-line signal, we shall make use of Fig. 9.5. An ultrasonic pulse of amplitude A_0 is transmitted at time t_0 into the first medium, commonly gel or water, for which the acoustic impedance is Z_0. The distance to the first target is S_0. For convenience, let us mark the transmission coefficient from the transducer into the first target as $T_{(-1,0)}$, and let us denote the transmission coefficient from the first target back into the transducer as $T_{(0,-1)}$. The values of these coefficients can be determined as explained in Chapter 6. (Note that if the matching layer for the transducer is optimally set, then one may consider these transmission values to equal $|1|$.)

For a target consisting of three layers, the corresponding acoustic impedances are $\{Z_1; Z_2; Z_3\}$, respectively. The thicknesses of the first two layers are $\{S_1; S_2\}$, respectively. Based on the above assumptions, reflections will occur only at the boundaries of these layers. The propagating wave will start its

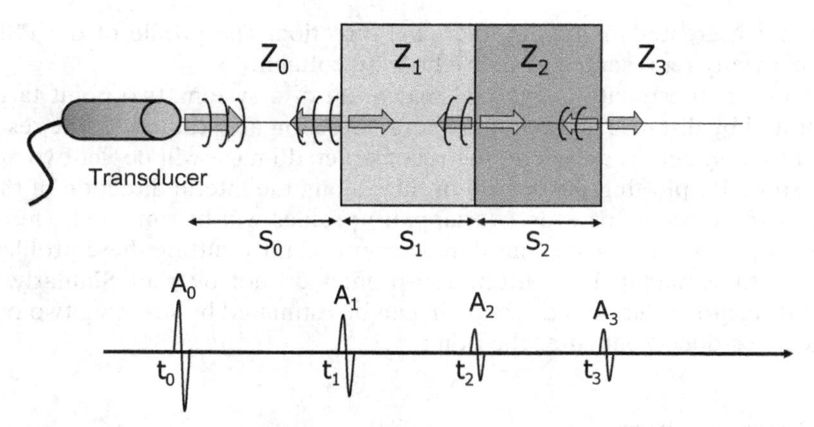

Figure 9.5. (Top) Schematic depiction of an ultrasonic pulse transmitted into a medium comprising of several layer. The transmitted and reflected waves are represented by the arrows (reverberations are excluded). **(Bottom)** Schematic representation of the corresponding A-line signal.

journey by traveling along the distance S_0, and its amplitude will decay exponentially according the corresponding attenuation coefficient μ_0 as explained in Chapter 5. The first echo will be obtained, as explained in Chapter 6, from the boundary between the first medium and the target where the acoustic impedance changes from Z_0 to Z_1. The amplitude of this echo will be determined by the reflection coefficient R_{01}. The echo will travel back toward the transducer, and its amplitude will decay exponentially according the corresponding attenuation coefficient μ_0. This first echo will be detected at time t_1, and its amplitude will be A_1. Without much difficulty, we can write the expression for the first echo signal $E_1(t)$ as

$$E_1(t) = A_0 \cdot T_{(-1,0)} \cdot T_{(0,-1)} \cdot e^{-2\mu_0 S_0} \cdot R_{01} \cdot \delta\left(t - \frac{2S_0}{C_0}\right), \qquad \text{where } R_{01} = \frac{(Z_1 - Z_0)}{(Z_1 + Z_0)} \tag{9.4}$$

Clearly, $E_1(t)$ has a nonzero value only when the argument for the δ function is zero—that is, when $t = t_1 = 2S_0/C_0$.

If the acoustic impedance of the first target layer is not too small or too large, relative to the first medium, then the echo will contain only part of the wave energy and the rest will be transmitted into this first target layer. The amplitude of this transmitted wave will be determined by the transmission coefficient T_{01}. This transmitted wave will propagate without reflections through distance S_1 until it will reach the boundary of layer number 2. Its amplitude will decay exponentially according to the attenuation coefficient μ_1.

At the boundary between the first and second target layers a second echo will be reflected. The amplitude of this second echo will be determined by the reflection coefficient R_{12}. This second echo will travel back along S_1, and its amplitude will again decay exponentially according to the attenuation coefficient μ_1. Part of its energy will be reflected, and the amplitude of the continuing wave will be determined by the transmission coefficient T_{10}. This part of the echo will continue to travel along distance S_0, and its amplitude will decay exponentially according the corresponding attenuation coefficient μ_0. The corresponding second echo signal $E_2(t)$ is thus given by

$$E_2(t) = A_0 \cdot T_{(-1,0)} \cdot T_{(0,-1)} \cdot e^{-2\mu_0 S_0} \cdot T_{01} \cdot e^{-2\mu_1 S_1} \cdot R_{12} \cdot T_{10} \cdot \delta\left(t - \frac{2S_0}{C_0} - \frac{2S_1}{C_1}\right)$$
$$\text{where} \quad T_{01} = \frac{(2Z_1)}{(Z_1 + Z_0)}, T_{10} = \frac{(2Z_0)}{(Z_1 + Z_0)}, \quad \text{and} \quad R_{12} = \frac{(Z_2 - Z_1)}{(Z_2 + Z_1)} \tag{9.5}$$

Again it can be noted that $E_2(t)$ has a nonzero value only when the argument for the δ function is zero—that is, when $t_2 = 2S_0/C_0 + 2S_1/C_1$.

In order to extend the derivation to the more general case, let us assume that the target is comprised of N layers. The corresponding echo received from the Nth target layer $E_N(t)$ is thus given by

$$\begin{cases} E_N(t) = A_0 \cdot \left[\prod_{n=0}^{N-1} e^{-2\mu_n S_n} \cdot T_{(n-1,n)} \cdot T_{(n,n-1)}\right] \cdot R_{(N-1,N)} \cdot \delta\left(t - \sum_{n=0}^{N-1} \frac{2S_n}{C_n}\right) \\ \text{where} \quad T_{(n-1,n)} = \frac{(2Z_n)}{(Z_{n-1} + Z_n)}, T_{(n,n-1)} = \frac{(2Z_{n-1})}{(Z_{n-1} + Z_n)}, n > 0; \\ \text{and} \quad R_{(N-1,N)} = \frac{(Z_N - Z_{N-1})}{(Z_N + Z_{N-1})} \end{cases} \tag{9.6}$$

where the transmission coefficients for $n = 0$, (i.e. $T_{(-1,0)} \cdot T_{(0,-1)}$) designate the losses occurring at the transducer's surface.

In conclusion, it can be stated that using this simple model, the overall A-line signal $E(t)$ which is comprised of all the received echoes is given by

$$E(t) = \sum_{N=1}^{\infty} E_N(t) \tag{9.7}$$

9.2.2 Extending the Model

The basic model developed above can be extended (and become more complicated) by avoiding some of the simplifying assumptions listed above. Nevertheless, it is highly recommended that one should always weigh the benefit of each added complication versus the resulting complexity. Indeed, one should seek to have the most accurate representation of the actual physical processes. However, a too complicated model may render the equations practically useless.

One simple and useful extension is to replace the delta function pulse by a more realistic pulse profile. As explained in Chapter 5, a good approximation of an actual ultrasonic pulse is the Gaussian pulse (as recalled the spectral profile of this pulse has a Gaussian profile), which is given by

$$p(t) = e^{(-\beta \cdot t^2)} \cdot \cos(2\pi \cdot f_0 \cdot t) \quad (9.8)$$

where the parameter β defines the bandwidth and f_0 sets the central frequency of the transmitted pulse.

The simple model given by Eqs. (9.6) and (9.7) can be upgraded by convolving the Gaussian pulse profile with the set of delta functions representing the echo train. Hence, we can write

$$S(t) = p(t) \otimes E(t) = p(t) \otimes \sum_{N=1}^{\infty} E_N(t) \quad (9.9)$$

This simple operation provides a much more realistic description of the A-line signal. As demonstrated in Fig. 9.6, the simulation results obtained for the echoes reflected from a plastic plate immersed in water calculated using Eqs. (9.9) and (9.8) realistically describes the actual A-line signal obtained by measurements. Indeed, the measured A-line signal is more noisy and distorted, but the good resemblance of the calculated signal is evident.

Another extension of the model which may be beneficial is to replace the planar wave with a more realistic model (as explained in Chapter 8). A precise calculation of the effects stemming from the angle of incidence between the wave and the targets may be too demanding in terms of computation complexity. However, as recalled from Chapter 8, the amplitude of a nonfocused acoustic beam typically decays according to wave travel distance in addition

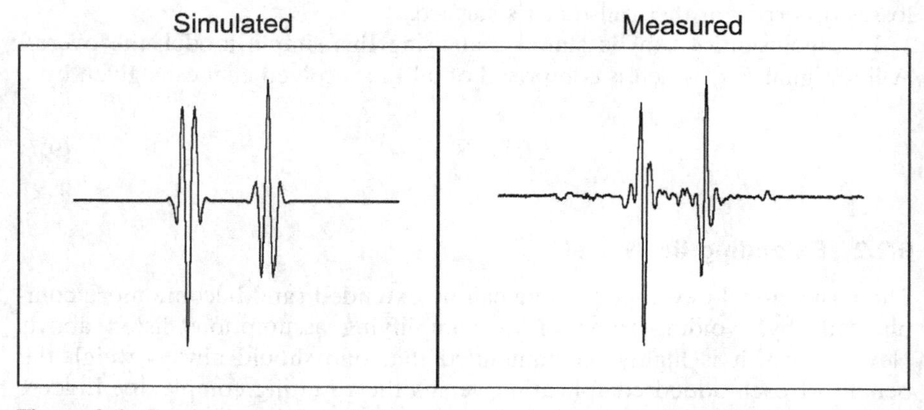

Figure 9.6. Comparison of an A-line signal depicting the reflected echoes obtained from a plastic plate immersed in water **(right)** and the corresponding calculated signal obtained by combining Eqs. (9.8) and (9.9) **(left)**.

to the attenuation effect. Thus, as a first-order approximation one can account for this effect using the following model:

$$E_N(t) = A_0 \cdot \frac{B_0}{\sum_{n=0}^{N-1} 2S_n} \cdot \left[\prod_{n=0}^{N-1} e^{-2\mu_n S_n} \cdot T_{(n-1,n)} \cdot T_{(n,n-1)} \right] \cdot R_{(N-1,N)} \cdot \delta\left(t - \sum_{n=0}^{N-1} \frac{2S_n}{C_n} \right) \tag{9.10}$$

where B_0 is a constant set to match the amplitude to a reference amplitude at a reference distance. Combining this equation with Eq. (9.9) may be beneficial in certain cases.

9.3 SCATTER MODEL FOR SOFT TISSUES

The model used above for deriving the mathematical expression for the A-line signal was based on the simplifying assumption that the target was comprised of many planar layers that are perpendicular to the acoustic beam (as shown in Fig. 9.5). However, in reality the situation is naturally more complicated. Living organs and tissues are comprised of a complicated assembly of cells and fluids of different physiological and mechanical properties. Moreover, even when limiting our region of interest to the size of a single PSF spot (a few wavelengths), the biological texture may still be complicated because it may contain different types of cells, fibers, connective tissues, and blood vessels. This kind of an assembly is therefore a dense cloud of discontinuity points in the acoustic impedance. Each point serves as a small speck for wave scatter. These scattering points are commonly smaller than the traveling wavelength. Therefore, when transmitting an ultrasonic wave toward a soft tissue target, many scattered waves in all direction may be obtained. This is shown schematically in Fig. 9.7.

Let us define the intensity of the incident waves as I_0 and the intensity of the reflected waves along direction θ as $I_s(\theta)$. As in the configuration shown in Fig. 9.7, the angle $\theta = \pi$ designates back-scatter (i.e., waves reflected back toward the transmitting transducer). The following relation can be defined:

$$I_S(\theta) = R(\theta) \cdot I_0 \tag{9.11}$$

where $R(\theta)$ designates the effective reflection factor along angular direction θ for a bulk scattering medium. It is important to recall that the scatter is in fact three-dimensional. However, for simplicity we shall assume symmetry around the beam axis and limit our discussion to two dimensions.

The solution of the problem for many small spherical scatterers of radius a for which the condition $ka << 1$ is valid (k is the corresponding wave number) was derived in the nineteenth century by the famous scientist Raleigh. The solution assumes that all the energy impinging upon a single particle is

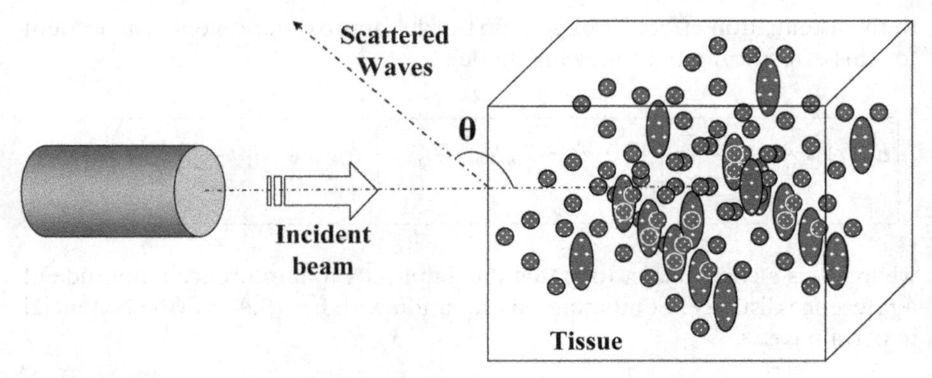

Figure 9.7. A more realistic model for describing the scatter of waves from soft tissues. The scattering points are randomly arranged within the tissue. Consequently, waves may be scattered in all directions.

scattered. This implies in our context that the acoustic impedance of the reflecting particle is much greater than that of the surrounding medium. Raleigh used this model to explain why the sky is blue despite the fact that the sunlight is white. For this model, Rayleigh found that the reflection factor can be described by (see Szabo [3])

$$R(\theta) = \frac{k^4 a^6}{9r^2} \cdot \left(1 - \frac{3\cos\theta}{2}\right)^2 \tag{9.12}$$

where r is the distance from the scatterer. Using this equation, the corresponding reflection factor for back scatter is obtained by substituting $\theta = \pi$:

$$R(\pi) = \frac{25k^4 a^6}{36r^2} \tag{9.13}$$

Considering the fact that the wave number fulfills the relation $k = 2\pi f/C$, it can be easily conceived that this reflection factor depends on the fourth power of the frequency.

It is important to recall that the basic assumption of Rayleigh's model is that the acoustic impedance of the scattering points is much greater than the surrounding medium. However, this is not the case for soft tissues. In fact the acoustic properties of different soft tissues are quite similar to each other (see Appendix A at the end of the book). It is more realistic to consider the scattering regions as locations with small perturbation in properties relative to the surrounding medium. For this purpose, two dimensionless parameters can be defined:

$$\gamma_\rho = \frac{(\rho_s - \rho_0)}{\rho_0}, \qquad \gamma_\kappa = \frac{(\kappa_s - \kappa_0)}{\kappa_0} \tag{9.14}$$

where ρ_s and ρ_0 are the densities of the scatterer and the surrounding medium, respectively. The term $\kappa = 1/\beta$ is the reciprocal value of the compressibility coefficient (see Chapter 3) and is also known as the *bulk modulus*. The terms κ_s and κ_0 are the corresponding bulk modulus for the scatterer and the medium, respectively. Therefore, the parameters γ_ρ and γ_κ describe the normalized changes in the density and compressibility, respectively.

Next, let us define another parameter $\sigma(\theta)$ by

$$\sigma(\theta) = \lim_{d\Omega \to 0} \frac{dW(\theta)_s}{I_0 d\Omega} \tag{9.15}$$

where $dW(\theta)_s$ is the power scattered along solid angle $d\Omega$ as a result of an impinging wave with an intensity of I_0. The parameter $\sigma(\theta)$ has the physical units of area. Thus, it is called the "differential cross section." The value of this parameter can be estimated by various ways (see, for example, Dickinson [4]),

For the case of a spherical scatterer and a cylindrical one, Nassiri and Hill [5] calculated the differential cross section using the Born approximation (see Chapter 1).

For a small spherical scatterer of radius a the following expression was obtained:

$$\sigma_{\text{sphere}}(\theta) = \frac{k^4}{k_s^6} \cdot (\gamma_\kappa + \gamma_\rho \cos\theta)^2 \cdot (\sin k_s a - k_s a \cdot \cos k_s a)^2 \tag{9.16}$$

where $k_s = 2k\sin(\theta/2)$.

For a small cylindrical disc-shaped scatterer of radius a and thickness h the following expression was obtained (*Note*: The original reference by Nassiri and Hill [5] seems to have an error in their Eq. (7). This I believe is the right expression.):

$$\sigma_{\text{disc}}(\theta) = \frac{h^2 k^4 a^4}{4} \cdot (\gamma_\kappa + \gamma_\rho \cos\theta)^2 \cdot \left(\frac{J_1(k_s a)}{k_s a} \right)^2 \tag{9.17}$$

In order to use these expressions, one may assume a certain distribution of discrete scatterers in space. The scatterers can be from both types. Their spatial arrangement may attempt to mimic the structure of the studied tissue by combining regular and random distributions. In order to simplify the model, one has to assume that each wave is scattered only once. In other words, a scattered wave from an individual scatterer does not encounter any more scatterers along its way to the receiver. Thus, the sum of all the contributions of all scattered waves along a specific direction will provide the overall effective differential cross section. It should be also pointed out that in order to make the model more realistic, attenuation of the beam must also be accounted for.

9.3.1 The Speckle Texture

When the transmitted ultrasonic beam encounters the boundary between two organs or tissue regions, strong and spatially ordered echoes are commonly detected. These well-defined echoes enable us to map the general anatomy. However, as explained above, the structure of soft tissues is very heterogeneous. Even within "homogeneous" tissue regions, many points of acoustic impedance discontinuities are spread. Consequently, reflections and scatter will occur almost throughout the entire path of the propagating acoustic beam.

The reflecting points located within the tissue are typically smaller than the wavelength, and they scatter the acoustic energy in almost every direction. These individual scatters are commonly very weak, and we wouldn't have been able to detect them if not for the fact that there are so many reflectors in the medium. This results in constructive interferences of waves almost everywhere. These points are detected and appear as speckle texture. Although this texture may seem like random noise (part of it is indeed image noise), its structure is not random. The spatial arrangement of the scattering points (stemming from the tissue structure) and the relative orientation of the transducer define this texture. The image is therefore reproducible if the tissue orientation relative to the transducer's acoustic field is kept the same.

An exemplary ultrasonic image obtained with the pulse-echo technique for a commercial phantom is given in Fig. 9.8. The transducer was placed on the top side of the image. The phantom is comprised of many small reflectors

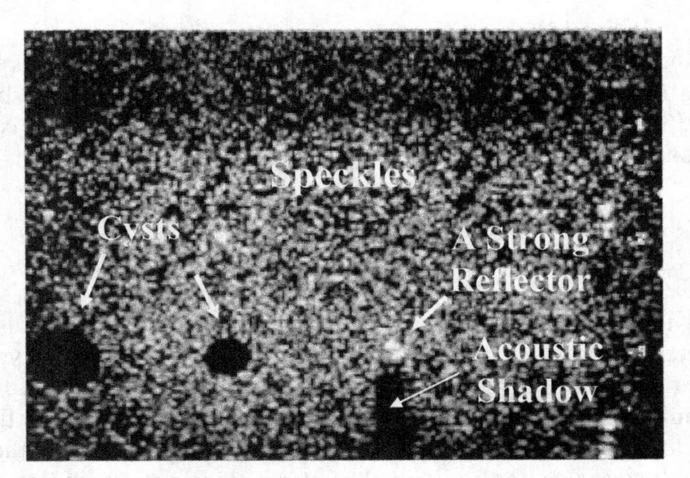

Figure 9.8. A pulse-echo image of a commercial phantom. The transducer was positioned at the top side of this picture. Notice the typical speckle texture. The dark circles are regions containing fluids (simulating cysts) which are naturally echoless. There is one strong reflector in the phantom. Notice the acoustic shadow behind this strong reflector.

embedded in a homogeneous medium. It contains two circular regions containing fluids which do not provide any echoes (simulating cysts), and a strong reflector. The typical speckle structure of the obtained image is clearly visible. It is also worth noting the strong acoustic shadow behind the strong reflector.

9.4 TIME GAIN COMPENSATION

Because it is difficult to use elaborated models of scattering tissue properties, we shall revert to the simple multilayered model. Studying Eq. (9.10), one can note that the "interesting" information, consisting of the reflectivity indices $R_{(N-1,N)}$, is masked by the term in the square brackets. This term contains all the cumulative effects of attenuation and transfer coefficients. Furthermore, the speed of sound varies from one location to another in a manner that is not known to the operator. Thus, we cannot determine the exact location from which the echo has been reflected. These facts render even Eq. (9.10) impractical for quantitative use.

A partial remedy for overcoming these difficulties is provided by the time gain compensation (TGC). The basic idea is to apply a gain function that will amplify the echoes received from deeper layer in a manner that will at least partially compensate for the term in the square brackets of Eq. (9.10). In order to build the TGC function, two basic assumptions are made:

I. The attenuation coefficient is the same throughout the acoustic path, that is $\mu(x) = \bar{\mu}$. Naturally, this assumption does not really match the facts (see Appendix A at the end of the book). However, as long as there are no bones or air along the acoustic path, this assumption is tolerable.

II. The speed of sound is the same throughout the medium, that is, $c(x) = \bar{C}$. This is of course not true. But since the changes in the speed of sound are on the order of several percent, this assumption is also reasonable.

Using these two assumptions, we can rearrange Eq. (9.10) to obtain

$$E_N(t) = A_0 \cdot R_{(N-1,N)} \cdot \delta\left(t - \frac{1}{\bar{C}} \sum_{n=0}^{N-1} 2S_n\right) \cdot \left[e^{-\bar{\mu} \cdot \sum_{n=0}^{N-1} 2S_n} \cdot \prod_{n=0}^{N-1} T_{(n-1,n)} \cdot T_{(n,n-1)} \right] \tag{9.18}$$

The corresponding time–range relation can be written as

$$Xn = \sum_{n=0}^{N-1} S_n = \bar{C} \cdot \frac{t}{2} \tag{9.19}$$

where Xn is the range to the nth reflection and t corresponds to the time elapsed from transmission. The division by 2 stems from the fact that the transmitted wave has to travel all the way to the target and the echo has to travel this distance back to the transducer. Substituting Eq. (9.19) into Eq. (9.18) yields

$$E_N(t) = A_0 \cdot R_{(N-1,N)} \cdot \delta\left(\frac{\bar{C} \cdot t}{2} - Xn\right) \cdot \left[\prod_{n=0}^{N-1} T_{(n-1,n)} \cdot T_{(n,n-1)}\right] \cdot e^{-\bar{\mu} \cdot \bar{C} \cdot t} \tag{9.20}$$

As can be noted, the relevant information, which is the range to the reflector and its corresponding reflectivity index, is given by the third and second terms, respectively. The last two terms—that is, the term in the square brackets and the exponential decay—still mask the information we need. However, since they are now more simplified, we can partially reduce their effect by multiplying the echo signal by a time-dependent gain compensation function, the TGC. Thus, ideally we can define the following TGC function,

$$\boxed{TGC_N(t) = \frac{1}{\left[\prod_{n=0}^{N-1} T_{(n-1,n)} \cdot T_{(n,n-1)}\right]} \cdot e^{+\bar{\mu} \cdot \bar{C} \cdot t}} \tag{9.21}$$

and its resulting effect by

$$\boxed{\Rightarrow S_N(t) = E_N(t) \cdot TGC_N(t) = A_0 \cdot R_{(N-1,N)} \cdot \delta\left(\frac{\bar{C} \cdot t}{2} - Xn\right)} \tag{9.22}$$

where $S_N(t)$ is the TGC corrected signal, and $\mathrm{TGC}_N(t)$ is the gain correction needed for the nth target.

The TGC is obtained typically by applying a dedicated hardware component in the ultrasonic scanner. The problem, however, is that we would like to correct the entire A-line, and we do not know a priori the required ideal TGC parameters. As a compromise, the TGC is set empirically at real time. This can be done automatically by the system or manually by the operator. Thus, we can write the general expression for the TGC as

$$\boxed{TGC(t) = \Pi(t) \cdot e^{+\bar{\mu} \cdot \bar{C} \cdot t}} \tag{9.23}$$

where $\Pi(t)$ is some empirically set continuous function.

9.5 BASIC PULSE-ECHO IMAGING (B-SCAN)

The most popular ultrasonic imaging method is based on the pulse-echo technique. With this method a short ultrasonic pulse is transmitted into the medium

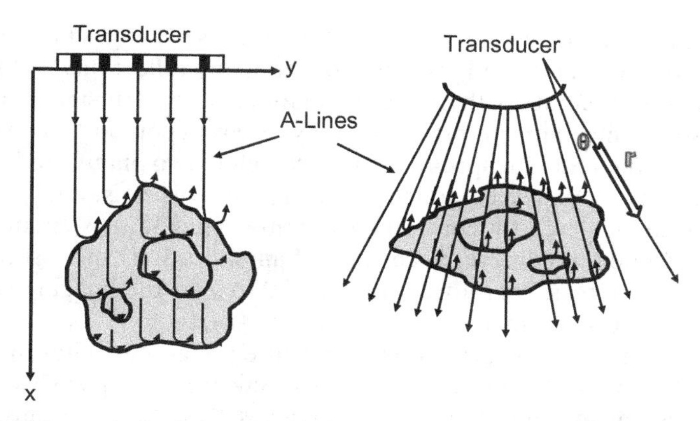

Figure 9.9. Schematic depiction of the pulse-echo imaging (B-scan) strategies. **(Left)** Cartesian scan. **(Right)** Sector scan. The straight arrows represent the A-lines, and the curved arrows represent the reflections (echoes).

and the resulting wave propagates as explained above. The intensity of the wave at any location is determined by the corresponding acoustic field (see Chapter 8), and reflections occur at every discontinuity in the acoustic impedance within the medium. In most cases the same transducer is used both as a transmitter and as a receiver. Thus, the echoes reflected at an angle of 180°; that is, backscattered waves are sampled.

Two-dimensional imaging is obtained by steering the acoustic field using one of the methods described in Chapter 8. The scanned area is "covered" by collecting many A-lines along each direction toward which the beam is aimed. After processing the A-line signals as will be explained in the following, a cross-sectional image is obtained. Basically, there are two main scanning strategies as depicted in Fig. 9.9. With the first approach a Cartesian strategy is utilized and the scanned area is rectangular in shape. With the second strategy a polar scan is applied, and the obtained image is shaped as a circular sector. Naturally, these two approaches can be combined or alternative approaches can be utilized.

After completing the data collection stage as explained above, two more steps are needed to obtain an image:

I. *Contrast Generation.* By conversion of each A-line signal into a vector of gray levels.
II. *Spatial Mapping.* Each obtained gray level value must be allocated to a specific position in the image space.

9.5.1 Conversion to Gray Levels

As explained in the previous sections and demonstrated in Fig. 9.6, the signal detected by the ultrasonic transducer (commonly referred to as the RF signal) may have many rapid changes in its amplitude. These changes may stem from the profile of the transmitted wave and/or the many reflections that the wave may encounter along its path. These fluctuations in the RF signal are typical to soft tissues, which are comprised of many different components. Moreover, in addition to the inherent speckle texture, which is also manifested in the RF signal, electric noise may be added to the signal. This noise may stem from the electric circuits used for controlling and processing the system or from the cables. In addition, nonlinear processes and amplifications can distort the RF signal and induce bias. Therefore it is essential to start the imaging procedure by improving the RF signal.

The first element of the processing procedure is high-pass filtering. Because in most of the medical applications the ultrasonic wave frequencies are in the range of several megahertz, lower frequencies than the transmitted signal band may be attributed solely to noise and hence can be removed. (It is important to note, however, that in some applications such as monitoring cavitation bubbles or using contrast materials, one may be interested in detecting the subharmonic signal.)

The other element is low-pass filtering. This element is needed to further reduce the noise and prepare the signal for digitization by the A/D converter. As recalled from the Shannon–Whittaker sampling theorem [6], the highest frequency that can be used is half the sampling frequency. This frequency is sometimes referred to as the Nyquist frequency. Thus, it is essential to limit the maximal frequency in the RF signal before digitizing in order to avoid aliasing.

The two filters can be applied consecutively or simultaneously using a band-pass filter. Following the filtering stage, the time gain compensation (TGC) should be applied as explained above. And then the signal is digitized using an A/D converter.

At this point it is recommended to apply a median filter to remove isolated spikes from the signal. A two-dimensional median filter is also recommended at the end of the process. This filter effectively removes speckle noise from the image while retaining the border and without smearing the picture.

After cleaning the signal and adjusting its amplitudes, the envelope of the signal needs to be extracted. This stage allows us to display the changes stemming from the texture and not from the wave profile. The envelope of the signal can be obtained using the Hilbert transform $H\{\}$. This transform is defined as [7]

$$\begin{aligned} H\{S(t)\} &= \frac{1}{\pi}\int_{-\infty}^{\infty}\frac{S(\tau)}{(t-\tau)}\,d\tau, \\ H(\omega) &= -j\cdot\operatorname{sign}(\omega) \equiv \begin{cases} -j & \omega > 0, \\ j & \omega < 0 \end{cases} \end{aligned} \tag{9.24}$$

This transform changes the phase of all the negative frequencies by 90° and the phase of all the positive frequencies by −90°. With the Hilbert transform we can generate a signal that is known as the "analytic signal" (see Papoulis [7]). The analytic signal $\hat{S}(t)$ is defined as

$$\hat{S}(t) \triangleq S(t) + j \cdot H\{S(t)\} = S(t) + j\frac{1}{\pi}\int_{-\infty}^{\infty}\frac{S(\tau)}{(t-\tau)}d\tau \tag{9.25}$$

The required envelope of the signal is given by the absolute value of the analytic signal,

$$\mathbb{S}(t) = \text{Envelope}\{S(t)\} = \left|\hat{S}(t)\right| = \left|S(t) + j\frac{1}{\pi}\int_{-\infty}^{\infty}\frac{S(\tau)}{(t-\tau)}d\tau\right| \tag{9.26}$$

A demonstrative example of this "envelope peeling" process is depicted in Fig. 9.10.

After extracting the envelope $\mathbb{S}(t)$ of the signal, a decimation procedure is needed. This is done in order to match the time scale with the distance scale, as will be explained in the following. Also, it is sometimes required to apply log compression of the signal. This procedure is recommended if there is a very large difference in the amplitudes of echoes within the scanned range (or if the TGC is not effective enough). This operation diminishes the gaps in contrast between different regions of the image and allows us to visualize weak and strong reflections in the same image.

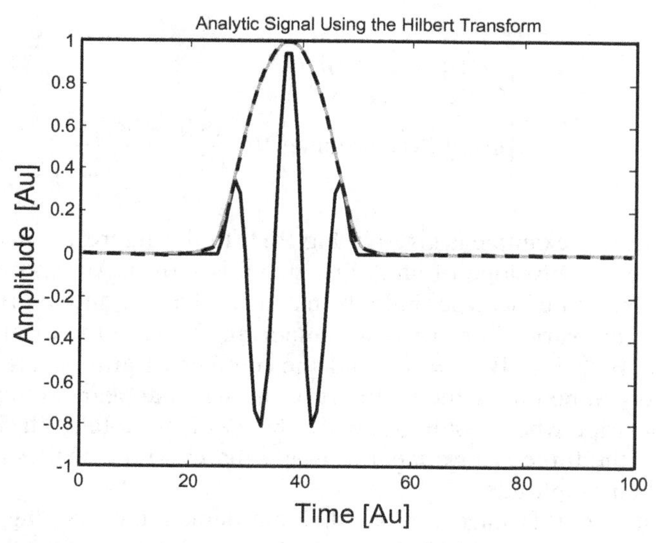

Figure 9.10. A demonstrative example of the envelope extraction procedure obtained using the Hilbert transform. The sampled signal $S(t)$ is plotted in black and its corresponding envelope (absolute value) is plotted in gray and dashed line.

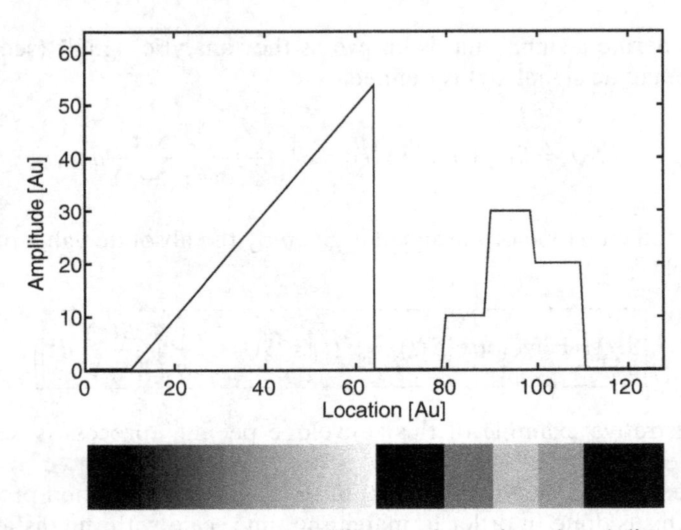

Figure 9.11. A demonstrative example of the transformation of an A-line signal **(top)** into a vector of gray levels **(bottom)**. High amplitudes are depicted in bright colors, and low amplitudes are depicted as dark colors.

The last stage of the procedure is the transformation of the signal into a vector of gray level values. In order to do that, we have to set a display window $\{W_{min}, W_{max}\}$ and the number of gray levels in the scale Ng. The transformation is given by

$$\breve{\mathbb{S}}(t) = \begin{cases} \mathbb{S}(t) < W_{min} \Rightarrow \text{Black} \\ \mathbb{S}(t) > W_{max} \Rightarrow \text{White} \\ \text{else} \quad \breve{\mathbb{S}}(t) = \text{round}\left\{ Ng \cdot \dfrac{(\mathbb{S}(t) - W_{min})}{(W_{max} - W_{min})} \right\} \end{cases} \tag{9.27}$$

A demonstrative example is given in Fig. 9.11. In this figure, a synthetic signal representing the envelope of an A-line signal is shown. As can be observed, the signal amplitude increases initially in a linear manner until it drops to zero. Then the signal varies in a stepped manner. In this case the display window was set to $[W_{min} = 0, W_{max} = 54]$ and the number of gray levels was set to $Ng = 54$. As can be noted, the first part of the signal appears in the gray level vector as a stripe whose color gradually changes from totally black to totally white while the three steps appear as gray squares whose shades are proportional to their amplitude.

To summarize this stage, a block diagram depicting every stage along the process is shown in Fig. 9.12. It should be clarified that not all stages are necessary in every application and the first two filters can be combined into a single band-pass filter. Also, postprocessing stages are not shown.

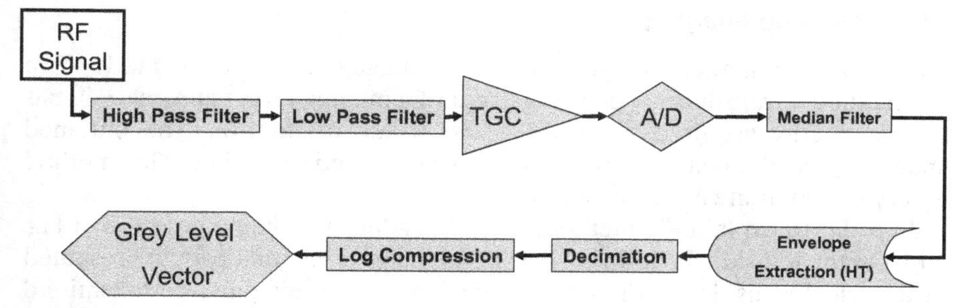

Figure 9.12. A block diagram of the process applied in order to transform the A-line signal into a gray level vector. Some of the stages are optional, and of course the order of the stages may be slightly changed.

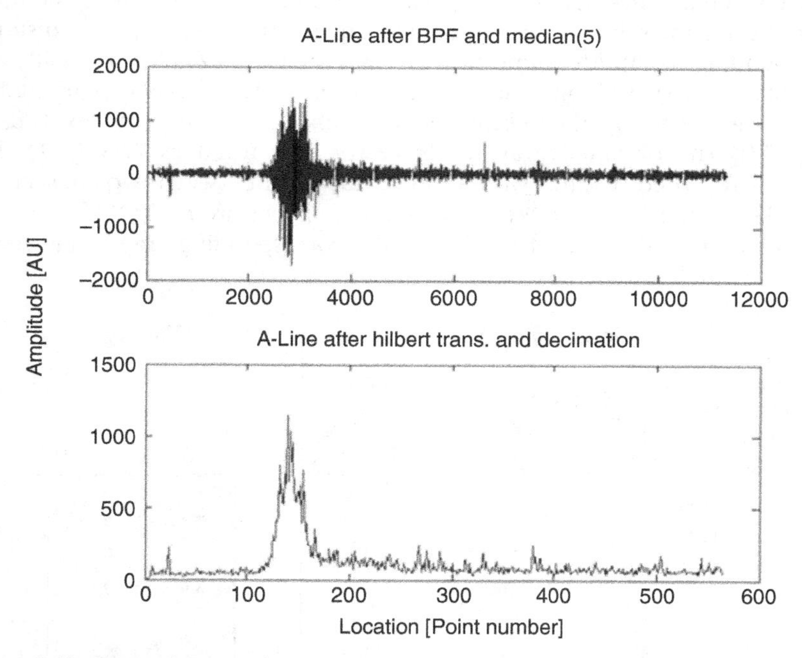

Figure 9.13. Exemplary results of a processed A-line signal. **(Top)** After band-pass filtering and median filtering. **(Bottom)** The corresponding extracted envelope obtained by applying the Hilbert transform and decimation, as explained in the text.

To further clarify the procedure, the results obtained at two stages are depicted in Fig. 9.13. On the top panel, an A-line signal following band-pass filtering (BPF) and median filtering is depicted. The bottom panel depicts the extracted envelope obtained using the Hilbert transform, as explained above and after decimation.

9.5.2 M-Mode Imaging

The most basic method that generates a 2D image is called the M-mode. With this method a transducer is placed in front of a moving target and echo signals are repeatedly acquired along the same A-line orientation. The obtained image depicts the distances to the targets as a function of time. This method is very common in cardiac echography.

In order to explain the method, let us consider the scheme depicted in Fig. 9.14. On the left side the left ventricle of the heart is schematically represented by a circle. Let us define the time at which an ultrasonic pulse is transmitted as $T(i) = i \cdot \Delta t$, where $i = 1{:}\infty$ is an index designating the number of pulse transmitted. The time elapse between two consecutive pulses is Δt. At time $T(1)$ a pulse is transmitted and two echoes are detected corresponding to the anterior and posterior walls of the heart (represented by the circle). The obtained signal is marked as $s(1, t)$. This temporal signal is transformed into distance using the relation $y(j) = j \cdot \Delta y$ and Eq. (9.19), where $y(j)$ is the distance corresponding to the jth temporal point (the range of which is $j = 1{:}m$) and Δy is the pixel size along that direction. The displayed range is limited to $0 < y(j) < y_{\max}$. The signal amplitude is then transformed into gray levels using Eq. (9.27). The obtained gray levels vector is marked as $M(y(j), 1)$. The process is repeated N times with a time delay of Δt between two consecutive pulses. This yields a matrix $\bar{\bar{M}}(j, i)$ whose dimensions are $i = 1{:}N$ and $j = 1{:}m$. Every column in this matrix represents the corresponding gray levels vector

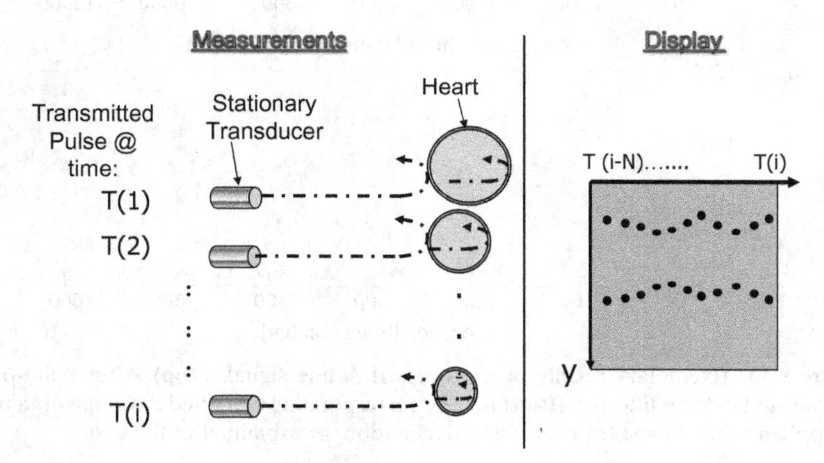

Figure 9.14. Schematic representation of the M-mode technique. **(Left)** A stationary transducer transmits pulses toward the heart, which is represented schematically by the circle. The heart contracts and reduces its diameter during systole. The pulses are transmitted at times $T(i) = i \cdot \Delta t$, and the echoes are transformed into a gray levels vector. **(Right)** The echo matrix is displayed. With every additional pulse, the left-hand column is removed, the columns are shifted to the left, and the new vector is placed at the first column on the right.

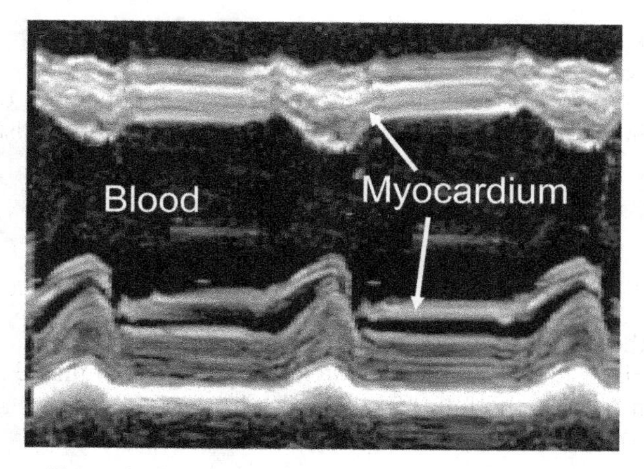

Figure 9.15. M-mode image of a contracting heart.

for a single A-line signal. Every line in the matrix corresponds to the range that is given by $y(j) = j \cdot \Delta j$. This matrix is displayed continuously on the scanner's screen as shown schematically on the right-hand side of Fig. 9.14.

After completing the first full display of the filled echo matrix $\bar{\bar{M}}(j, i)$, new A-lines are acquired and their corresponding gray levels vector is calculated. From here on, the indices are marked as $i > N$. With every additional pulse the first column on the left-hand side, which contains the oldest information, is removed. The columns are shifted to the left and the new vector is placed at the first column on the right. The new matrix refreshes the older one on the screen, and the procedure is repeated continuously.

A demonstrative M-mode image of a contracting heart is shown in Fig. 9.15. The upper band of echoes corresponds to the myocardial wall closer to the transducer. The black zone corresponds to the blood from which the reflections are very weak. The bottom band of echoes corresponds to the distal myocardial wall. During systole the distance between the walls is reduced, and this is manifested by the small "bumps" shown.

9.5.3 Spatial Mapping—The Simple Model

The simplest model that can be implemented for image reconstruction ignores the width of the acoustic beam and assumes that each transmitted pulse travels along a very narrow beam (similar to a laser ray) that can be well-approximated by the A-line signal. Also, assuming that the speed of sound is constant within the medium and that the TGC was effectively applied, we can use Eq. (9.22) to describe this acoustic ray:

$$\boxed{S(t) = \sum_N S_N(t) = \sum_N A_0 \cdot R_{(N-1,N)} \cdot \delta\left(\frac{\bar{C} \cdot t}{2} - Xn\right)} \tag{9.28}$$

where $S(t)$ is the TGC corrected signal.

In order to obtain an image, we need to transform the time scale into distance. For that purpose we shall use the assumption made above of a constant speed of sound and set the time-scaled gray levels vector into a distance scaled vector by

$$\breve{\mathbb{S}}(x') = \breve{\mathbb{S}}\left(\frac{\bar{C} \cdot t}{2}\right) \tag{9.29}$$

where x' is the distance along the ray. If a polar typed scan (e.g., sector scan) is applied, this distance can be translated into a Cartesian image according to the direction θ, using the following equations:

$$\begin{cases} F(i,j) = \breve{\mathbb{S}}(x') = \breve{\mathbb{S}}\left(\frac{\bar{C} \cdot t}{2}\right) \\ \text{where} \begin{cases} i = \text{round}\left[\frac{1}{\Delta} \cdot \left(\frac{\bar{C} \cdot t}{2}\right) \cdot \cos\theta\right] \\ j = \text{round}\left[\frac{1}{\Delta} \cdot \left(\frac{\bar{C} \cdot t}{2}\right) \cdot \sin\theta\right] \end{cases} \end{cases} \tag{9.30}$$

where $F(i, j)$ is the matrix representing the reconstructed image, $\{i, j\}$ represents the address of a specific pixel, and Δ is the size of each pixel in the reconstructed image.

If the number of acoustic rays (A-lines) used is large and the chosen pixel size is relatively big, then certain overlap between A-lines can occur in the reconstructed image. This is quite common in a sector scanning mode, particularly near the transducer. If such overlap occurs, then it is advisable to use an averaging procedure for pixels through which more than one acoustic ray has passed. In order to do that, let us define an "occurrence matrix" $\mathbb{N}(i, j)$ that counts the number off A-lines passing through each pixel. The corresponding average image value for each pixel is therefore given by

$$\bar{F}(i,j) = F(i,j)/\mathbb{N}(i,j) \qquad \forall \mathbb{N}(i,j) > 1 \tag{9.31}$$

Another operation which may be needed is decimation. As can be noted, the continuous A-line signal has been transformed into a digitized vector. It is therefore advisable to decimate the signal according to the corresponding pixel size. The decimation window can be set by the ratio between the pixel size and the average speed of sound in the medium using the following relations:

$$\left\{ S(\tau) = \text{Decimation}\,\{S(t), \tau\}, \qquad \text{where} \quad \tau = \frac{2 \cdot \Delta}{\bar{C}} \right. \tag{9.32}$$

As recalled, decimation implies reduction in the sampling frequency. Consequently, the corresponding Nyquist frequency is reduced. Hence, in order to avoid aliasing of the signal, it is advisable to low-pass filter it prior to the decimation operation. For example, if a signal is sampled at a frequency of *fs*, the corresponding Nyquist frequency is *fs*/2. On the other hand, after decimation the new Nyquist frequency will be only $1/2\tau$. Thus, the frequency band between the two values has to be removed.

Another operation that might improve the signal is the replacement of the down-sampling operation with window averaging. Thus, one can use the average value within the temporal decimation interval τ as the new image value, that is, $F(i, j) = \langle \breve{\mathbb{S}}(\tau) \rangle$. Naturally, averaging is inherently a low-pass filter. Thus, further filtering might not be required.

To conclude this section, three demonstrative figures are depicted: In Fig. 9.16 a B-scan image obtained for a tissue-mimicking phantom is shown. The transducer was placed on the left side and the scan was done along the vertical coordinate. The phantom was immersed in water (the black region on the left). Several targets with a relatively high reflectivity are imbedded in this phantom. These targets that simulate tumors are depicted in the image as bright spots. TGC was not implemented in this case. Thus, the attenuation effect is clearly noted; that is, the echo intensity is reduced with the distance (the horizontal coordinate). The A-line signal shown in Fig. 9.13 was taken from this image and is marked on the figure.

A typical B-scan image of a baby in her mother's womb is depicted in Fig. 9.17. This image was acquired using a sector scan protocol such as the one shown schematically in Fig. 9.9 (right).

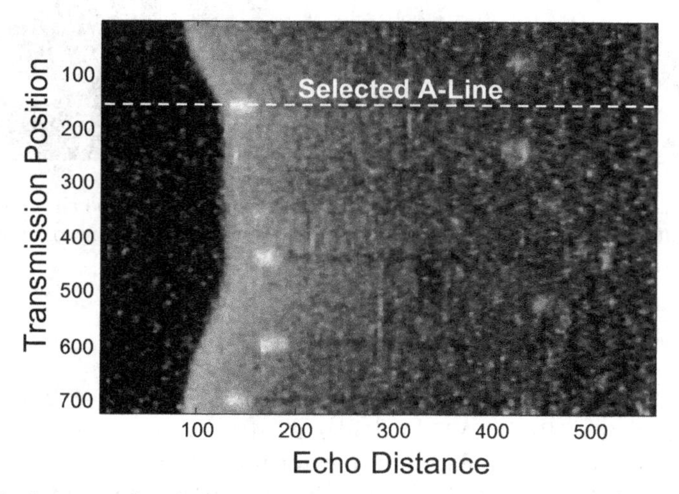

Figure 9.16. An example of a B-scan image obtained by scanning a phantom containing several tumor-mimicking targets (bright spots). The phantom was immersed in water (black zone on the left). The processed A-line depicted in Fig. 9.13 is marked on this image.

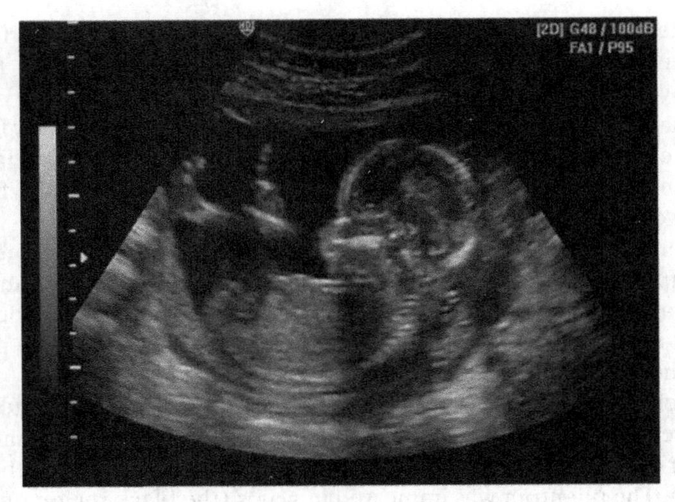

Figure 9.17. Typical B-scan image of a baby in her mother's womb. The image was acquired using a sector scan protocol.

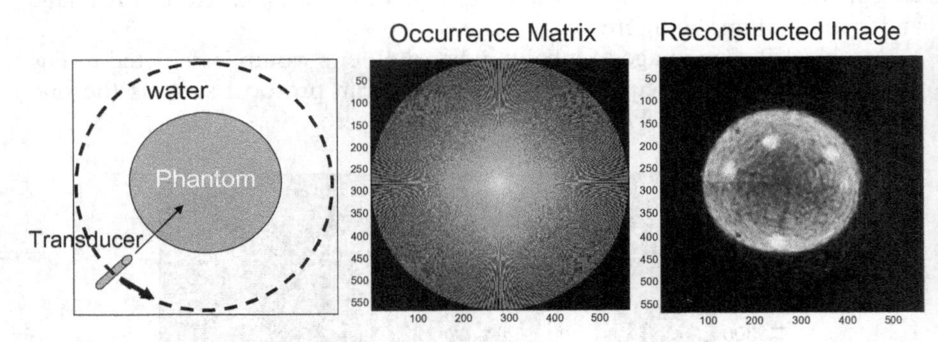

Figure 9.18. Circular B-scan procedure. **(Left)** Schematic of the the circular trajectory used. **(Middle)** Occurrence matrix designating the number of A-line crossings per pixel. **(Right)** Corresponding reconstruction.

A demonstrative polar-circular B-scan is shown in Fig. 9.18. A commercial breast phantom (*BB1* by ATS®) that is used for training the procedure of image-guided biopsy was placed in a water tank. The phantom was scanned using a circular trajectory suitable for automated breast imaging. The scanning trajectory is shown on the left side. The occurrence matrix designating the number of A-line crossings per pixel (i.e., $\mathbb{N}(i, j)$ in Eq. (9.31), is shown in the middle. And the final reconstruction is shown on the right.

9.5.4 Deconvolution Methods

The simple model suggested above ignores the pulse profile (the wavelet) and the acoustic field of the transmitted beam. Many techniques have been suggested, and many attempts have been made to improve the images by applying a deconvolution procedure. Some of the methods have focused on the wave profiles, while others have dealt with the actual 2D acoustic filed.

In this section we shall briefly present the basic 1D deconvolution approach. For that purpose we recall the improved model of the A-line, which was given in Eq. (9.9). Considering that in reality noise is added to the signal, the mathematical description of the A-line is given by

$$S(t)=[p(t)\otimes E(t)]+noise(t) \tag{9.33}$$

If the noise can be ignored, then the transform Fourier of the signal $\mathbb{S}(f)$ is approximately given by

$$\mathbb{S}(f)=\mathbb{P}(f)\cdot\mathbb{E}(f) \tag{9.34}$$

where $\mathbb{P}(f)$ and $\mathbb{E}(f)$ represent the Fourier transforms of the wave profile and the echo train, respectively. Seemingly, in order to remove the convolution effect from the process, all we need to do is to divide both sides of Eq. (9.34) by $\mathbb{P}(f)$ and take the inverse Fourier transform, that is,

$$E(t)=\mathfrak{F}^{-1}\left\{\frac{\mathbb{S}(f)}{\mathbb{P}(f)}\right\} \tag{9.35}$$

However, in most cases this approach is impractical. There are two main reasons for that. The first is that the transmitted pulse is of a limited spectral band. Therefore there are many frequencies for which the spectral power is small or even negligible (in addition to the off-band frequencies which are equal to zero). These components will be more susceptible to noise effects and inaccuracies in our parameter estimation. And since these components appear in the denominator, their effect on the solution will be considerably large. As a result, the obtained solution might be inaccurate and instable.

The second source of difficulty stems from the fact that the pulse profile is not known throughout the wave propagation path. As explained in Chapter 5, the pulse is attenuated with the range unevenly, that is, higher frequencies decay faster than the lower frequencies. As a result, the spectrum of the wavelet changes with the propagation distance. Hence, in order to implement a realistic deconvolution procedure, one needs to estimate the pulse profile in time and in space—that is, to determine $\mathbb{P}=\mathbb{P}(x,f)$. And this task is not trivial at all. There are suggested techniques that attempt to overcome this problem by applying blind-deconvolution algorithms. With such algorithms an approximation of the convolution kernel is derived based on several assumptions and

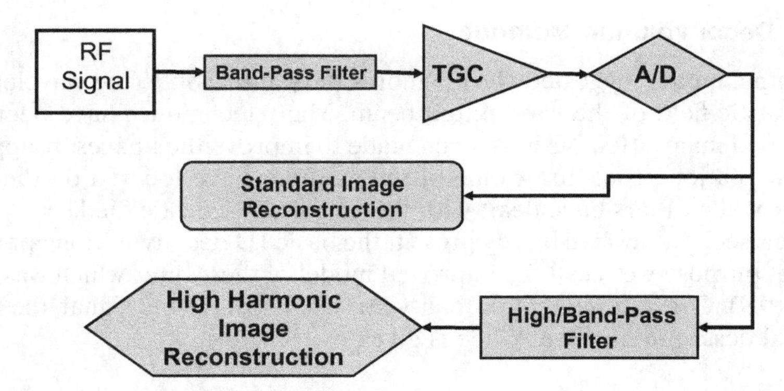

Figure 9.19. Block diagram for the high harmonic imaging procedure.

some optimization procedure. Few example references are listed herein [8–10].

9.6 ADVANCED METHODS FOR PULSE-ECHO IMAGING

The methods presented above constitute the fundamental layer of ultrasonic imaging. However, there are numerous ideas and techniques that have been suggested to improve image quality in terms of resolution, contrast, and SNR. Indeed, the scope is too wide to be entirely covered. In the following sections and in Chapter 10, some notable techniques are presented.

9.6.1 Second Harmonic Imaging

One of the techniques that have become a standard in almost all current ultrasonic scanners is the second harmonic imaging technique. This technique is based on a nonlinear phenomenon that occurs when the passing waves interact with the tissue. As a result of these nonlinear effects, new waves whose frequency is an integer multiple of the fundamental frequency of the transmitted wave are generated in the medium. In this book we shall not explain the reasons for this effect. The interested reader is referred to an excellent book by Beyer [11]. Using mathematical terms, it is stated that if the fundamental frequency is f_0, the waves generated due to nonlinearity in the tissue have the frequencies given by

$$f' = N \cdot f_0, \quad \text{where} \quad N = 2, 3, \ldots \tag{9.36}$$

It is important to note, as demonstrated in Chapter 10, that waves with additional frequencies are also generated. The amplitudes of these new waves are by order of magnitude smaller than the amplitude of the fundamental wave. Nevertheless, the second harmonic (i.e., $N = 2$) does provide a sufficiently high amplitude that can be useful.

The basic idea of second harmonic imaging is to transmit waves with a relatively low fundamental frequency and detect the echoes stemming from the generated second harmonic. The advantage is that the transmitted low frequency allows the waves to penetrate deep into the tissue. As explained in Chapter 5, the attenuation coefficient increases exponentially as the frequency increases. Hence, high frequencies are absorbed within a relatively short range. Although low-frequency waves penetrate deeper, their wavelength is longer. Thus, the corresponding image resolution is poor. High frequencies, on the other hand, have shorter wavelength, thus providing better resolution. By transmitting at low frequency and detecting at high frequency, we try to gain from both sides. The penetration is deep and the resolution is improved. Nevertheless, because the amplitude of the second harmonic is smaller than the fundamental harmonic, the corresponding SNR of this method is poorer. However, because the contrast is based on the nonlinearity of the tissue, it may provide an added value for diagnosis.

The implementation of this technique is quite simple. All we have to do is to band pass or high pass the detected echo signal. In practice, there is no reason why we should lose the information provided by the fundamental frequency, which also has better SNR. Thus, the two-image reconstruction process can be done consecutively or in parallel (depending on the available hardware). First the procedure described in Fig. 9.12 is implemented to yield the fundamental harmonic image. Then, the raw data are reprocessed by the high harmonic filters and the high harmonic image is reconstructed. To clarify the technique, a block diagram of the process is depicted in Fig. 9.19.

To conclude this section, an exemplary image is shown in Fig. 9.20. This image demonstrates the advantage of second harmonic imaging in terms of

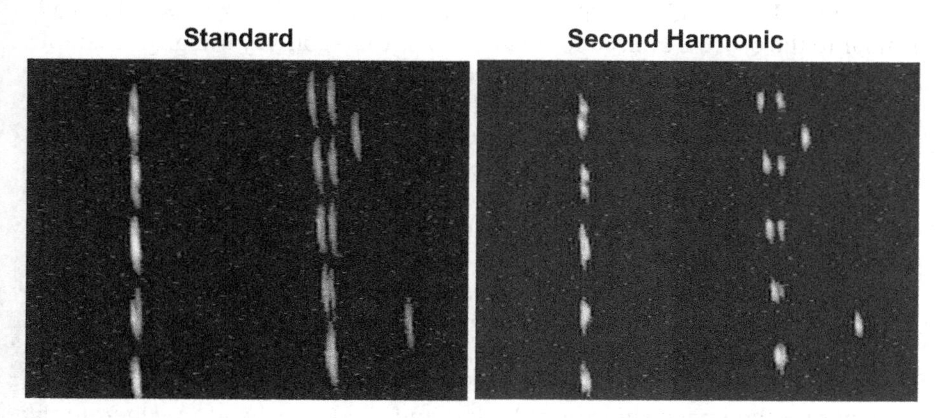

Figure 9.20. Demonstration of the improved resolution obtained by second harmonic imaging **(right)** relative to the standard imaging **(left)**. The transducer was placed on the left side, and the scan was done along the vertical axis. The target was comprised of pairs of nylon strings with a laterally (left column) or axially (right column) varying gap between them. Note the improved PSF and the improved resolution for the second harmonic image.

resolution. A specially built resolution phantom made of thin (0.1 mm in diameter) nylon strings was B-scanned using a Cartesian schema. The transducer was placed on the left side and the scan was done along the vertical axis. The phantom was comprised of two sets of pairs of strings separated by an increasing distance. In the left set, the gap between each pair of strings increases laterally, and in the right set the gap increases axially. As can be noted, the PSF for the second harmonic image is better (i.e., smaller) and the resolution (manifested by the separation between two adjacent spots) is improved.

9.6.2 Multifrequency Imaging

The intensity of the reflected echoes commonly serves as the source of image contrast as explained above. In order to accurately map the location of these sources of echoes, a very short pulse in time is preferable. Of course, such a pulse has a broad spectral band. This implies that the reflected echoes are actually a blend of waves having quite different frequencies. As explained in Section 9.3, each wave has its own reflectivity properties. To express this mathematically, we can rewrite Eq. (9.11) to include the frequency related scatter as

$$I_S(\theta, f) = R(\theta, f) \cdot I_0(f) \tag{9.37}$$

where $I_0(f)$ is the intensity of the impinging wave at frequency f, $I_S(\theta, f)$ is the intensity of the waves with the same frequency scattered toward direction θ, and $R(\theta, f)$ is the corresponding reflection coefficient.

This fact can be used to generate a few (at least two) images corresponding to echoes of different spectral bands. The discrepancy between the images can be exploited as an additional source of information. The process is basically similar to the process utilized for high harmonic imaging. By using, for example, two band-pass filters, two images can be reconstructed, with one image depicting the high-frequency band and the other the low-frequency band. As recalled, the scatter depends on both the wavelength and the tissue properties. This dependency may be different for different tissue types. Thus the discrepancy between the two images contains tissue-related information. This information may be used for tissue characterization (see, for example, references 12 and 13).

As a demonstration, two B-scan images obtained using a very broad band phased array transducer (5–12 MHz) are depicted side by side in Fig. 9.21. In these images a solid breast tumor and an adjacent cyst (a pathologic structure containing mainly fluids) are shown. The upper band of frequencies was used to generate the image shown on the left and the lower band to generate the image shown on the right. All other acquisition parameters were identical. As can be observed, there is a significant change in the intensity of the echoes from within the solid tumor. The change within the cyst is very minor. These changes were quantified and used as an index for tissue characterization.

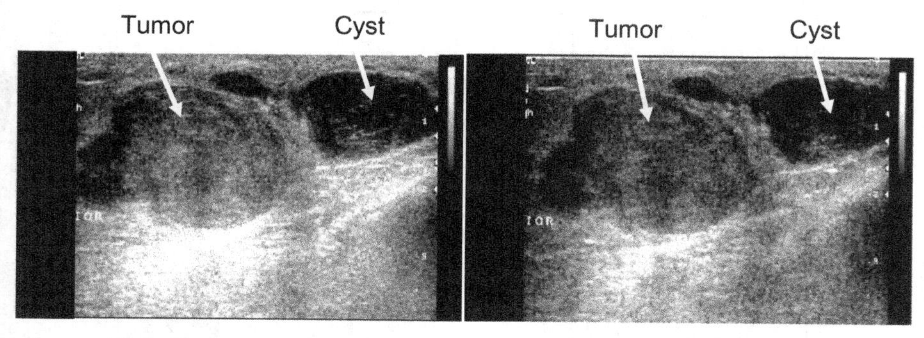

Figure 9.21. Demonstrative example of using two frequency bands for characterizing breast tumors are shown. In this image a solid tumor and an adjacent cyst are shown. **(Left)** Image obtained using the high-frequency band. **(Right)** Image obtained using the low-frequency band. As can be noted, the echo pattern within the solid tumor changes significantly.

9.6.3 Image Compounding

One characteristic problem of ultrasonic pulse-echo imaging is its poor lateral resolution. As can be noted in Fig. 9.20 (second harmonic imaging), the PSF is much more smeared along the vertical direction which is perpendicular to the acoustic beam (lateral direction) than along the horizontal direction which is parallel to the beam (axial direction). Also, as can be appreciated from Fig. 9.8, the speckle texture of the B-scan image may mislead the image reader, because some of the speckles in the image depend on the direction of sonication.

These two problems may at least partially be reduced by adding images of the same field of view acquired from different directions. This process is called *image compounding*. Let $\{I_k\}$ be a set of B-scan images of the same object acquired from different directions, and let T_k designate the spatial image transformation (e.g., rotation and translation) needed to align a certain image I_k from this set into a reference image. The compound image F_0 is therefore given by

$$F_0 = \sum_k T_k \{I_k\} \tag{9.38}$$

The above operation represents a simple image addition. However, this operation is not good enough in many cases because some of the images may not overlap. As a result, the gray levels at image zones where the overlap is maximal may present unrealistic high values. In order to correct for this effect, one must calculate the corresponding occurrence matrix and normalize the corresponding gray level values, as given by Eq. (9.31):

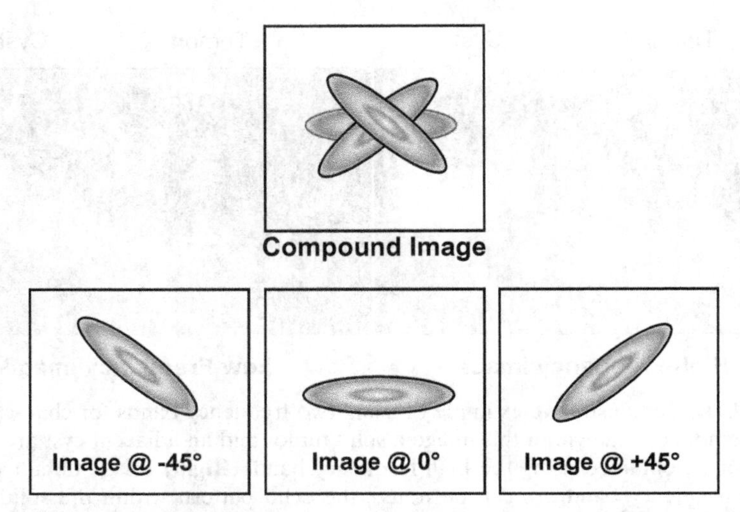

Figure 9.22. (Top) Schematic depiction of the compounding effect on the PSF obtained by adding the three images depicted at the bottom. **(Bottom)** Note that these images have been rotated to match the reference coordinate system.

$$\bar{F}(i,j) = F(i,j)/\mathbb{N}(i,j) \qquad \forall \mathbb{N}(i,j) > 1 \tag{9.39}$$

In order to graphically explain the image compounding process let us consider the schematic drawing shown in Fig. 9.22. The image obtained for a single-point target is schematically represented by an ellipse. This ellipse is in fact the corresponding PSF of the system. Commonly, as explained above, the PSF is more smeared along the lateral direction. Hence the resolution will be poorer along that direction. By acquiring two additional images from two different directions (e.g., ±45° relative to the basic image), we shall obtain two additional ellipses. Then, by rotating these images—that is, by applying the corresponding transformation T_k—the images shown at the bottom of Fig. 9.22 will be obtained. Accurate alignment and adition of these images will yield the image schematically shown at the top of Fig. 9.22. The central point of the compound image will be sharpened yielding better PSF. Moreover, because each rotated image actually contains the same information, the corresponding SNR will also improve by

$$\mathrm{SNR}_{\mathrm{compound}} = \mathrm{SNR}_0 \cdot \sqrt{M} \tag{9.40}$$

where M is the number of images compounded, and SNR_0 and $\mathrm{SNR}_{\mathrm{compound}}$ are the corresponding SNR for a single and compounded images, respectively. (It should be noted that it is assumed here that the noise is "white." If the noise is "colored," then alternative estimation may be needed.)

The compounding process can be obtained by physically moving the transducer relative to the target. However, this process is technically inconvenient because the transducer and the body must be in good physical contact to allow smooth passage of waves into the body. The motion of the transducer can reduce the quality of the contact and hence the quality of the obtained images. This problem can be solved by three alternative manners. The first solution is to use a water bath or a rubber bag filled with water. In this manner the water will serve as a coupling layer between the transducer and the body. In this case the spatial motion of the transducer will not affect the position of the studied organ. This type of solution can be utilized for imaging the breast or the hands and legs. However, in many abdominal applications this method may be too cumbersome and impractical.

The second solution is to attach special position detection sensors with a dedicated tracking system. There are several commercially available systems today. One type determines the position of each sensor by measuring the magnetic field at the sensor's location. The other system utilizes optical (IR) measurements. Using such a special tracking system, the images are acquired by free hand motion of the operator. The coordinates of the plane imaged at each acquisition are registered and stored in the computer memory. Compounding is obtained by combining all the obtained images into a single 3D matrix. This approach has two major limitations: First of all it requires a special system and dedicated software. Secondly, each time the transducer is pressed against the body, it can induce deformations in the studied soft tissue. Hence, the compounding process can be inaccurate. This problem is less significant in cardiac imaging where the transducer is attached to the relatively rigid chest wall. Indeed there were several works who have applied this method in echocardiography to obtain 3D images of the heart.

Another solution that allows compounding without moving the transducer is commercially available (SonoCT® by Philips). The corresponding process is schematically shown in Fig. 9.23. The idea is based on the abilities offered by a phased array transducer. Using electronic beam steering as explained in Chapter 8, a set of images are acquired. At each acquired image, the A-lines are slightly slanted relative to the other images. Thus, in fact one obtains several images acquired from different irradiation directions without physically moving the transducer. Furthermore, the images are continuously acquired yielding real-time imaging.

9.6.4 Three-Dimensional Imaging

Three-dimensional (3D) imaging using pulse-echo techniques is usually obtained by simply filling a scanned volume with multiple planar B-scan images. The acquired B-scan planes are spatially moved mechanically or electronically. Two simple examples (there are alternative configurations) are schematically shown in Fig. 9.24. In the first configuration a sector scan is rotated around the transducer's axis of symmetry. The outcome is a cone-

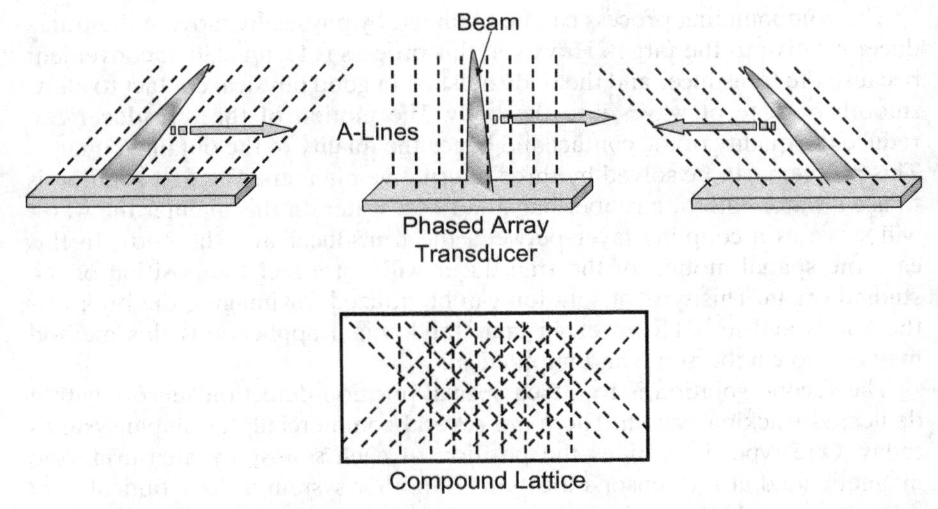

Figure 9.23. Schematic illustration of the compounding process obtained using a phased array transducer. Using electronic beam steering, several images at different angular orientations such as the images shown at the top row are acquired. The images are then combined into a single image using a common lattice as shown at the bottom.

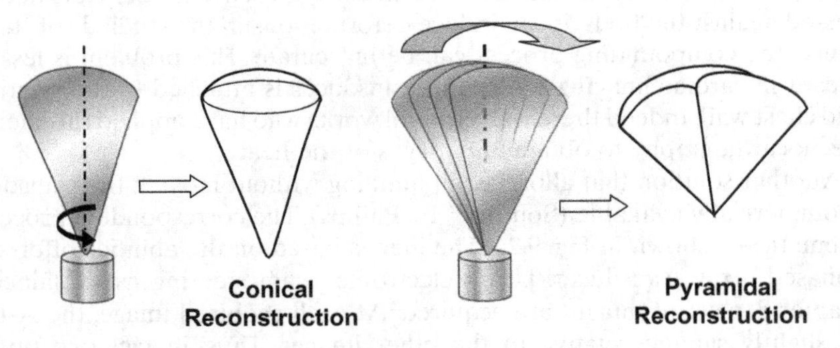

Figure 9.24. Schematic depiction of two 3D imaging configurations. **(Left)** A cone-shaped volume is scanned by rotating the B-scan plane around an axis of symmetry. **(Right)** A pyramidal-shaped volume is scanned by tilting the B-scan planes. B-scan plane steering can be obtained mechanically or electronically.

shaped scanned volume. With the second configuration the planes are swept like a fan yielding a pyramid-shaped scanned volume.

Three exemplary 3D images are shown in Fig. 9.25. The images were obtained by scanning a baby inside his mother's womb. The images depict mainly reflections from the baby's skin which have relatively strong amplitude and are received first.

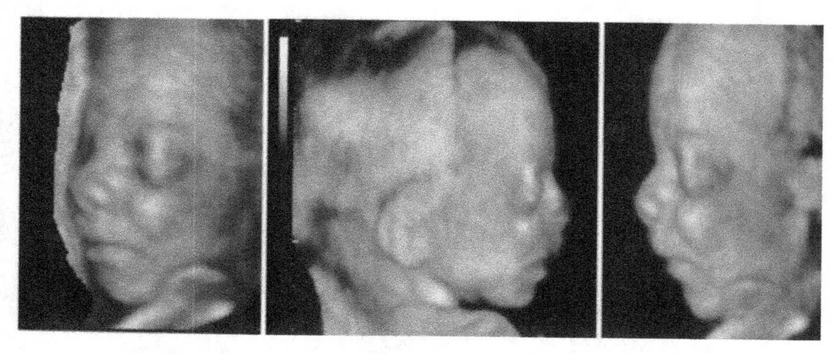

Figure 9.25. Three exemplary 3D scans of a baby's face.

9.6.5 Semi-invasive Imaging

Medical ultrasonic imaging is popular among other things because it is considered hazardless and convenient to use (as was explained in the Introduction chapter). However, not all the organs are adequately accessible to allow proper imaging. Consequently, in order to obtain the needed information and in a suitable quality, it is sometimes required to give up patient comfort and insert probes into his/her body. There are basically two intrusive techniques: (i) invasive imaging, where the probes are inserted into the body through a specially made cut, and (ii) semi-invasive imaging, where the probes are inserted into the body without wounding the patient. The three most popular semi-invasive imaging techniques are described in the following.

9.6.5.1 Trans-esophageal Echo. Trans-esophageal imaging (TEE) is usually used to overcome the acoustic obstacles present in the chest. The lungs are naturally filled with gas. Their acoustic impedance is therefore negligible relative to the surrounding soft tissues. Consequently, the borders between the tissues and the gas will have a reflection coefficient of about 1. This implies that through transmission through the lungs is very difficult (though not impossible). The other acoustic obstacle is set by the ribs. As recalled, bones have high acoustic impedance. As a result, most of the energy of waves propagating from soft tissues into the bones will be reflected. Furthermore, the attenuation coefficient of bones is also very high. These two facts make the transmission of ultrasonic waves through the ribs quite difficult. Moreover, reflected echoes that will be generated behind the ribs will be significantly attenuated. On the other hand, imaging the heart, which is located within the chest cage, is clinically extremely important. Thus, in order to solve this problem a semi-invasive procedure is used. A small transducer (commonly a phased array), which is connected to a cable, is inserted into the esophagus through the patient's mouth, as described schematically in Fig. 9.26.

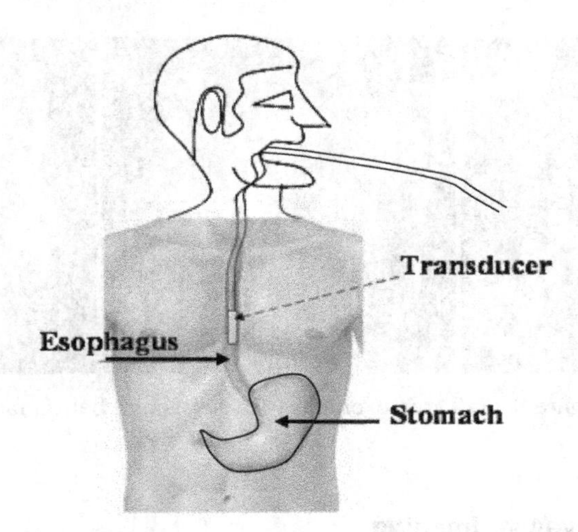

Figure 9.26. Schematic illustration of the ultrasonic trans-esophageal echo (TEE) imaging procedure. The electric cable, which is inserted through the mouth, is connected to the scanner system.

With TEE a sector B-scan is steered mechanically or electronically to provide 3D imaging as explained above. Since the esophagus is adjacent to the heart and there are no acoustic obstacles between the two, TEE can provide high-resolution and high-SNR images of the heart.

9.6.5.2 Intra-vaginal Imaging. Intra-vaginal imaging is usually applied in order to allow better gynecological diagnosis and to scan the fetus systems during the early stages of pregnancy. In these applications it is important to visualize fine anatomical details, and for that purpose particularly high resolution is needed. In order to obtain high image resolution, short wavelengths are used. This implies that the transmission is done at high frequencies (commonly 7–12 MHz). Nevertheless, as recalled from Chapter 5, the attenuation coefficient increases with the frequency, and these waves decay vary rapidly with the range. If transmitted externally through the abdomen, these waves will reach the imaged target with a very low amplitude. Consequently, the SNR might be too low. In order to solve this problem, it is advisable to minimize the range between the transducer and the target. (Using high amplitudes at transmission may be unsafe.) The common solution is to utilize a thin and long probe with a small phased array transducer at its tip. The probe is inserted into the vagina to the cervical canal. The range to the imaged targets is thus substantially shortened. The obtained images are typically of high resolution and very good quality.

9.6.5.3 Trans-rectal Imaging. Trans-rectal imaging is typically done for urological diagnosis and for imaging the prostate in men. There are also com-

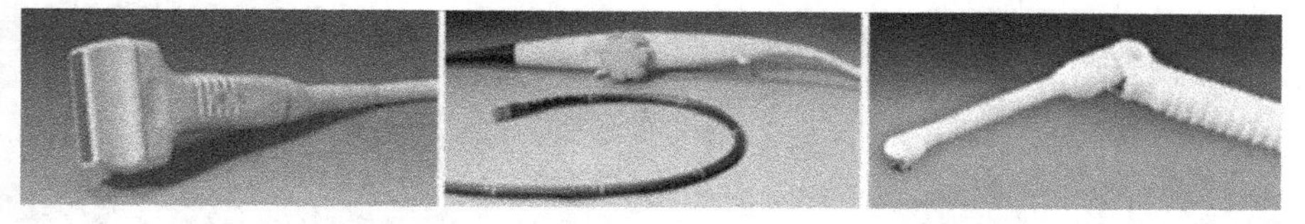

Figure 9.27. Three commonly used ultrasonic probes. **(Left)** A regular transducer. A special device for guiding a biopsy needle can be attached to this transducer. **(Middle)** TEE probe. The black cable is the part that is inserted into the esophagus. **(Right)** An intra-vaginal probe. GE Healthcare, a division of General Electric Company.

mercial systems that utilize trans-rectal imaging to guide a high-intensity focused ultrasonic (HIFU) treatment of the prostate (see Chapter 12). The motivation is similar to that of intra-vaginal imaging and the uncomfortable acoustic windows at the pelvis. By inserting a long probe into the rectum, the range to the target is substantially shortened and fine details can be visualized.

To conclude this section, three images of commercial transducers are depicted side by side in Fig. 9.27.

9.6.6 Invasive Imaging

9.6.6.1 Intravascular Ultrasound. Invasive imaging is commonly utilized in cardiovascular applications. The most popular system is called intravascular ultrasound (IVUS). With the IVUS system, the ultrasonic probe is attached to a dedicated catheter. The catheter is inserted into a blood vessel through a small cut made by the physician. The procedure has to be done under sterile conditions. In addition, an external "C-Arm" X-ray imaging system is used to provide real-time imaging for tracking the catheter. The IVUS images are needed to provide images of pathologic narrowing of blood vessels (stenosis). Although X-ray imaging of blood vessels (known as angiography) can demonstrate stenosis with the aid of an injected contrast material. The images provided by the X-rays are projection images. The IVUS, on the other hand, provides cross-sectional images from within the blood vessels, as explained in the following.

Pictures of an IVUS probe are shown in Fig. 9.28. A general overview of the IVUS catheter is shown in the top image. A zoomed image of the IVUS tip is shown in the bottom right picture. The rear part of the IVUS is shown in the bottom left picture. This part is plugged into the main system. It serves as an interface for signal transfers and for rotating the ultrasonic head of the IVUS. In this case the rotation is mechanical, and newer systems use a miniature circular phased array. In Fig. 9.29, three schematic drawings of the IVUS tip are shown.

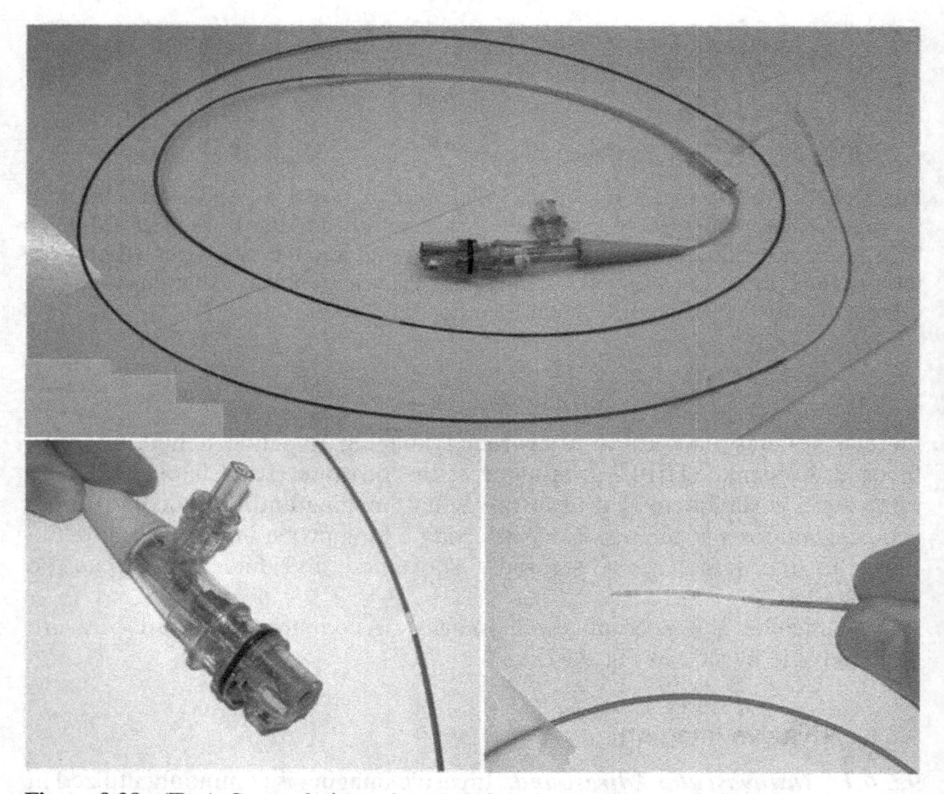

Figure 9.28. (Top) General view of an IVUS catheter. **(Bottom left)** Rear part of the IVUS, which connects to the main system. **(Bottom right)** Tip of the IVUS catheter, which contains the ultrasonic transducer. The diameter of the IVUS is about 1 mm and its length is about 1.5 m.

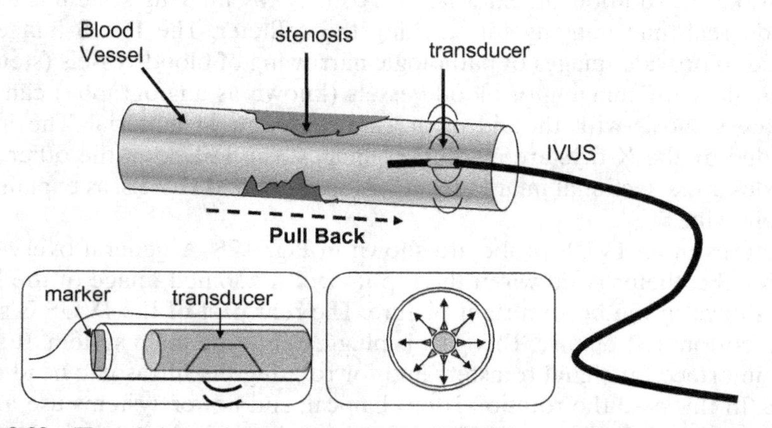

Figure 9.29. (Top) Schematic description of an IVUS front end. **(Bottom left)** Cross-sectional view through the IVUS tip. **(Bottom right)** Illustration of the transmission configuration.

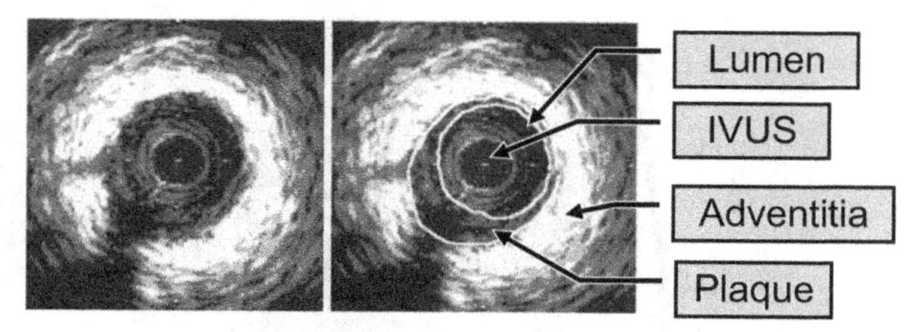

Figure 9.30. (Left) Typical IVUS image depicting a cross section through a coronary plaque. **(Right)** Same image with the markings corresponding to the different regions.

The IVUS catheter is actually made of two sleeves, one within the other. The outer sleeve is opened to the blood at its tip, and at the rear end it has an opening for injecting fluids (see bottom left image of Fig. 9.28) as in a conventional catheter. The inner sleeve contains the ultrasonic unit. The inner sleeve is sealed at its front end. It contains a cable that leads to the piezoelectric elements located at the tip. Small markers are placed near the tip to allow better localization in the X-ray images. Transmission of acoustic waves and detection of echoes are done perpendicular to the catheter axis and through the walls of the sleeves. The typical frequency range is 20–50 MHz. The imaged field of view is only a few millimeters in diameter because the attenuation at such high frequencies is very strong. On the other hand, the wavelength is very small and the spatial resolution is good.

There are currently two main types of IVUS probes. The first type utilizes a single piezoelectric element that spins very rapidly around its axis. The second type contains a circular phased array that rotates the acoustic beam electronically. With both types a large number of radial A-lines are acquired as shown schematically at the bottom right of Fig. 9.29. The IVUS image is therefore circular as shown in Fig. 9.30. At its center a circular shape corresponding to the IVUS catheter wall appears.

9.6.6.2 Intraventricular Echo. In a manner similar to that used by IVUS, intraventricular imaging of the heart can be performed. In this case the catheter is inserted into one of the hearts ventricles as shown schematically in Fig. 9.31. The scan is circular or sector-shaped. Because the imaged field of view is much larger than that of IVUS, lower frequencies are utilized. The advantage of such imaging is that it allows unobstructed data acquisition and that it can be combined with minimal invasive therapeutic procedures. Combining the system with a magnetic tracking system, such as the one provided by Biosense-Webster®, allows precise 4D (space and time) imaging.

9.6.6.3 Laparoscopic Ultrasonic Imaging. Ultrasonic imaging is also used in many minimal invasive (laparoscopic) surgical procedures. Basically, there are two modes to incorporate ultrasonic imaging in a laparoscopic procedure. With the first mode, ultrasound is used for guiding a laparoscopic device that

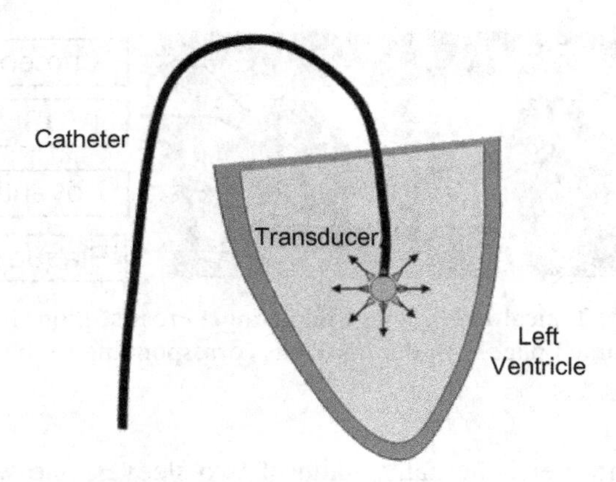

Figure 9.31. Schematic illustration of the intracardiac imaging principle.

contains a miniature camera and optic fibers. The insertion of the device is guided to the target by the external acoustic scanner, and the inner optical system is used for monitoring the fine surgical motions. With the other mode, an ultrasonic probe is attached to a rigid device that is inserted into the body (commonly abdominal). The operator moves the device manually or robotically inside the body to form good contact with the target tissue. The acoustic waves transmitted into the tissue and the detected echoes are used to provide real-time internal B-scan images. Due to the proximity to the tissue, high frequencies can be used to provide good spatial resolution. These images can be used to monitor the ablation or resection of tumors and for collecting biopsy samples. The difficulty in such procedures stems from the need to understand the images in real time while steering the device. This naturally requires good skills of the operator.

REFERENCES

1. Gonzales RC and Wintz P, *Digital Image Processing*. Addison-Wesley, Reading, MA, 1977.
2. Shepp LA and Logan BF, The Fourier transform of a head section, *IEEE Trans Nucl Sci* **NS-21**:21–43, 1974.
3. Szabo TL, *Diagnostic Ultrasound Imaging: Inside Out*, Elsevier Academic Press, Burlington, MA, 2004.
4. Dickinson RJ, Reflection and scattering, in *Physical Principles of Medical Ultrasonics*, Hill CR, editor, Ellis Horwood, Chichester, 1986, Chapter 6.
5. Nassiri DK and Hill CR, The use of angular acoustic scattering measurements to estimate structural parameters of human and animal tissues, *J Acousti Soc Am (JASA)* **79**:2048–2054, 1986.

6. Bellanger M, *Digital Processing of Signals*, 2nd edition, John Wiley & Sons, Tiptree, Essex, GB, 1989.
7. Papoulis A, *Probability, Random Variables and Stochastic Processes*, 2nd edition, McGraw-Hill, New York, 1984.
8. Adam D and Michailovich O, Blind deconvolution of ultrasound sequences using non-parametric local polynomial estimates of the pulse, *IEEE Trans Biomed Eng* **42**(2):118–131, 2002.
9. Michailovich O and Adam D, A novel approach to 2-D blind deconvolution of medical ultrasound images, *IEEE Trans Med Imag* **24**(1):86–104, 2005.
10. Kaaresen KF and Bolviken E, Blind deconvolution of ultrasonic traces accounting for pulse variance, *IEEE Trans Ultrason Ferroelectr Freq Control* **46**:564–574, May 1999.
11. Beyer RT, *Non-Linear Acoustics*, Naval Systems Command, Department of the Navy, 1974.
12. Sommer G, Olcott EW, and Tai L, Liver tumors: Utility of characterization at dual-frequency ultrasoud, *Radiology* **211**:629–636, 1999.
13. Topp KA, Zachary JF, and O'Brien WD, Quantifying B-mode images of in vivo mammary tumors by the frequency dependence of backscatter. *J Ultrasound Med* **20**:605–612, 2001.

CHAPTER 10

SPECIAL IMAGING TECHNIQUES

Synopsis: In this chapter the basic principles of several special ultrasonic imaging techniques are introduced. These methods can provide additional information to that obtained by the popular pulse-echo technique. The first technique, which is called "impediography" attempts to map the acoustic impedance of the tissue. The second technique, which is called "elastography," provides information on the elastic properties of the tissue. Few techniques that utilize through transmitted waves are then presented, and finally the principles of ultrasonic computed tomography (CT) are explained.

10.1 ACOUSTIC IMPEDANCE IMAGING—IMPEDIOGRAPHY

Thus far, our analysis of the signals received using the pulse-echo technique was aimed at providing images that map the reflectivity of the scanned tissues. The analysis was rather qualitative in nature. However, as explained in previous chapters, the cause for the generated echoes is a change in the acoustic impedance that the transmitted beam encounters along its path. Thus, it seems logical to attempt to map the acoustic impedance profile along the path and get quantitative information on the tissue. This idea was suggested by J. P. Jones and H. Wright [1, 2] The full mathematical derivation was later given by H. Wright [3].

In order to explain the principles of this technique in mathematical terms, let us consider the following model: An ultrasonic planar wave is transmitted into a medium that is comprised of many planar layers perpendicular to the

Basics of Biomedical Ultrasound for Engineers, by Haim Azhari

acoustic beam, as depicted in Fig. 9.5. Each layer has homogeneous acoustic impedance that is different from its adjacent neighbors. As explained in the previous chapter, the reflection coefficient between layers number n and layer number $n-1$ is given by

$$R_{(n-1,n)} = \frac{Z_n - Z_{n-1}}{Z_n + Z_{n-1}} \tag{10.1}$$

At this point we make a significant approximation and assume that all the reflection coefficients between each pair of layers have the same value of R, that is,

$$R_{(n-1,n)} = R \qquad \forall n \tag{10.2}$$

Next, we ignore the attenuation of the medium and the corresponding transmission coefficients $T_{(n-1,n)}$. Also, it is assumed that the transmitted pulse is short in time. Under these assumptions it can be assumed that the impulse response of a medium which contains M layers is given by

$$\begin{aligned}\Psi(t) = R\cdot\delta(t) + R\cdot(1-R^2)\cdot\delta(t-2t_2) + R\cdot(1-R^2)^2\cdot\delta(t-2t_3) + \\ \cdots + R\cdot(1-R^2)^{M-1}\cdot\delta(t-2t_M)\end{aligned} \tag{10.3}$$

This impulse response can be estimated by deconvolution of the echo signal using the transmitted pulse as the kernel. Assuming that the number of layers is infinite (i.e., $M \to \infty$) and integrating over the entire echo train duration (i.e., $t = 0 : 2t_M$) yields

$$\lim_{M\to\infty} \int_{t=0}^{2t_M} \Psi(t)\cdot dt = \frac{1}{2}\,\mathrm{lan}\left(\frac{Z_M}{Z_0}\right) \tag{10.4}$$

By accounting also for higher-order reflections, this can be modified to

$$\lim_{M\to\infty} \int_{t=0}^{2t_M} \Psi(t)\cdot dt = \tanh\left\{\frac{1}{2}\,\mathrm{lan}\left(\frac{Z_M}{Z_0}\right)\right\} \tag{10.5}$$

This implies that if the function $\Psi(t)$ is properly estimated, then the corresponding acoustic impedance of layer number M (i.e., Z_M) can be determined from

$$Z_M = Z_0\cdot\exp\left[2\cdot\tanh^{-1}\left\{\int_{t=0}^{2t_M} \Psi(t)\cdot dt\right\}\right] \tag{10.6}$$

Hence, in order to map the acoustic impedance along the A-line path, we need to start the transmission in a medium that has a known acoustic impedance

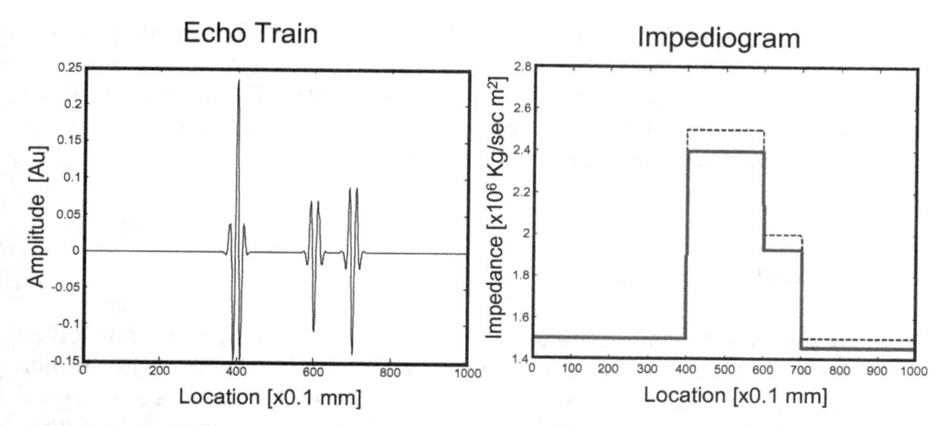

Figure 10.1. (Left) Simulated echo train signal obtained by simulating the encounter of a Gaussian pulse of a central frequency of 5 MHz with two adjacent plates immersed in water (reverberations were excluded). **(Right)** Corresponding acoustic impedance profile reconstructed using Eq. (10.4). The accurate profile is depicted by the dashed lines for comparison.

Z_0 (e.g., water). Then, by integrating in time along consecutive short intervals and by assuming that the distance x is given by $x = \tilde{C} \cdot \Delta t/2$, a profile of the acoustic impedance $Z(x)$ can be obtained.

As was shown by J. P. Jones [1], the acoustic impedances of various phantoms can be obtained. He has also shown a 2D map of the acoustic impedance for a rabbit leg. In order to demonstrate this technique, the results obtained using the simulated echo signal for two plates immersed in water are shown in Fig. 10.1. The reference impedance for water was $Z_0 = 1.5 \times 10^6$ [kg/sec·m²]. The acoustic impedance of the first target was $Z_1 = 2.5 \times 10^6$ [kg/sec·m²], and for the second target it was $Z_2 = 2 \times 10^6$ [kg/sec·m²]. The attenuation of the waves in water was considered negligible and for the targets the value of $\alpha = 0.01$ [1/mm] was used. Every pixel along the path was taken to be of 0.1 mm in length. As can be observed despite the fact that the simulation included no noise or distortions, the reconstructed acoustic impedance profile was not correct. This stems from the fact that the attenuation was ignored in the process.

This method has several drawbacks that one needs to overcome. First, the assumption that the target is comprised of parallel plates that are perpendicular to the beam is not correct in many applications. Second, the attenuation reduces the amplitudes of the echoes with the range, and the accurate attenuation coefficients in the tissues are not known a priori. The correction by the TGC (as explained in Chapter 9) is commonly not precise enough. Hence, the error is accumulated with the range. Third, in order to evaluate the function $\Psi(t)$, a deconvolution procedure is applied. This poses additional problems, since as explained in the previous chapter, the wave profile is distorted due to the frequency-dependent attenuation. Thus, blind deconvolution and iterative estimations are needed.

Some of these problems are discussed in the work of B. Perelman [4]. In this work a new approach for mapping the 2D acoustic impedance by using a computed tomography method is suggested.

10.2 ELASTOGRAPHY

One of the oldest methods used by physicians to detect tumors and other pathologies in tissues is palpation. By applying a slight pressure with his/her fingers, the physician senses the stiffness of the examined tissue. A stiff region within a soft tissue will attract the physician's attention. Commonly, a pathologic region is characterized by lower elasticity (high stiffness) as compared to normal soft tissues. Palpation is very useful and commonly applied to detect breast tumors. The ultrasonic elastography methods aim to imitate the palpation procedure by using acoustic waves.

The reflected echo signals that provide the basis for the imaging methods described above were used to map reflectivity properties and geometry. This was done by assuming that the examined organ is stationary or that it moves only due to internal physiological changes. However, if a known external mechanical load is applied to the tissue, a new kind of contrast that depicts the elastic properties of the tissue—for example, Young's modulus or the shear modulus (see Chapter 4)—can be obtained. The basic idea behind the elastographic imaging approach is to acquire a series of consecutive images while pressing or vibrating the studied organ in a precisely known manner. By processing the acquired information, the corresponding geometrical changes or the regional acoustic properties can be quantified. Using these measured values, the corresponding strain field can be determined; and using the stress–strain relationship (see Chapter 4), the regional elastic properties of the studied tissue can be determined. Finally, an image (a map) that depicts the spatial distribution of the elastic coefficients is generated.

In order to clarify the process, let us start with a very simple example. Consider a soft tissue that is represented by an isotropic block, as described schematically in Fig. 10.2 (Left side). If pressure is applied to the top plane of this block and if no friction between the bottom plane and the supporting surface exists, then the block's height will be reduced and its width will increase. By plotting the corresponding deformations as a function of height, a diagonal line is obtained.

At this point let us recall the definition introduced in Chapter 4 for the elastic strains:

$$U_{ik} = \frac{1}{2}\left[\frac{\partial \xi_i}{\partial x_k} + \frac{\partial \xi_k}{\partial x_i}\right] \tag{10.7}$$

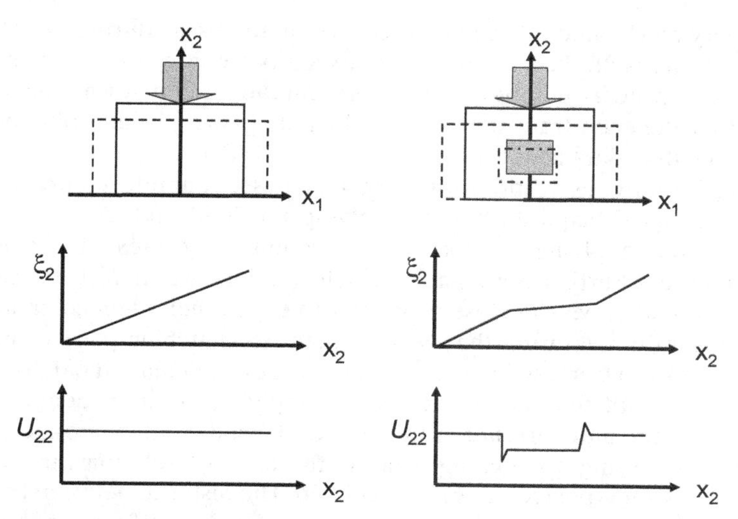

Figure 10.2. (Left top) Schematic depiction of a loaded homogeneous block. **(Left middle)** Vertical displacement as a function of height. **(Left bottom)** Corresponding strain along the vertical direction. **(Right side)** Same as on the left for a block containing a small stiff cube.

where ξ_i is the deformation along direction i and U_{ik} is the corresponding strain. For the simple case introduced above this term reduces to,

$$U_{22} = \frac{\partial \xi_2}{\partial x_2} \tag{10.8}$$

Consequently, the resulting strain will be described by the deformation slope and its value will be constant along the x_2 axis.

Next, let us consider the case depicted on the right side of Fig. 10.2. In this case a small cube made of a stiffer material is placed within the larger, softer cube. In this case the slope of the deformation curve will be different for the range corresponding to the inner cube. The strain, which is given by the derivative of this curve, will depict two distinct regions: (a) the region outside the inner cube, which has higher strain, and (b) the strain within the inner cube, which is smaller. Thus, by measuring with ultrasonic waves the deformations and by calculating the corresponding strain field, we will be able to map regions of different elasticity.

The first problem that we have to solve is how to measure accurately the deformations with the ultrasonic waves. One commonly used approach is to use the cross-correlation function of the echo train signal. The basic assumption is that the deformations are very small. This can be achieved by applying a small load on the tissue and acquiring the required data and then incrementally increasing the load and repeating the process. Alternatively, a scanner with a very high frame rate can be used to scan the tissue during a continuous loading. As a result, the deformation between two consecutive frames will be small. The second assumption that stems from the small deformation assumption is that the echo train signal approximately preserves its profile between two consecutive data acquisitions.

Using the above assumptions, it is expected to see common features between the A-line signal acquired before applying the load and the A-line signal acquired after applying the load. These common features can be used as virtual markers. Furthermore, since the deformations are small, these common features should appear in close proximity to each other when comparing the two A-lines. By measuring the distance between matching pairs of markers, the local deformation can be estimated. In order to calculate the deformation, a small section of the first A-line (before applying the incremental load) is selected and the autocorrelation function of a small window around it is calculated. Then, using a matching window, the cross-correlation function with the second A-line (post loading) is calculated. The distance between the peaks of the auto-correlation function and the cross-correlation function is the estimated migration distance of this selected section. Naturally, since the location of the auto-correlation peak always appears at the same point, it is enough to calculate only the cross–correlation function.

In order to clarify the procedure, let us examine Fig. 10.3. In this figure a simulated A-line signal before the load is applied (solid line) and after the load is applied (dashed line) are shown. In order to estimate the deformation of point A (i.e., ΔA) the autocorrelation function of the pre-load A-line signal was initially calculated and then the corresponding cross-correlation between the two A-lines was calculated. The distance between the two function peaks designates the displacement distance (as indicated in Fig. 10.3).

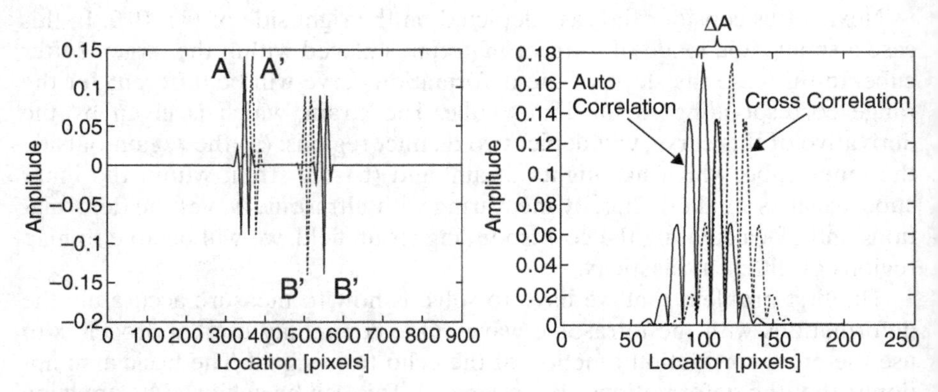

Figure 10.3. (Left) Simulated A-line signal before (solid line) and after (dashed line) deformation. **(Right)** The corresponding auto-correlation and cross-correlation functions of a small window around point A are shown. The displacement is estimated from the distance between the two peaks.

There are alternative methods for estimating the unidirectional deformations. For example: (i) *Phase-based estimators*—where the change in phase is related to deformation. The disadvantage is that 2π phase wrapping may occur under large deformations and a suitable unwrapping algorithm may be required. (ii) *Least-squares strain estimators*—where the local deformation that minimizes the variance between the two signals is estimated. It was reported that better SNR is obtained with this method.

After calculating the displacements, the local unidirectional strain can be calculated by selecting a section AB defined by two proximal features. (In practice, the division into sections is commonly arbitrary.) If the time elapse between the two features was Δt_{AB} before applying the load and $\Delta t_{A'B'}$ after applying the load, the corresponding strain can be estimated by

$$U = \frac{\Delta t_{A'B'} - \Delta t_{AB}}{\Delta t_{AB}} \tag{10.9}$$

Assuming that the medium is isotropic and that the stress is unidirectional σ_1 and known (as in Fig. 10.2), then the strains can be related to the elastic properties of the medium (Young's modulus E and Poisson's ratio υ) by the following relations (see, for example, Shigley [5]):

$$\begin{aligned} U_1 &= \frac{\sigma_1}{E} \\ U_2 &= -\upsilon U_1 = -\upsilon \frac{\sigma_1}{E} \\ U_3 &= -\upsilon U_1 = -\upsilon \frac{\sigma_1}{E} \end{aligned} \tag{10.10}$$

As can be noted, the strains along the loading direction and along one perpendicular direction suffice to calculate the elastic constants. However, if three principal normal stresses σ_1, σ_2, and σ_3 are applied, a method that measures the strains along all three directions is needed. After measuring the local strain tensor, the three local principal strains U_1, U_2, and U_3 can be estimated by calculating the corresponding eigenvalues. Using these principal strains, the following relations can be used to calculate the corresponding elastic coefficients (see Chapter 4):

$$\begin{aligned} U_1 &= \frac{\sigma_1}{E} - \frac{\upsilon\sigma_2}{E} - \frac{\upsilon\sigma_3}{E} \\ U_2 &= \frac{\sigma_2}{E} - \frac{\upsilon\sigma_1}{E} - \frac{\upsilon\sigma_3}{E} \\ U_3 &= \frac{\sigma_3}{E} - \frac{\upsilon\sigma_1}{E} - \frac{\upsilon\sigma_2}{E} \end{aligned} \tag{10.11}$$

It is important to remember that these relations are based on the not-so-reliable assumption of an isotropic medium.

After obtaining the corresponding elastic coefficients, it is useful to depict them as a colored graphic overlay atop the regular B-scan image. This provides the additional information to the physician that may allow him/her to reach the correct clinical diagnosis. However, the calculation of the principal stresses is also a complicated task. In order to do so, an exact geometrical model and exact loading conditions need to be known. This is impractical or too complicated in many cases. Hence, simple models and qualitative depiction of the elastic coefficients are commonly used.

There are other alternative elastography techniques that utilize more robust approaches. One such technique is based on estimation of the spatial frequency shift resulting from the deformation. Assuming that the scatterers in the scanned tissue region have some characteristic periodicity stemming from the tissue texture, then the corresponding spatial spectral signature of the echo train will be centered around some spectral frequency f_{C0}. After applying the mechanical load to that tissue (e.g., compression), the characteristic distance between the scattering points will be changed due to the resulting deformation (e.g., it will be shortened). As a result, the spatial spectral signature will be changed and the so will be the characteristic central frequency (it will be higher for shortening and lower for elongation). Designating the new characteristic central frequency as f_{C1}, it can be assumed that the spectral shift is correlated to the corresponding strain, that is,

$$D \cdot U = \frac{f_{C1} - f_{C0}}{f_{C0}} \tag{10.12}$$

where D is some linearity coefficient.

Another approach that utilizes spectral analysis is based on applying a periodical shear stress to the tissue as schematically depicted in Fig. 10.4. Resulting from the application of these stresses, deformations which are parallel to the applied force will be generated in the tissue. When the actuating

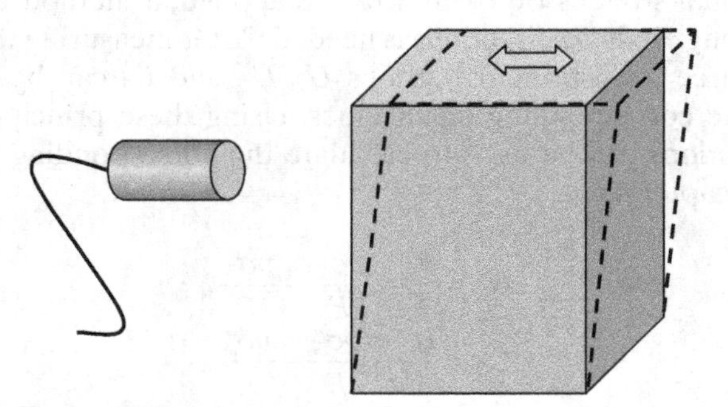

Figure 10.4. Applying a periodical shear stress to the top wall of the specimen while keeping its opposite bottom wall fixed will induce a periodical shear deformation such as the one depicted by the dashed lines. This will generate shear waves in the specimen. The deformations are measured using the M-mode technique.

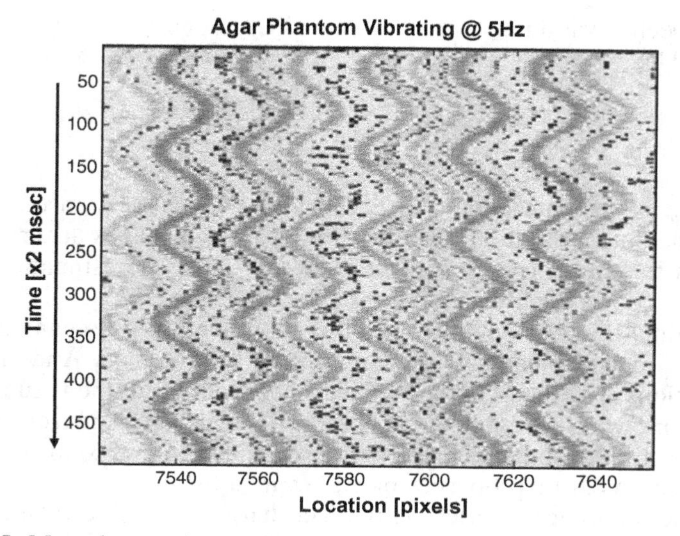

Figure 10.5. M-mode-type image depicting the deformations occurring in an agar phantom containing random scatterers. The phantom was vibrated at a frequency of 5 Hz. The image was reconstructed by accumulating 500 consecutive A-line signals using a PRF of 500 Hz. The gray levels correspond to the echo amplitudes. The periodical deformations are clearly noted.

force is periodical (e.g., sinusoidal), the different tissue layers will vibrate with different amplitudes according to their location and elastic properties. In some cases, standing waves may appear as explained in Chapter 1. An exemplary M-mode-type deformation image is depicted in Fig. 10.5. This image was reconstructed by accumulating 500 consecutive A-line signals of an agar phantom. The pulse repetition frequency (PRF) was 500 Hz. Acquisition time was 1 sec. The phantom was vibrated by a piezoelectric actuator at a frequency of 5 Hz. The gray levels correspond to the echo amplitude of each scattering point. The periodical deformation is clearly noted.

The amplitudes of the vibrations in the studied tissue can be assessed in various manners. One suggested technique utilizes the Doppler effect (see Chapter 11). For example, Huang et al. [6] have related the spectral broadening of the resulting Doppler shift to the deformation amplitude.

Expressing the spectral broadening by its variance σ^2 the following relation was derived:

$$\sigma = \frac{\varepsilon_m \cdot \sqrt{2} \cdot \omega \cdot \omega_V}{C} \tag{10.13}$$

where ε_m is the deformation amplitude, C is the speed of sound in the medium, ω is the angular frequency of the transmitted wave, and ω_V is the vibration frequency.

The spectral variance σ^2 was estimated using the following relation (see Chapter 11):

$$\sigma^2 = \frac{2}{T_{prf}^2} \cdot \left(1 - \frac{|R(T_{prf})|}{R(0)}\right) \tag{10.14}$$

where T_{prf} is the time interval between two consecutive transmitted pulses, $R(0)$ is the autocorrelation function of the received echoes at zero shift, and $R(T_{prf})$ is the autocorrelation function at T_{prf} shift. The estimation of these functions is done using the equations given in Chapter 11.

By estimating the regional vibration amplitude ε_m and scanning different regions, a map depicting it distribution can be obtained. And since these deformations are related to the local elastic properties of the tissue, it follows that regions with different elastic properties will have different ε_m values. Consequently, the obtained map provides an image whose contrast represents the changes in elastic properties of the scanned tissue.

Another approach is based on the fact that soft tissues can be considered incompressible. Therefore their corresponding Poisson's ratio (see Chapter 4) is in the range of $\upsilon \approx 0.49$–0.5. As recalled from Chapter 4, the speed of shear waves in a homogeneous solid medium is given by

$$C_{\text{shear}} = \sqrt{\frac{\mu}{\rho_0}} \tag{10.15}$$

where μ is the shear modulus and ρ_0 is the density of the medium at rest.

As recalled from Chapter 4, the shear modulus is related to Young's modulus E through the relation

$$\mu = \frac{E}{2(1+\upsilon)} \tag{10.16}$$

By substituting a Poisson's ratio of $\upsilon \approx 0.5$, we obtain

$$\mu \approx \frac{E}{3} \tag{10.17}$$

Hence, the relation between Young's modulus and the velocity of the shear waves in the tissue is given by

$$\boxed{C_{\text{shear}} \approx \sqrt{\frac{E}{3\rho_0}}} \tag{10.18}$$

Using Eq. (10.18), one can measure the speed of the shear waves in different locations of the tissue and map the corresponding values of Young's modulus.

And since pathologies (e.g., tumors) are characterized by high values, this map can be utilized to detect them.

Shear waves can be generated in the tissue by applying intrinsic or continuous deformations at one of the studied organ boundaries. This is typically done with special piezoelectric actuators. The mapping of C_{shear} can be done by simultaneous imaging of the tissue using the pulse-echo technique. The resulting deformations in the tissue are visualized in the obtained image as demonstrated in Fig. 10.5. This stems from the fact that the speed of shear waves in soft tissues is on the order of a few meters per second, while the speed of longitudinal waves that are used for imaging is on the order of 1500 m/sec.

For further information on ultrasonic elastography the interested reader is referred to references 7–10.

A new recent development in this context is a method which utilizes an acoustic radiation force impulse (ARFI) [11, 12]. By focusing an acoustic beam on a point within the tissue and transmitting a strong ultrasonic impulse, local force is generated at the focal zone. The radiation force (as explained in Chapter 3) can noninvasively indent a selected region within the tissue. By mapping the reaction of the tissue to this rapid indentation, an image mapping the stiffness of the tissue can be generated.

10.3 TISSUE SPECKLE TRACKING

As a complementary or alternative method to elastography, a method that analyzes tissue motion and deformation by tracking its texture has been suggested. This method is in fact an image processing technique, which utilizes the speckle texture obtained by the pulse-echo imaging method for tracking specific tissue points. The method is based on the assumption that each region within the tissue has a characteristic 2D signature in the echo image. Using high-frame-rate B-scan imaging, it can be further assumed that each tissue region has moved only a small distance between two consecutive frames and that its characteristic features are retained. This implies that the new position of the region of interest can be located and the distance can be calculated.

In order to better understand the principle of tissue tracking, let us observe Fig. 10.6. In this figure, two schematic B-scan frames are shown side by side. In these images, one can observe four notable textural elements (the dark regions). Since the two frames were acquired within a short time, the distance that each textural element has moved to is small. The task is to recognize each feature in these two images and represent its translation by a vector as depicted schematically at the right-hand side of Fig. 10.6.

In order to accomplish this goal, a region of interest (ROI) of size $N \times M$ pixels is chosen in the first frame (see Fig. 10.7). Around the center coordinates of the ROI a search window of size $L_1 \times L_2$ is chosen. This window

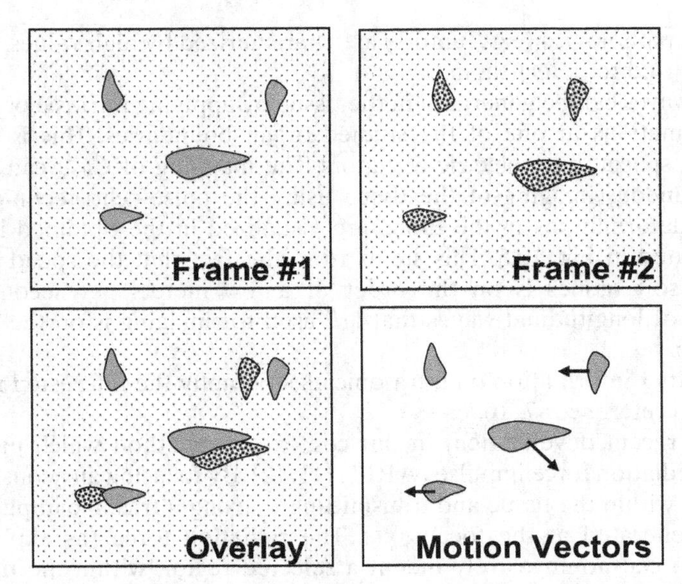

Figure 10.6. **(Top Row)** Two schematic images representing two consecutive images (frames) of the tissue, acquired within a short time. **(Bottom Left)** A graphic overlay indicates that the three out of the four notable features (dark regions) have moved. The task is to quantify and display this motion as shown on the bottom right image.

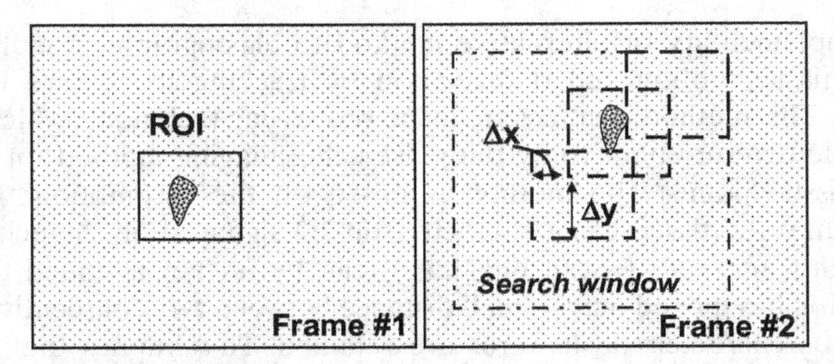

Figure 10.7. **(Left)** It is required to know the new location of a region of interest (ROI) marked in the first frame. **(Right)** A search window is chosen in the second frame, and quantitative comparison is done for a set of optional new ROI regions within this window. The ROI in the second frame which yields the maximal resemblance to the ROI in the first frame is assumed to be the new location.

defines the search area around the ROI and naturally has to be bigger than the ROI. Within this search window a temporary ROI of size $N \times M$ is chosen in the second frame. Then, a quantitative comparison between the two regions is done and the results are registered. Next, a new temporary ROI is chosen in the second frame and again a quantitative comparison between the two regions is done and the results are registered. The process is repeated several times for different regions, and finally the region that has yielded the maximal similarity is assumed to be the new location of the original ROI.

The optimal quantitative comparison of two ROI regions taken from two consecutive frames is naturally the correlation coefficient that is calculated by

$$\eta(\Delta x, \Delta y) = \frac{\sum_{i=1}^{N}\sum_{j=1}^{M}[s_1(i,j)-\bar{s}_1]\cdot[s_2(i+\Delta x, j+\Delta y)-\bar{s}_2]}{\sqrt{\sum_{i=1}^{N}\sum_{j=1}^{M}[s_1(i,j)-\bar{s}_1]^2\cdot[s_2(i+\Delta x, j+\Delta y)-\bar{s}_2]^2}} \tag{10.19}$$

where $\eta(\Delta x, \Delta y)$ is the correlation coefficient for a shift of distance $(\Delta x, \Delta y)$, relative to the location of this region in the first frame. $S_1(i, j)$ and $S_2(i, j)$ are the gray levels of the original and new ROI, respectively. $\bar{S}_1$ and $\bar{S}_2$ are the mean gray level values in these two regions, respectively. The summation is carried out on all pixels located in the two ROIs.

The drawback of this estimator is the relatively long computation time needed. As a faster alternative, it was suggested (see, for example, reference 13) to use the difference as a resemblance estimator. The suggested index is called "SAD," an acronym for sum of absolute differences. This index is calculated by

$$\varepsilon(\Delta x, \Delta y) = \sum_{i=1}^{N}\sum_{j=1}^{M}|s_1(i,j) - s_2(i+\Delta x, j+\Delta y)| \tag{10.20}$$

where $\varepsilon(\Delta x, \Delta y)$ is the SAD index for a shift of $(\Delta x, \Delta y)$. Naturally, this index has the smallest value when the two compared regions resemble the most. And this region is assumed to be the new location of the selected ROI.

There are alternative indices and methods for tissue tracking. Currently, the main application is for studying the myocardial motion, in echocardiography. One such example is an integrated software package developed at the Technion IIT and installed in the scanners of GE health care systems. The algorithm is described in [14–16]. To conclude this section, two images obtained by this algorithm and system are shown in Fig. 10.8. The top image was acquired during the left ventricular contraction (see the directions of the motion-indicating arrows) and the second during diastolic relaxation.

10.4 THROUGH-TRANSMISSION IMAGING

The pulse-echo imaging method is very convenient because it allows access to almost any part of the body. However, this method has several significant drawbacks. First of all, the quantitative assessment of distances and geometry

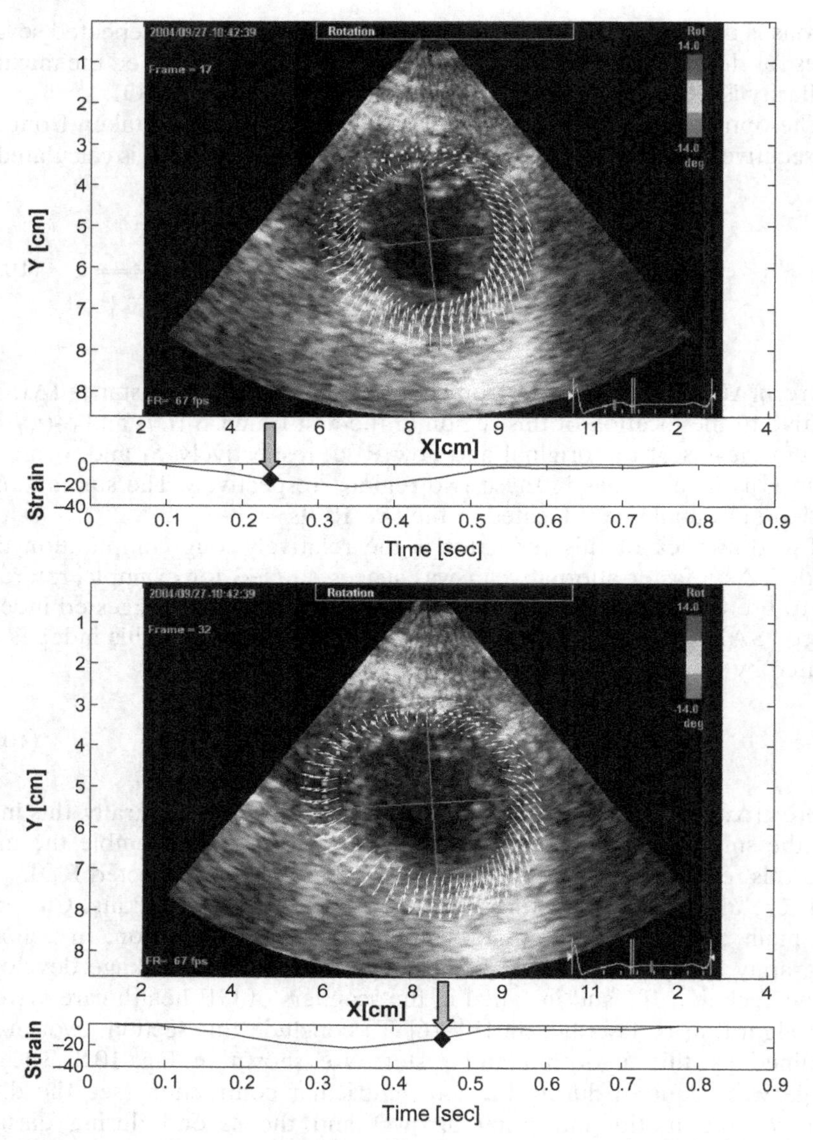

Figure 10.8. Cross-sectional B-scan images of a heart's left ventricle during contraction **(top)** and relaxation **(bottom)**. The arrows indicate graphically the direction and velocity of the instantaneous motion. The graphs at the bottom of each image designate the global strain, and the temporal location is indicated by the dot. (Courtesy of: Noa Bachner and Dan Adam—Technion, and Zvi Friedman—GE Healthcare). See insert for color representation of this figure.

is based on the assumption that the speed of sound is uniform for all tissues. This of course is not true. Secondly, stemming from the fact that the changes in the acoustic impedances of soft tissues are small (see Appendix A at the end of the book), the corresponding reflection coefficients are small as well. Hence, the corresponding SNR is poor and high amplifications and extensive filtering is needed. Thirdly, since the attenuation and transfer coefficients along the acoustic path are not known, it is almost impossible to quantify the acoustic properties of the scanned tissues.

Through-transmission imaging provides a solution for these three problems, as will be explained in the following. However, this method also has two significant drawbacks: (i) A receiving transducer (or a reflector) needs to be positioned on the other side of the examined organ, and (ii) the method is ineffective for scanning organs containing bones and air because these elements screen the beam and cast an acoustic shadow. Consequently, through transmission is useful mainly for scanning the breast and partial scanning of limbs.

10.4.1 Acoustic Projection Imaging

With this method an organ is scanned by transmitting waves from one side of the organ to the other, while transmission and reception are done on two parallel planes positioned from both sides of the organ. The commonly used protocol is called a raster scan where parallel lines are scanned as schematically shown in Fig. 10.9.

The transducers are usually coupled to the organ by using a water container. The time of flight (TOF)—that is, the traveling time of the waves from one side to the other—provides information on the speed of sound through the medium. The amplitude and the spectral signature of the waves detected on the other side provide information on the attenuation propertied of the scanned tissues.

Using a simple model for the transmitted waves, while ignoring diffraction and the Fermat's principle (see Chapter 7), it is assumed that each wave trans-

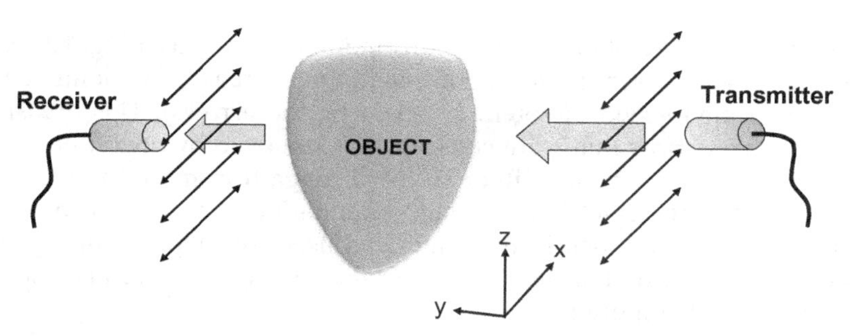

Figure 10.9. Schematic depiction of a raster scan utilizing through transmission to obtain an acoustic projection image.

mitted from one side will travel along a straight line and reach the receiver located along a light of sight. Consequently, the measured time of flight $\Delta t(x, z)$ can be related to the medium's speed of sound $C(x, y, z)$ by

$$\Delta t(x,z)=\int\frac{1}{C(x,y,z)}dy \tag{10.21}$$

As recalled from Chapter 7, the reciprocal ratio for the speed of sound is defined as the acoustic refractive index. Hence, the obtained image depicts the sum of all the refractive indices along each acoustic ray. This type of image is actually a projection image. The problem with this type of image is that the geometrical information along the integrated axis (in this case the y axis) is lost. Thus, in order to retrieve this information, two alternative options may be used. The first option is to combine this method with the pulse-echo method. This will provide an estimate of the geometry along the projected axis. The second option, as will be explained in the following, is to collect projections from various directions and reconstruct the image using computed tomography methods.

Assuming that the distance between the transmitting and receiving transducers L is fixed and known and that the dimensions of the studied organ $D_y(x, z)$ are known, and so is the speed of sound in water C_{water} (which serves as a coupling medium), one can calculate the average speed of sound in the studied organ $\bar{C}(x, z)$ using the following relation:

$$\Delta t(x,z)-\Delta t_{\text{water}}(x,z)=D_y(x,z)\cdot\left(\frac{1}{\bar{C}(x,z)}-\frac{1}{C_{\text{water}}(x,z)}\right) \tag{10.22}$$

where $\Delta t_{\text{water}}(x, z)$ is the TOF through water when no object is present in the scanned volume. After rearrangement, we obtain

$$\bar{C}(x,z)=\left(\frac{\Delta t(x,z)-\Delta t_{\text{water}}(x,z)}{D_y(x,z)}+\frac{1}{C_{\text{water}}(x,z)}\right)^{-1} \tag{10.23}$$

An exemplary image obtained using this method is shown in Fig. 10.10. The image depicts the average speed of sound of a breast containing several regions with high speed of sound (marked by the arrows). This breast had several benign tumors that were consequently removed by operation.

By measuring also the amplitude of the through-transmitted waves, a projection image that depicts the average attenuation coefficient can also be obtained (this can be done in the same scan used for mapping the speed of sound). In order to obtain the relevant projection, let us write first the term for the measured amplitude,

$$A(x,z)=A_0\cdot e^{-\int\mu(x,y,z)dy} \tag{10.24}$$

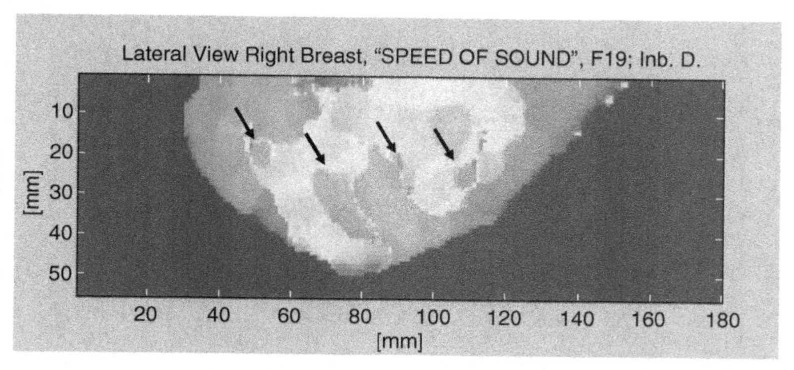

Figure 10.10. Speed-of-sound projection (lateral view) of a breast containing several benign tumors. Regions with extremely high speed of sound, which is characteristic to solid tumors, are marked by the arrows. See insert for color representation of this figure.

where A_0 and $A(x, z)$ are the transmitted and received amplitudes, respectively, and $\mu(x, y, z)$ is the local attenuation coefficient. Dividing both sides of this equation by the amplitude of the transmitted wave and taking the logarithm from both sides yields

$$\mathbb{F}(x, z) \triangleq \operatorname{lan}\left(\frac{A(x, z)}{A_0}\right) = -\int \mu(x, y, z)\, dy \tag{10.25}$$

Assuming again that the distance L between the transducers and the dimensions of the studied organ $D_y(x, z)$ are known, then this expression can be replaced by

$$\begin{aligned}\mathbb{F}(x, z) &= -\bar{\mu}(x, z) \cdot D(x, z) - \mu_{\text{water}} \cdot [L - D(x, z)] \\ &= [\mu_{\text{water}} - \bar{\mu}(x, z)] \cdot D(x, z) - \mu_{\text{water}} \cdot L\end{aligned} \tag{10.26}$$

where $\bar{\mu}(x, z)$ and μ_{water} are the average attenuation coefficient of the tissue and the attenuation coefficient for water respectively.

If a reference scan is performed through water only (i.e., without the object), then the following reference value $\mathbb{F}_{\text{water}}(x, z)$ is obtained:

$$\mathbb{F}_{\text{wwater}}(x, z) = \operatorname{lan}\left(\frac{A_{\text{water}}(x, z)}{A_0}\right) = -\mu_{\text{water}} \cdot L \tag{10.27}$$

By subtracting Eq. (10.27) from Eq. (10.26), we obtain

$$\mathbb{F}_{\text{water}}(x, z) - \mathbb{F}(x, z) = \bar{\mu}(x, z) \cdot D(x, z) - \mu_{\text{water}} \cdot D(x, z) \tag{10.28}$$

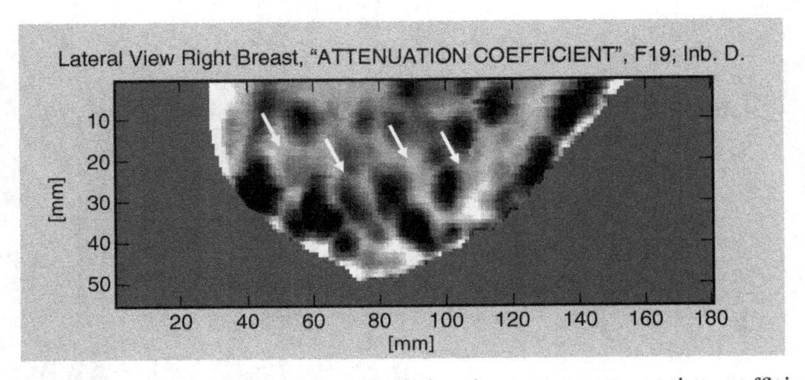

Figure 10.11. Lateral projection image depicting the average attenuation coefficient of the same breast shown in Fig. 10.10. The arrows indicate the same regions depicted there.

And after rearrangement the average attenuation coefficient for the tissue is given by

$$\bar{\mu}(x,z) = \frac{[\mathbb{F}_{\text{water}}(x,z) - \mathbb{F}(x,z)]}{D(x,z)} + \mu_{\text{water}} \tag{10.29}$$

An exemplary image depicting the average attenuation coefficient for the same breast shown in Fig. 10.10 is shown in Fig. 10.11.

10.5 VIBRO-ACOUSTIC IMAGING

The models presented above which were utilized for imaging presuppose a linear process. However, as was discussed in previous chapters, waves passing through tissues are also subjected to nonlinear processes. These nonlinear processes may lead to the generation of new waves within the medium. These waves have a frequency that differs from the transmitted one. By applying proper filtering, the information contents of these waves may be retrieved and incorporated in the imaging process to yield new source of contrast or to improve the spatial resolution as was demonstrated in Chapter 9 for the second harmonic imaging technique (Fig. 9.20). The disadvantage of these waves is that their amplitude is substantially smaller than the amplitude of the fundamental frequency waves. Another drawback of high harmonic waves is that their attenuation coefficient is also much higher than the attenuation coefficient of the fundamental frequency waves. Resulting from these facts the corresponding SNR is poor and penetration is limited.

The technical challenge is to obtain high-resolution images with low-frequency waves that have low attenuation and better penetration. A solution to this challenge was suggested by Fatemi and Greenleaf [17, 18]. The suggested new technique is called "vibro-acoustic imaging." The technique is based on the fact that when two high-intensity waves—one having a frequency of f_1 and the other having a frequency of f_2—intersect near a reflector, new waves that have frequencies that consist of linear combinations of the two fundamental frequencies are generated. This can be expressed by

$$\hat{f} = n \cdot f_1 \pm m \cdot f_2, \qquad \text{where} \quad n = 0, 1, 2, \ldots; m = 0, 1, 2, \ldots \tag{10.30}$$

where $\hat{f}$ is the frequency of the new generated wave.

To demonstrate this phenomenon, we have taken two high-intensity transducers and have focused their beams on a small metal target immersed in water. The frequency of the first wave was $f_1 = 0.986$ MHz and the frequency of the second wave was $f_2 = 3.125$ MHz. The scattered and generated waves were detected by a third broad-band transducer. The obtained spectrum of the detected waves is depicted in Fig. 10.12. The new generated frequencies that are manifested by the spectral peaks are marked by the arrows. Their corresponding values are given by

$$\begin{array}{ll} @a & \hat{f} = 2 \cdot f_1 \\ @b & \hat{f} = 1 \cdot f_2 - 1 \cdot f_1 \\ @c & \hat{f} = 1 \cdot f_2 + 1 \cdot f_1 \\ @d & \hat{f} = 1 \cdot f_2 + 2 \cdot f_1 \\ @e & \hat{f} = 2 \cdot f_2 - f_1 \\ @g & \hat{f} = 2 \cdot f_2 \\ @h & \hat{f} = 2 \cdot f_2 + 1 \cdot f_1 \end{array}$$

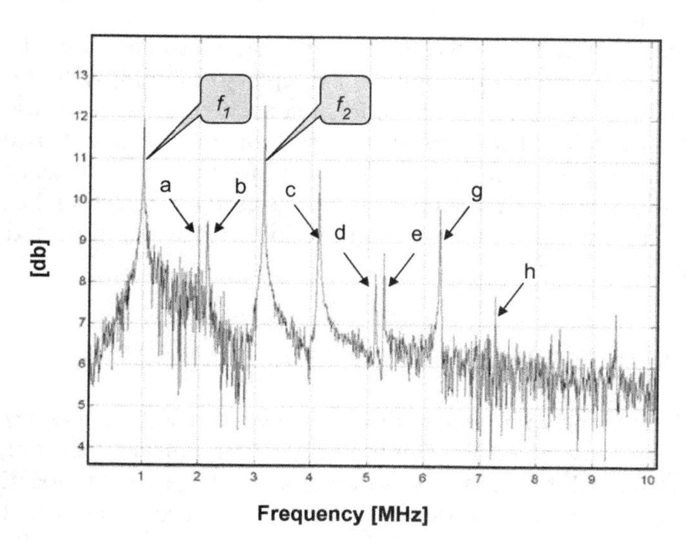

Figure 10.12. Spectrum obtained for the waves detected after focusing two strong beams on a small metal target immersed in water. The fundamental frequencies were f_1 and f_2. The spectral peaks, which are marked by the arrows, correspond to the linear combination frequencies generated at the focal point (see text).

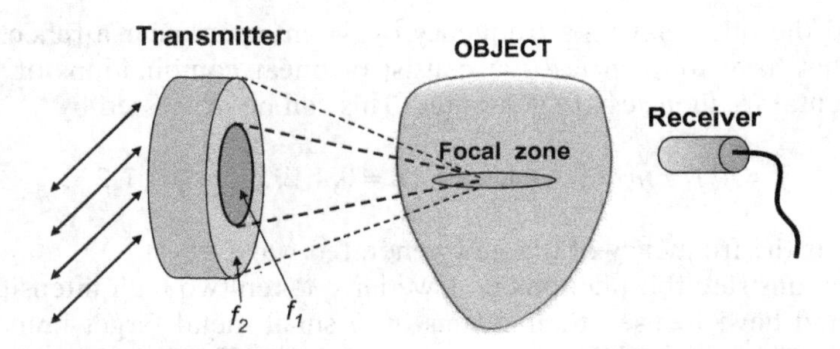

Figure 10.13. Schematic depiction of a vibro-acoustic Imaging system.

Of all the obtained combinations, let us consider the generated wave that has a frequency of $\hat{f} = (f_2 - f_1)$ (spectral peak **b** in Fig. 10.12). The basic idea of the vibro-acoustic method is to use two fundamental high frequencies (in the range of several megahertz) that differ (Δf) by only a few tens of kilohertz. Clearly, if the two frequencies are close to each other, then the generated wave will have a very low frequency $\hat{f} = \Delta f$. This provides three advantages: First of all, the two high-frequency beams can have a sharp focal point (see Chapter 7) and hence can provide high-resolution images. Secondly, the generated new wave which has a very low frequency is almost not attenuated in the tissue. Thirdly, the effect itself can be used as a new source for image contrast. The drawbacks of this technique are that the transmitted waves must have high intensity and that the target should be highly reflective in order to provide decent signals.

In reference 19 it was suggested to apply the vibro-acoustic imaging technique for acoustic mammography. An exemplary scanning system is depicted schematically in Fig. 10.13. Two confocal concentric focused transducers are used for transmission. The transmitting assembly is moved to provide a raster scan. The fundamental frequencies f_1 and f_2 are very close to each other. A hydrophone combined with a low-pass filter is used for detecting the generated low-frequency waves. A lateral resolution of 0.1 mm was reported with this technique.

10.6 TIME REVERSAL

Another interesting approach to utilize through transmitted waves is a method that is called "time reversal." The method is based on the fact that if a function $P(\bar{r}, t)$ is a solution to the wave equation (where $\bar{r}$ is the location vector and t is time), then it can be shown that $P(\bar{r}, -t)$ is also a solution of the same wave equation. This implies that if time could be moved backwards, we would have seen the wave traveling backwards (as though it were played in a reverse mode on a video display). Of course we cannot reverse the time coordinate

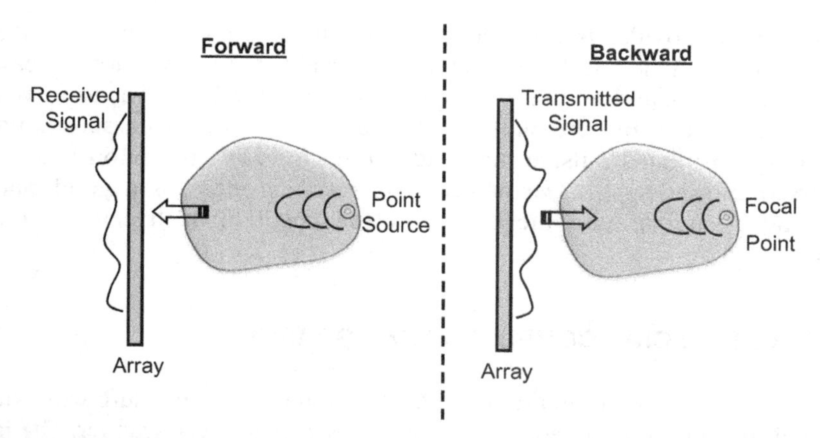

Figure 10.14. Schematic depiction of the "time-reversal" procedure. **(Left)** An ultrasonic pulse is transmitted from a point source within the target. The through-transmitted waves are detected and digitized by a transducer array. **(Right)** The detected signals are time-reversed and transmitted back toward the medium. As a result, the energy is focused at the target despite the distortions induced by the medium.

(at least presently), but we can make a wave travel backwards. This can be obtained by digitizing a through transmitted wave at the opposite location and then re-transmit from the receiver side to the transmitted side with a time-reversed profile.

In order to clarify the idea, let us consider the schematic drawing shown in Fig. 10.14. Suppose that we have a point source that transmits an acoustic pulse into the medium. The resulting waves will be transmitted through the medium and naturally their shape will be distorted. On the opposite side, an array of receivers is placed. Each element i in the transducer array receives a signal $S_i(t)$. After registering (i.e., digitizing) the entire profile, the receiver array is switched into transmit mode and the same profile but is time-reversed—that is, $S_i(-t)$ is transmitted backwards. The new wave front will pass through the same regions in the medium, but this time in the opposite direction. As a result, the refractions and reverberations will be reversed and the reversed wave will be refocused at the point originally used for transmission.

Using time reversal, high-quality focusing can be obtained despite the distortions induced by the medium. Possible applications in medicine are suggested for imaging and therapy. Several applications and algorithms for using time reversal are outlined, for example, in references 20–22. For example, time reversal can be used to focus high-intensity focused ultrasonic beam (HIFU; see Chapter 12) in the skull. Stemming from the unknown and strong distortions induced by the skull bones and the brain tissue, a simple transmission of a focused beam is practically useless. In many cases the beam will be too distorted to provide useful focusing. One suggested solution is to utilize time reversal. A needle transducer can be inserted into the target location within the brain, and an ultrasonic pulse is transmitted. The waves traveling though the brain and the skull bones are naturally distorted substantially. An array positioned outside the head is used for wave reception and time-reversed transmission, as explained. The resulting effect is a good-quality focusing at and around the region of the needle (which can of course be removed).

10.7 ULTRASONIC COMPUTED TOMOGRAPHY

Ultrasonic computed tomography (UCT) is an imaging procedure with which through-transmitted acoustic waves replace X rays in a CT scanner. The idea was first suggested (to the best of my knowledge) by James F Greenleaf [23]. Initial works have mapped the attenuation coefficients, and later works (e.g., reference 24) have also mapped the speed of sound. The main suggested application of UCT is breast imaging for tumor detection [25, 26]. In this section the basic principles of UCT in 2D and 3D are presented.

10.7.1 Basic Computed Tomography Principles

As was explained in the previous sections, projection images of certain acoustic properties of tissues can be obtained. For example, in Eq. (10.21) it was shown that by measuring the time of flight (TOF), the obtained signal is in fact an integral of the refractive index along the beam path. Similarly, it was shown via Eq. (10.25) that an integral of the acoustic attenuation coefficient can be obtained by measuring the amplitude of the through-transmitted wave. Using general terms, it can be stated that any integrative property of the medium [whether the integration is direct as in Eq. (10.21) or nondirect as in Eq. (10.25)] can yield a projection image of the scanned object. Or using mathematical terms, a projection onto the $\{x, z\}$ plane is given by

$$P(x,z) \triangleq \int_{-\infty}^{\infty} f(x,y,z)\,dy \tag{10.31}$$

where $f(x, y, z)$ is the desired (scanned) object property and $P(x, z)$ is in this case the obtained projection image. The infinite borders are used for mathematical convenience, as will be shown in the following. In practice, it is sufficient that the borders are large enough to include the object. In the case of a 2D imaging process, we are interested in obtaining a cross-sectional image of the object at a specific height Z_0. This reduces both sides of Eq. (10.31) by one dimension. Using mathematical expressions, this leads to $f(x, y) \triangleq f(x, y, Z_0)$ and $P(x) \triangleq P(x, Z_0)$. Next let us perform a one-dimensional Fourier transform on the obtained projection. This will yield (up to a constant) [27]

$$\mathbb{P}(K_x) = \int_{-\infty}^{\infty} P(x) e^{(-jK_x \cdot x)}\, dx = \int_{-\infty}^{\infty} \left[\int_{-\infty}^{\infty} f(x, y) dy \right] \cdot e^{(-jK_x \cdot x)}\, dx \qquad (10.32)$$

where K_x is the horizontal coordinate (which is equivalent to the x axis) in the Fourier domain, and $\mathbb{P}(K_x)$ is the transform function of the corresponding projection. By rearrangement of Eq. (10.32), we obtain

$$\mathbb{P}(K_x) = \int_{-\infty}^{\infty} P(x) e^{-jK_x \cdot x}\, dx = \int_{-\infty}^{\infty} \left[\int_{-\infty}^{\infty} f(x, y) \cdot e^{-jK_x \cdot x} \right] dxdy \qquad (10.33)$$

Now, let us compare this expression to the 2D Fourier transform of the image $\mathbb{F}(K_x, K_y)$, which is given (up to a constant) by

$$\mathbb{F}(K_x, K_y) = \int_{-\infty}^{\infty} \left[\int_{-\infty}^{\infty} f(x, y) \cdot e^{-j(K_x \cdot x + K_y \cdot y)} \right] dxdy \qquad (10.34)$$

As can be noted, this expression is almost identical to the expression given in Eq. (10.33). There is only one discrepancy between the two terms. The exponential term $K_y \cdot y$ is zero in Eq. (10.33). Thus in order to equate the two equations, we should use $K_y = 0$ and then write the following:

$$\mathbb{F}(K_x, K_y = 0) = \mathbb{P}(K_x) = \int_{-\infty}^{\infty} \left[\int_{-\infty}^{\infty} f(x, y) \cdot e^{-j(K_x \cdot x + 0 \cdot y)} \right] dxdy \qquad (10.35)$$

This implies that the one-dimensional Fourier transform of the projection $P(x)$ is equal to the horizontal line in the 2D Fourier transform of the image.

Next, let us consider the rotational property of the Fourier transform. As recalled, the Fourier transform of a rotated object is equal to the rotated Fourier transform of the nonrotated object. This is expressed mathematically by [28]

$$f(\theta + \theta_0, \bar{r}) \Leftrightarrow \mathbb{F}(\theta + \theta_0, \bar{K}) \qquad (10.36)$$

where the vector $\bar{r}$ represents the location in the regular (i.e., $\{x, y, z\}$) space and the vector $\bar{K}$ represents the location in the Fourier domain (i.e., $\{K_x, K_y, K_z\}$ space). This property is demonstrated graphically in Fig. 10.15. In this figure an image of a white rectangle on a black background is depicted (top left). The corresponding 2D Fourier transform of this object is shown below it. On the top right-hand side of the figure an image of the same object rotated by an angle θ is shown. The corresponding 2D Fourier transform of the rotated object is shown below (bottom right). As can be observed, the 2D Fourier transform of the object is also rotated by the same angle. It follows from these two properties that the Fourier transform of the projection obtained after rotating the object yields a horizontal line in the *rotated* 2D transform of the

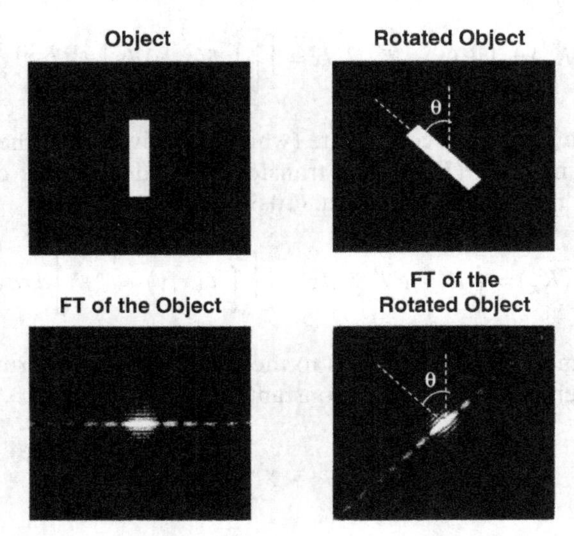

Figure 10.15. Graphical demonstration of the Fourier transform (FT) sensitivity to rotation. **(Top left)** Original object. **(Bottom left)** Corresponding FT of the object. **(Top right)** The object is rotated by angle θ. **(Bottom right)** Corresponding FT of the rotated object.

object. This property is called the "Projection Slice Theorem," which can be stated as follows: "The one-dimensional Fourier transform of a projection obtained at angle θ is equal to a line in the 2D Fourier domain (i.e., $\{K_x, K_y\}$) rotated by the same angle." Using this theorem, a tomography cross-sectional image of the object can be obtained, as explained in the following.

Next, let us consider a scanning system that can rotate around the object and acquire projections from various angles using through-transmitted ultrasonic waves. Each projection is designated as $P(\theta, x')$, where x' is the rotated horizontal axis (i.e., the rotated x axis). It is important to note that the rotation of the scanning system is equivalent to a rotation of the object. A schematic depiction of such system is shown in Fig. 10.16. The system consists of two array transducers parallel to each other which are positioned from both sides of the scanned object. One array transmits ultrasonic waves into the object. The waves travel through the object and are detected by the other array. The pair of transducers is then rotated to a new angle θ and the procedure is repeated.

By collecting projections around the object starting at an angle of $\theta = 0°$ up to $\theta = 180°$, a data matrix is obtained. In this matrix, each row represents a projection acquired at angle θ, that is, $P(\theta, x')$. Each column represents the projection values for a specific location along the x' coordinate. The size of this matrix is therefore the number of projections acquired times the number of acoustic rays acquired. This matrix is called a "sinogram" and is mathematically known as the Radon transform of the object [29].

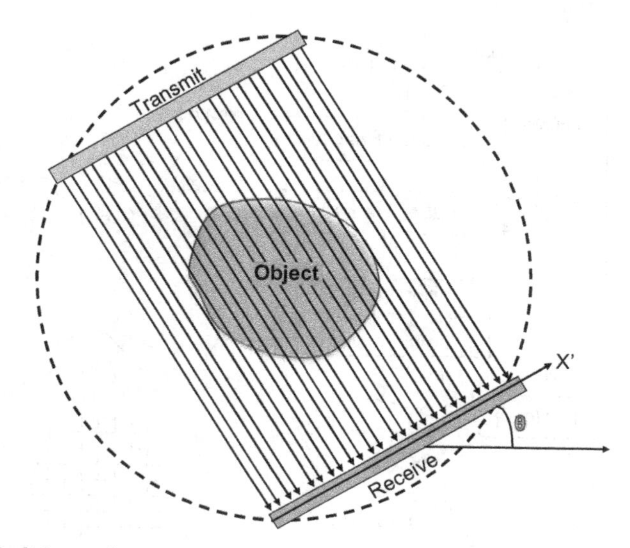

Figure 10.16. Schematic representation of a system for ultrasonic computed tomography (UCT). The system is comprised of a pair of parallel array transducers that are positioned on both sides of the object. Projections are acquired by transmitting waves from one side into the object and detecting them on the other side. The pair is rotated in order to collect data from various angles θ.

It follows from the "Projection Slice Theorem" that by applying a one-dimensional Fourier transform to a raw in the sinogram matrix, the value of a rotated line (by the same corresponding angle θ) in the 2D Fourier transform of the image is obtained. By repeating the procedure for all the sinogram rows and plotting the obtained lines in the 2D Fourier domain, a star-shaped coverage of this domain is obtained. Using interpolation, this polar arrangement of 2D Fourier values are rearranged into a Cartesian matrix. The required image is obtained by applying the inverse Fourier transform to the obtained matrix of 2D Fourier values.

The image obtained using the procedure above will depict the desired cross section of the object. However, due to the star-shaped coverage of the 2D Fourier domain, the lower spatial frequencies of the object will be enhanced. This will result in blur image similar to the effect of a 2D low-pass filter. In order to overcome this problem and obtain a proper reconstruction, a filter whose value in Fourier domain is given by $|\bar{K}|$ (known as the K-filter) is initially applied to every sampled projection line. This filter, which is shaped like a double ramp, creates a better balance between the high and low spatial frequencies in the image. This method is called the filtered back projection (FBP) method [29].

In order to further clarify the FBP method, a flow chart depicting the sequence of operations is shown in Fig. 10.17. It should be pointed out that

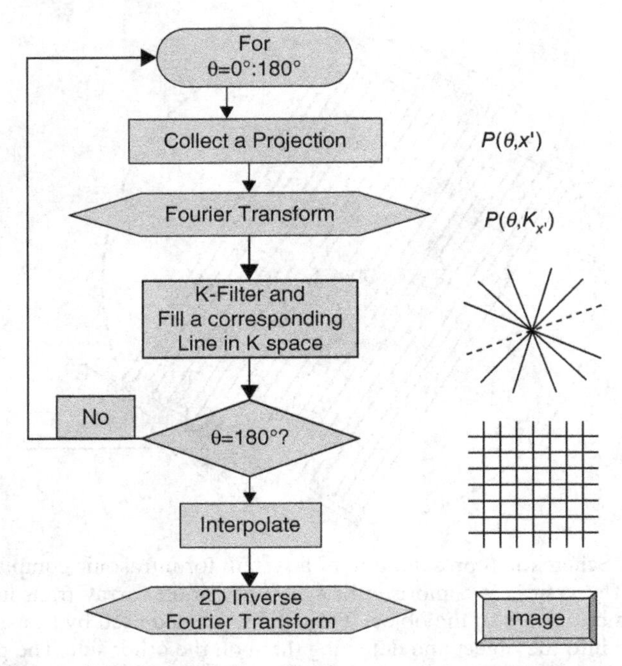

Figure 10.17. Flow diagram depicting the sequence of operations needed to reconstruct a tomographic image using the FBP method.

the filtering operation need not necessarily be carried out in the Fourier domain, but can be replaced by spatial filtering using a convolution operation with the inverse Fourier transform of the K-filter.

As a demonstration for the FBP method, the sinogram (radon transform) depicting the time of flight (TOF) of ultrasonic waves passing through a commercial breast phantom is depicted on the left-hand side of Fig. 10.18. The phantom was immersed in water. The sinogram was obtained by transmitting short ultrasonic pulses with a central frequency of 5 MHz. The through-transmitted waves were sampled with sampling frequency of 50 MHz. The TOF was calculated from the digitized signals. The projections were acquired every 10 degrees.

The corresponding tomographic cross-sectional image obtained with the FBP method is shown on the right-hand side of Fig. 10.18. The gray levels depict the refractive index for each pixel. The phantom contained tumor-mimicking targets. The targets are marked by the arrows. As can be noted, there are streak artifacts in the image (the straight lines passing from side to side in triangular patterns). These streaks are, of course, not part of the object (which is the rounded bright shape in the center). They stem from the fact that the number of projections used is too small. The number of projections needed in order to avoid these artifacts is given by

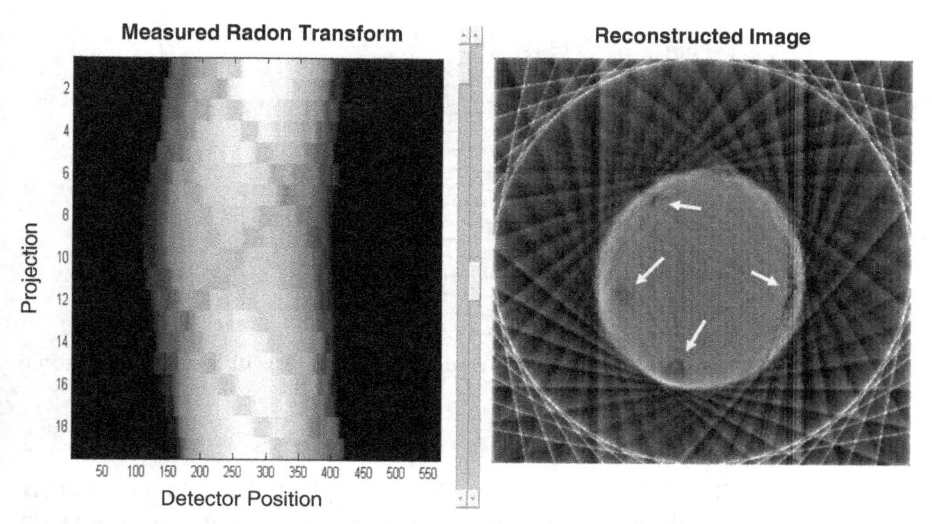

Figure 10.18. (Left) "Sinogram" (radon transform) depicting the time of flight of ultrasonic waves through a breast phantom immersed in water. Projections were acquired every 10 degrees. **(Right)** Cross-sectional image of the phantom depicting the refractive index. The reconstruction was obtained using the FBP method. The arrows indicate tumor-mimicking targets. Note the streak artifacts stemming from under sampling of the data.

$$\text{Number of projections} \geq \frac{\pi}{2} \cdot N_P \tag{10.37}$$

where N_P is the number of sampled points along each projection.

Although FBP is efficient and robust, it is important to note that there are alternative reconstruction methods to the FBP. Some of these methods are based on algebraic computations and are generally referred to as [29] algebraic reconstruction tomography (ART) methods. The reconstruction process in ART is iterative; thus their computation times are typically long. Nevertheless, when the number of projections is too small for the condition given by Eq. (10.37), or when the data are noisy, ART methods can be beneficial. Alternatively, a hybrid method that combines B-scan information (which is acquired anyway) with projection information has been suggested [30]. With this method, artifacts generated due to undersampling can be diminished.

10.7.2 Spiral Computed Tomography

In order to obtain a 3D reconstruction of an object, one can collect a large set of 2D cross-sectional images that are parallel to each other (i.e., by virtually slicing the object) and stack the images one atop the other. If the imaged slice

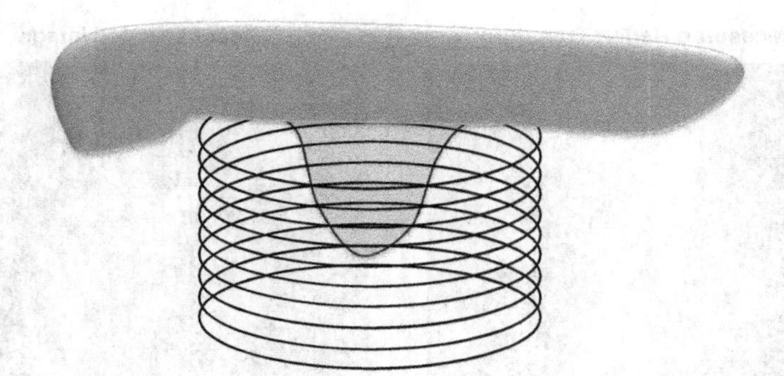

Figure 10.19. Schematic depiction of the scanning trajectory utilized for ultrasonic spiral CT imaging of the breast.

thickness is small and the gap between slices is negligible, a 3D image matrix can be obtained, with every element in this matrix corresponding to a voxel (3D pixel) in the object. This matrix can then be displayed in various modes using computer graphics. This approach has two major disadvantages: Firstly, the needed acquisition time will be long (unless special equipment that allows parallel imaging is used). Secondly, small objects that are located at the border between two consecutive slices might be missed because their signal is "diluted" between the two images.

An alterative imaging technique, which is currently used in modern X-ray CT, is spiral tomography. With spiral CT, projections are acquired along a helical trajectory as depicted in Fig. 10.19. After each projection acquisition, the system is rotated and translated to yield a projection at a slightly different height and angular position. Using suitable interpolation, a full 3D coverage of the object with a relatively small number of acquisitions (as compared to the multi-slice method) is possible. The application of spiral ultrasonic computed tomography was suggested by Azhari and Sazbon [31]. A 3D reconstruction of a commercial breast phantom obtained by this method is shown in Fig. 10.20.

10.7.3 Diffractive Tomography

The model used in the above-presented tomographic reconstructions is inaccurate. Medical ultrasonic imaging utilizes the lower megahertz frequency band. Considering the fact that the speed of sound in soft tissues is about 1500 m/sec, it follows that the wavelengths are commonly in the range of 0.1–1 mm. These wavelengths are too large to assume nondiffractive propagation, considering the typical image resolution. Hence, wave diffraction needs to be accounted for if a better theoretical model is desired.

As shown in the literature (e.g., references 29 and 32), the projection slice theorem, which assumes that the Fourier transform of a projection is a straight line in the 2D Fourier domain, is not valid when wave diffraction is significant.

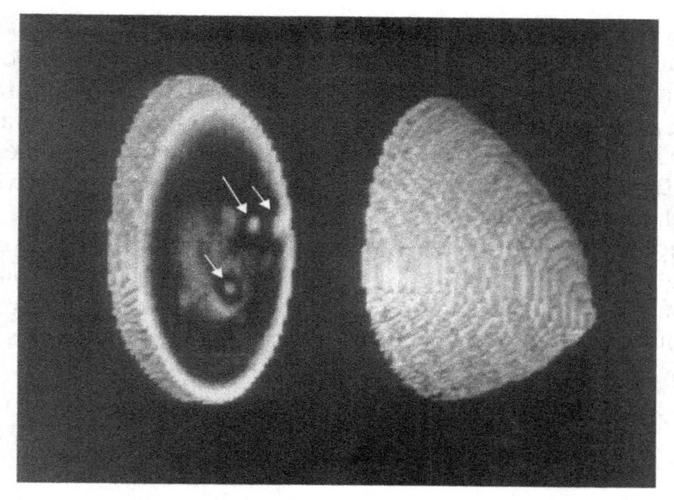

Figure 10.20. A 3D reconstruction of a commercial breast phantom obtained using ultrasonic spiral CT imaging. The arrows designate tumor mimicking targets. (Reprinted with permission from: Azhari and Sazbon [31], *Radiology* **212**(1):270–275, 1999 © Copyright by Radiological Society of North America (RSNA).

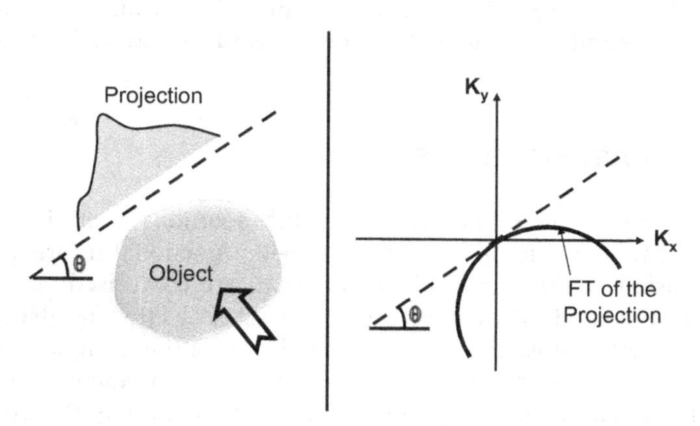

Figure 10.21. When wave diffraction is accounted for, the Fourier transform (FT) of a projection at angle θ **(left)** is mapped onto an arc in the 2D Fourier domain of the image.

Instead, it can be shown that it describes an arc in the 2D Fourier domain as shown schematically in Fig. 10.21. This arc is tangent to the straight line oriented at the projection angle θ. The radius of curvature of this arc increases as the wavelength decreases. Its value is given by [27]

$$K_0 = \frac{2\pi}{\lambda} \tag{10.38}$$

where λ is the corresponding wavelength.

When using a broadband pulse for imaging, a set of arcs can be obtained from a single transmission. By proper filtering of the through-transmitted ultrasonic pulse, a set of projection signals each corresponding to a different frequency is first generated. Then by applying FT to each signal, the matching arc in the 2D Fourier domain is derived. These arcs are imbedded within each other with one common point at $\{K_x = 0, K_y = 0\}$. The most inner arc corresponds to the lowest frequency, whereas the most outer arc corresponds to the highest frequency. In order to be able to reconstruct a tomographic image, a sufficient number of arcs has to be collected so as to cover a circle in the 2D Fourier domain whose center is at the axes origin and whose radius is given by

$$K_{\max} = \frac{\pi}{\Delta x} \tag{10.39}$$

where Δx is the required spatial resolution (i.e., the image pixel size).

The reconstruction process in diffractive ultrasonic computed tomography is a little more complicated than the FBP. Devaney [32] has suggested an algorithm that was called *filtered back-propagation*. Explicit algorithms for diffraction UCT are presented in Kak and Slaney [29]. An alternative approach that applies nonuniform Fourier transform was suggested by Bronstein et al. [33].

10.8 CONTRAST MATERIALS

Contrast-enhancing materials are injectable solutions used in ultrasonic imaging and therapeutic procedures to provide functional information (e.g., blood perfusion). Currently available materials are comprised mainly of gas micro-bubbles coated with a polymeric shell [(e.g., Polyethylene glycol (PEG)] or some biological substance such as albumin, galactose, or lipids. The diameter of these bubbles should be small enough to allow them clear passage through the capillaries and the pulmonary system. Commonly, the size of these bubbles is on the order of 3–5 μm. A microscopic photo of such a contrast material (Definity™ by Bristol–Myers Squibb Medical Imaging) is shown in Fig. 10.22.

The gas for which the acoustic impedance is practically negligible compared to that of blood or soft tissues reflects acoustic waves almost entirely. Although the reflection from a single micro-bubble is too small to be detected by conventional imaging equipment, a cluster or a "cloud" containing a very large number of such bubbles provides very strong echoes that are very clearly visible (see, for example, Fig. 10.23).

The shell, which is made of a biocompatible material, retains the encapsulated gas in the body for a sufficiently long time, ranging from a few minutes

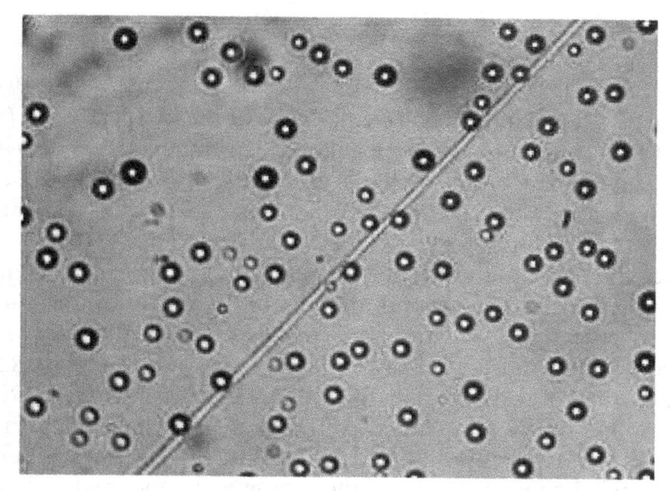

Figure 10.22. Microscopic image of a contrast material suspension (Definity™ by Bristol–Myers Squibb Medical Imaging, N. Billerica, MA, USA). Note the variability in the bubbles size. (Courtesy of Assaf Hoogi.)

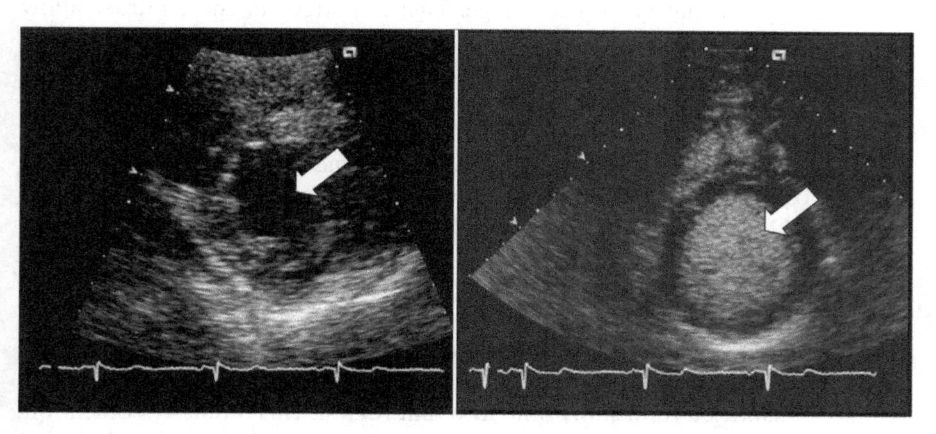

Figure 10.23. (Left) Short axis echocardiography (B-scan) image of the left ventricle prior to contrast material injection (the cavity is marked by an arrow). **(Right)** Same cross section after injecting the contrast material. Note the whitening of the cavity as a result of the bubbles arrival. (Courtesy of Jonathan Lessick, Rambam Health Care Campus, Cardiology Department.)

to several tens of minutes, to allow the completion of the needed data collection. The bubbles dissolve naturally after a while or can be actively destroyed by exposing them to a strong acoustic beam. These micro-bubbles may collapse when exposed to the ultrasonic beam of the imaging system if the beam is intensive enough or its "mechanical index" (MI) is large. The "mechanical index" is defined by

$$MI = \frac{P_{\text{negative}}}{\sqrt{f}} \tag{10.40}$$

where P_{negative} is the maximal negative pressure of the incident wave (measured in MPa), and f is the frequency in megahertz. This index serves also as a safety index (see Chapter 12), and its value is automatically displayed on the screen of modern imaging systems. Commonly, a transmitted acoustic beam with an MI > 1 will cause the bubbles to collapse.

Contrast-enhancing materials are used today in imaging and image-guided therapeutic procedure. The second application is mostly in a research mode today. The most popular imaging application is echocardiography. One possible protocol is to acquire a set of reference images of the heart. Then inject intravenously the contrast-enhancing solution. The material is carried by the blood system and reaches the myocardium. A myocardial region that is well-perfused will receive many bubbles. As a result, the echoes reflected from these regions will be augmented and the region will appear whiter in the post-injection image as compared to the pre-contrast reference image. On the other hand, regions with reduced perfusion or that are non-perfused will retain the same level of reflectivity and the same range of gray levels in the post-injection image relative to the reference image. These changes in the contrast allow better detection of ischemic regions.

Using a more quantitative approach, the rate by which the contrast-enhancing material is removed from the tissue is analyzed. Regions of interest (ROI) in the image are marked by the operator. Then a table containing the average gray levels in each ROI as a function of time is registered by analyzing a set of post-injection frames. Using these data, a "washout curve" is provided for each ROI. This curve can be characterized by a mathematical model. The estimated parameters of this model are then used to characterize the tissue in this ROI.

Another method is to study the wash in curve. With this method the contrast-enhancing material is injected continuously into the body. When the gray levels in the target organ are stabilized (more or less), a very intensive beam is transmitted for a short time into that region. This intensive sonication, which is called "flash," causes the collapse of the irradiated micro-bubbles. Consequently, the echoes are reduced and therefore the gray levels become darker. New bubbles that are washed into this region restore the high reflectivity and the high gray levels. The rate by which the gray levels are restored serves as a tissue characterization parameter.

It is important to note two additional facts related to these micro-bubbles. First, the micro-bubbles have a specific resonance frequency that stems from their size and the mechanical properties of their shell. Second, these bubbles have a characteristic spectral signature that is half the transmitted frequency, known as subharmonic frequency. This allows one to map the spatial distribution of these bubbles by band-pass filtering the signal around this subharmonic frequency.

10.9 CODED EXCITATIONS

Recently, the use of coded excitations for ultrasonic applications in medicine has been suggested (see, for example, references 34–36). The idea is to enrich the information contents of the transmitted wave, by embedding frequency as well as temporal-related coded information. Such transmitted waves can be of longer duration than Gaussian pulses, for example. Hence, they can carry more acoustic energy, yet be still within the safety limits (see Chapter 12). The potential advantages are improvement in signal-to-noise ratio (SNR), better penetration, and better characterization of the tissue properties, and recently it was also suggested to utilize such pulses for measuring the velocity of moving reflectors with a single transmitted wave [37]. The disadvantages are the need for suitable dedicated hardware and the need for a relatively cumbersome signal processing.

One of the commonly used coded pulses is known as a "chirp" pulse. This pulse changes its instantaneous frequency contents linearly with time. Mathematically, the chirp pulse is defined by

$$f(t) = f_0 + S_0 t, \qquad 0 \le t \le T_m, \qquad \text{where} \quad S_0 = \frac{f_1 - f_0}{T_m} \tag{10.41}$$

where f_0 and f_1 are the starting and ending instantaneous frequencies (which in fact define the frequency band), S_0 defines the frequency sweeping rate, and T_m is the transmission duration. The signal itself, $\psi(t)$, is given by

$$\boxed{\psi(t) = A(t) \cdot e^{2\pi j \cdot \left(f_0 t + S_0 t^2\right)}, \qquad 0 \le t \le T_m} \tag{10.42}$$

where $A(t)$ is the amplitude modulation function, which can be, for example, $A(t)$ = const. In practice, however, the signal is also modulated by the transducer's transfer function $H(f)$. Thus, the actual transmitted function is given by $\psi'(t) = \psi(t) \cdot H(f \Rightarrow t)$, where the arrow indicates the relation between time and frequency.

An exemplary chirp signal is given in Fig. 10.24. The lowest frequency for this signal was $f_0 = 1$ MHz and the highest frequency $f_1 = 5$ MHz. The displayed signal was received by a broadband transducer for which the central frequency is 5 MHz. The originally transmitted chirp pulse had a uniform amplitude. The depicted increase in amplitude of the received wave stems from the transducer's transfer function, which responds better to frequencies closer to its central frequency.

Another useful coded pulse is a double inverted pulse, such as the one depicted in Fig. 10.25. This pulse is comprised of two pulses transmitted one after the other. The first pulse is a regular pulse (e.g., Gaussian pulse), whereas the second, which is transmitted after a slight delay, is inverted; that is, it has a phase of π relative to the first one. The resulting echo train obtained from

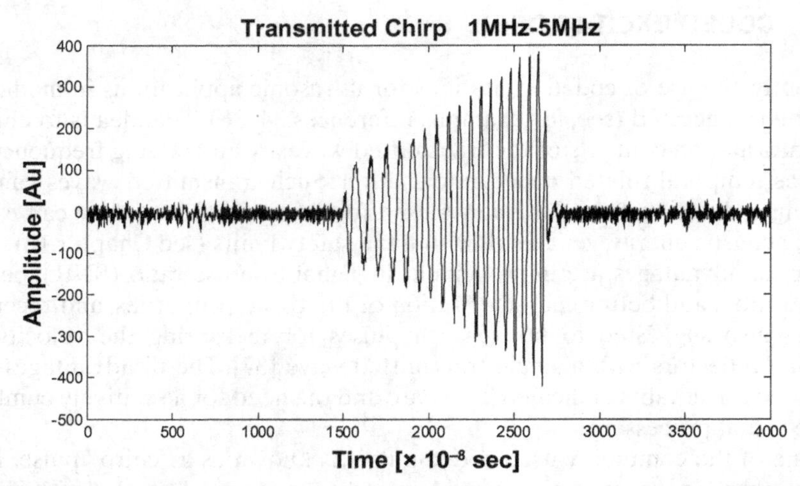

Figure 10.24. Chirp signal that sweeps the frequency range of 1–5 MHz. This signal was acquired by a broadband transducer for which the central frequency is 5 MHz.

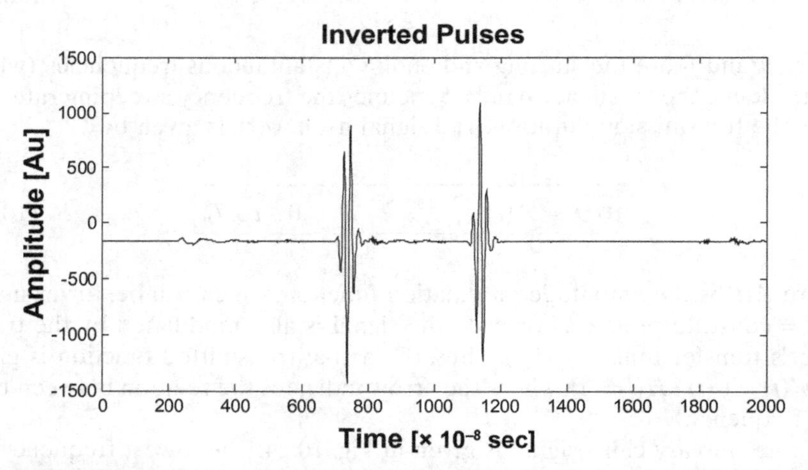

Figure 10.25. An exemplary double inverted coded pulse that can be utilized for nonlinear imaging. The two pulses are identical Gaussian pulses with a central frequency of 5 MHz. However, as can be noted, the second pulse is inverted relative to the first one.

the second pulse is therefore inverted relative to the echo train obtained from the first pulse.

By synchronizing the obtained pair of echo trains and adding them, the two signals will ideally cancel each other except for the additive noise. This is true as long as the medium is linear. However, if the medium is nonlinear, the signals will not cancel each other. This allows one to map regions with a strong nonlinear behavior.

One of the applications for such a technique is to image the distribution of injected contrast-enhancing materials that are comprised of gas bubbles (as explained above). Since the gas bubbles have strong nonlinear behavior, regions with high concentration of this contrast material will yield high signal even when the two echo trains are added. Consequently, functional images depicting strong nonlinearity in the tissue are obtained. It should be pointed out that anatomical images can be obtained as well from the very same data simply by using only one of the two echo trains or by inverting the second echo train and adding the two. (This operation is equivalent to subtraction of the two echo trains.)

REFERENCES

1. Jones JP and Wright H, A new board band ultrasonic technique with biomedical implications: I. Background and theoretical discussion, 83rd Meeting of the Acoustic Society of America. April 1972, *J Acoust Soc Am (JASA)* **52**:178, 1972, Abstract NN7.
2. Jones JP, Impediography: A new ultrasonic technique for non-destructive testing and medical diagnosis. Conference Proceedings, Ultrasonics International 1973, pp. 214–218.
3. Wright H, Impulse-response function corresponding to reflection from a region of continuous impedance change, *J Acoust Soc Am (JASA)* **53**(5):1356–1359, 1973.
4. Perelman B, An Investigation of Ultrasonic Impediography: Imaging Methods and Applications, M.Sc. dissertation, Faculty of Biomedical Engineering, Technion, 1997.
5. Shigley JE, *Mechanical Engineering Design*, 3rd edition, McGraw-Hill, New York, 1977.
6. Huang SR, Lerner RM, and Parker KJ, On the estimation of the amplitude of harmonic vibrations from Doppler spectrum of reflected signals, *J Acoustic Soc Am* **88**:310–317, 1990.
7. Ophir J, Alam SK, Garra B, Kallel F, Konofagou E, Krouskop T, and Vargehez T, Elastorgaphy: Ultrasonic estimation and imaging of the elastic properties of tissue, *Proc Inst Mech Eng* **213**(H):203–233, 1999.
8. Gao L, Parker KJ, Lerner RM, and Levinson SF, Imaging of the elastic properties of tissue—A review, *Ultrasound Med Biol* **22**(8):959–977, 1996.
9. Parker KJ, Taylor LS, and Gracewski S, A unified view of imaging the elastic properties of tissue, *J Acoust Soc Am* **117**(5):2705–2712, 2005.
10. Taylor LS, Porter BC, Rubens DJ, and Parker KJ, Three-dimensional sonoelastography: Principles and practices, *Phys Med Biol* **45**:1477–1494, 2000.
11. Palmeri ML, Frinkley KD, Zhai L, Gottfried M, Bentley RC, Ludwig K, Nightingale KR, Acoustic radiation force impulse (ARFI) imaging of the gastrointestinal tract, *Ultrason Imag* **27**(2):75–88, 2005.

12. Dumont D, Behler R, Nichols T, Merricks E, and Gallippi C, ARFI imaging for noninvasive material characterization of atherosclerosis, *Ultrasound Med Biol* **32**(11):1703–1711, 2006.
13. Hein A, O'Brien WD, Current time-domain methods for assessing tissue motion by analysis from reflected ultrasound echoes—A review, *IEEE Trans UFFC* **40**(2):84–102, 1993.
14. Rappaport D, Adam D, Lysyansky P, and Riesner S, Assessment of myocardial regional strain and strain rate by tissue tracking in B-mode echocardiograms, *Ultrasound Med Biol* **32**(8):1181–1192, 2006.
15. Adam D, Landesberg A, Konyukhov E, Lysyansky P, Lichtenstein O, Smirin N, and Friedman Z, Ultrasonographic quantification of local cardiac dynamics by tracking real reflectors: Algorithm development and experimental validation, *Comput Cardiol* **31**:337–340, 2004.
16. Bachner N, Friedman Z, Fehske W, and Adam D, Inhomogeneity of left ventricular apical rotation during the heart cycle assessed by ultrasound cardiography, *Comput Cardiol* **33**:717–720, 2006.
17. Fatemi M and Greenleaf JF, Ultrasound stimulated vibro-acoustic spectroscopy, *Science* **280**:82–85, 1998.
18. Fatemi M and Greenleaf JF, Vibro-acoustography: An imaging modality based on ultrasound stimulated acoustic emission, *Proc Nat-Acad Sci USA* **96**:6603–6608, 1999.
19. Fatemi M, Wold LE, Alizad A, and Greenleaf JF, Vibro-acoustic tissue mammography, *IEEE Trans Med Imaging* **21**(1):1–8, 2002.
20. Vignon F, Marquet F, Cassereau D, Fink M, Aubry JF, and Gouedard P, Reflection and time-reversal of ultrasonic waves in the vicinity of the Rayleigh angle at a fluid–solid interface, *J Acoust Soc Am* **118**(5):3145–3153, 2005.
21. Montaldo G, Aubry JF, Tanter M, and Fink M, Spatio-temporal coding in complex media for optimum beamforming: The iterative time-reversal approach, *IEEE Trans Ultrason Ferroelectr Freq Control* **52**(2):220–230, 2005.
22. Folegot T, de Rosny J, Prada C, and Fink M, Adaptive instant record signals applied to detection with time reversal operator decomposition. *J Acoust Soc Am* **117**(6):3757–3765, 2005.
23. Greenleaf JF, Johnson SA, Lee SL, Herman GT, and Wood EH, Algebraic reconstruction of spatial distributions of acoustic absorption within tissue from their two-dimensional acoustic projections, in *Acoustical Holography*, Vol. **5**, Plenum Press, New York, 1974, pp. 966–972.
24. Glover GH and Sharp JC, Reconstruction of ultrasound propagation speed distributions in soft tissue: Time-of-flight tomography, *IEEE Trans Sonics Ultrason* **SU-24**(4):229–234, 1977.
25. Greenleaf JF and Bahn RC, Clinical imaging with transmissive ultrasonic computerized tomography, *IEEE Trans Biomed Eng* **BME-28**(2):177–185, 1981.
26. Scherzinger AL, Belgam RA, Carson PL, et al. Assessment of ultrasonic computed tomography in symptomatic breast patients by discriminant analysis, *Ultrasound Med Biol* **15**(1):21–28, 1989.
27. Parker JA, *Image Reconstruction in Radiology*, second printing, CRC Press, Boca Raton, FL, 1991.

28. Gonzales RC and Wintz P, *Digital Image Processing*, Addison-Wesley, Reading, MA, 1977.
29. Kak AC and Slaney M, *Principles of Computerized Tomographic Imaging*, IEEE Press, New York, 1987, pp. 49–112.
30. Azhari H and Stolarski S, Hybrid ultrasonic computed tomography, *Comput Biomed Res* **30**:35–48, 1997.
31. Azhari H and Sazbon D, Volumetric imaging using spiral ultrasonic computed tomography, *Radiology* **212**(1):270–275, 1999.
32. Devaney AJ, A filtered backpropagation algorithm for diffraction images, *Ultrason Imag* **4**:336–350, 1982.
33. Bronstein MM, Bronstein AM, Zibulevsky M, and Azhari H, Reconstruction in diffraction ultrasound tomography using non-uniform FFT, *IEEE Trans Med Imag* **21**(11):1395–1401, 2002.
34. Misaridis T and Arendt Jensen J, Use of modulated excitation signals in medical ultrasound. Part I: Basic concepts and expected benefits, *IEEE Trans Ultrason Ferroelectr Freq Control* **52**(2):177–190, 2005.
35. Misaridis T and Arendt Jensen J, Use of modulated excitation signals in medical ultrasound. Part II: Design and performance for medical imaging applications, *IEEE Trans Ultrason Ferroelectr Freq Control* **52**(2):192–207, 2005.
36. Misaridis T and Arendt Jensen J, Use of modulated excitation signals in medical ultrasound. Part III: High frame rate imaging, *IEEE Trans Ultrason Ferroelectr Freq Control* **52**(2):208–219, 2005.
37. Levy Y and Azhari H, Velocity measurements using a single transmitted linear frequency modulated chirp, *Ultrasound Med Biol* **33**(5):768–773, 2007.

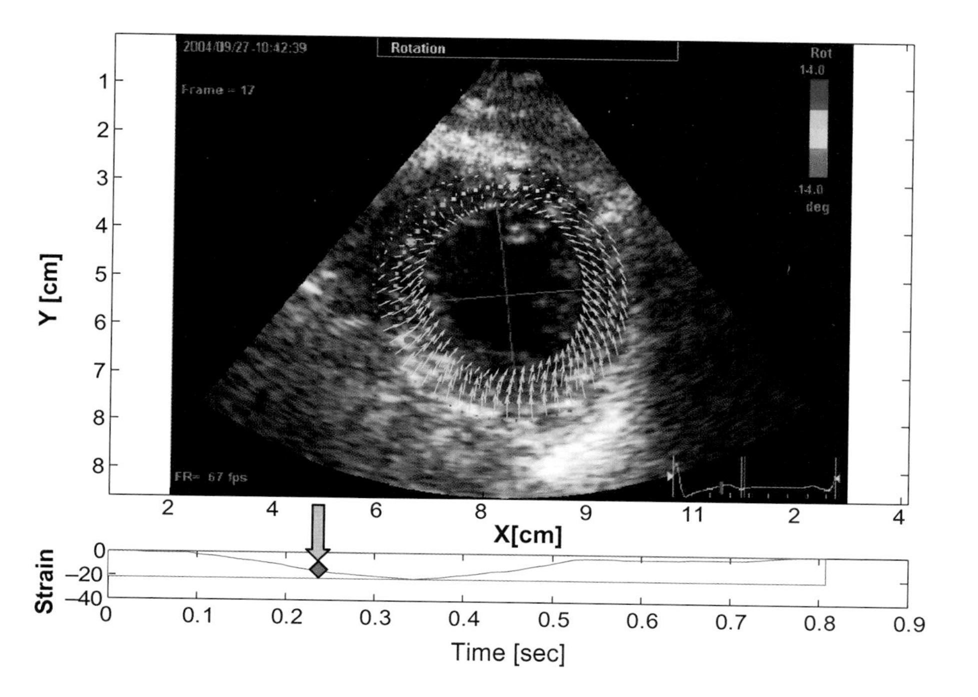

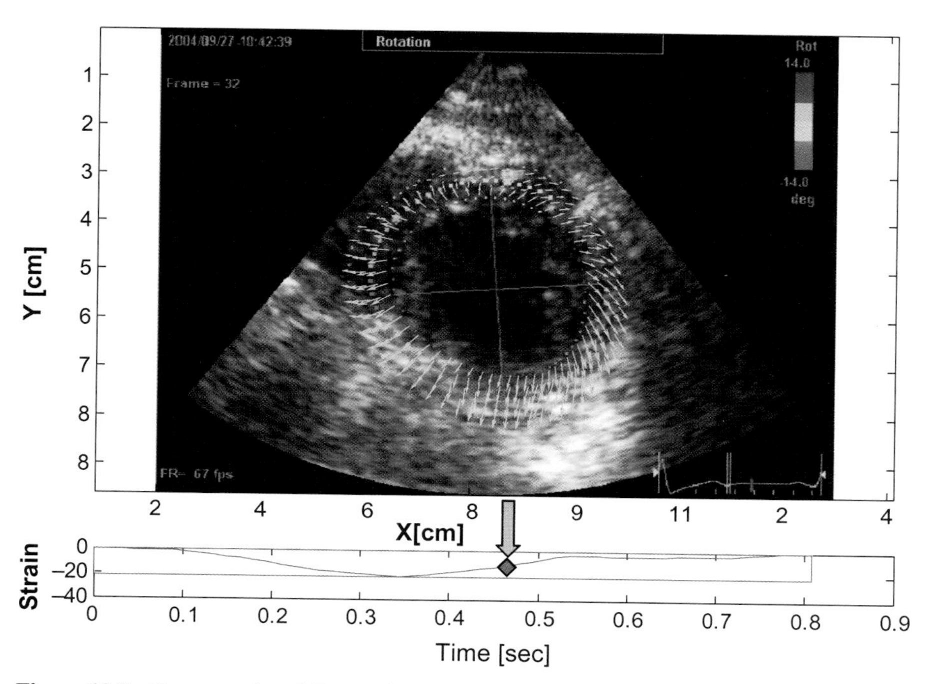

Figure 10.8. Cross-sectional B-scan images of a heart's left ventricle during contraction **(top)** and relaxation **(bottom)**. The arrows indicate graphically the direction and velocity of the instantaneous motion. The graphs at the bottom of each image designate the global strain, and the temporal location is indicated by the dot. (Courtesy of: Noa Bachner and Dan Adam—Technion, and Zvi Friedman—GE Healthcare). See page 245 for text discussion of this figure.

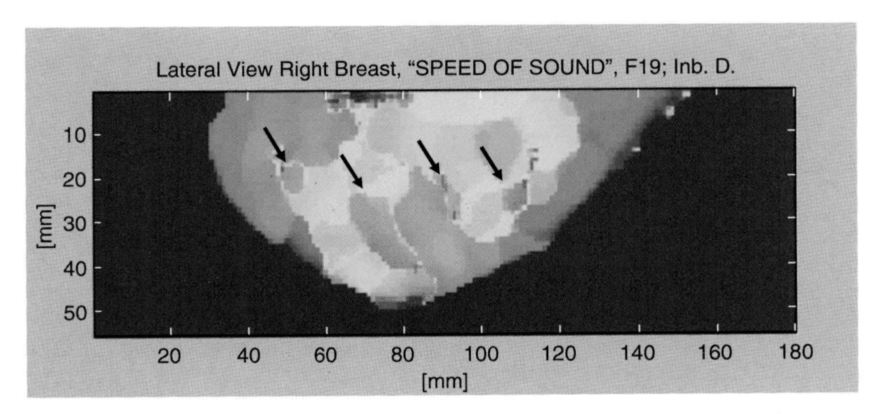

Figure 10.10. Speed-of-sound projection (lateral view) of a breast containing several benign tumors. Regions with extremely high speed of sound, which is characteristic to solid tumors, are marked by the arrows. See page 248 for text discussion of this figure.

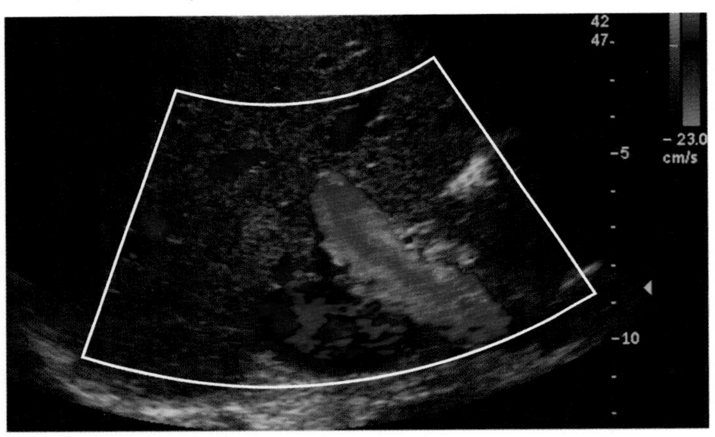

Figure 11.8. Duplex image (CFM) of the liver portal vein. The anatomy is depicted in shades of gray. The flow direction toward and away from the ultrasonic transducer is indicated by the blue and red colors. (Courtesy of Dr. Diana Gaitini.). See page 283 for text discussion of this figure.

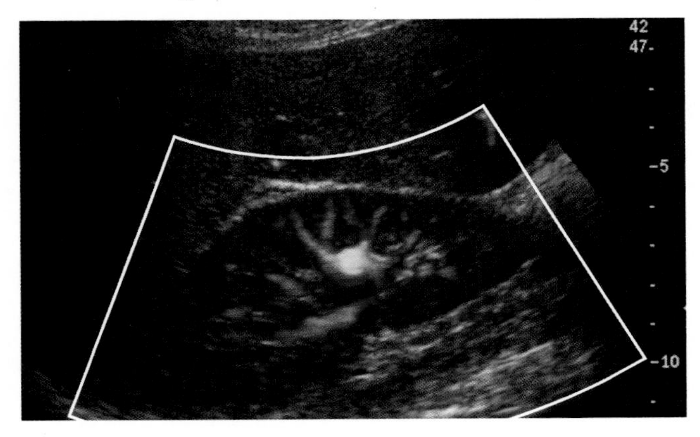

Figure 11.9. Another type of duplex imaging using the power Doppler technique. As can be noted, shades of only one color are used in this case because the flow mapping is nondirectional. (Courtesy of Dr. Diana Gaitini.). See page 283 for text discussion of this figure.

CHAPTER 11

DOPPLER IMAGING TECHNIQUES

Synopsis: In this chapter the basic principles used for measuring motion and flow with ultrasonic waves are presented. First, the Doppler effect is introduced. Then, it is explained how this effect can be utilized for measuring the temporal flow speed profile. The difficulties associated with the computation of the spectral Doppler shift are discussed, and numeric methods for rapid estimation of flow velocity and variance are presented. Finally, the principles of color flow mapping and duplex imaging are introduced.

11.1 THE DOPPLER EFFECT

One of the prominent advantages of ultrasonic imaging is its ability to combine anatomical imaging with flow or tissue velocity mapping. This is an essential tool in cardiovascular diagnosis. The basis for these imaging techniques is the Doppler effect.

Almost every one of us is aware of the fact that when a vehicle (e.g., a train) passes us by while blowing its horn, the sound of that horn seems to increase its pitch (its frequency). On the other hand, when that vehicle travels away from us, it appears as though its pitch is decreasing. However, for someone who is riding this vehicle the horn pitch seems constant. The change in pitch (frequency) is proportional to the relative velocity between the vehicle and the listener. The frequency shift trend—that is, an increase or a decrease—depends on whether this vehicle is moving toward us or away from us. This

Basics of Biomedical Ultrasound for Engineers, by Haim Azhari

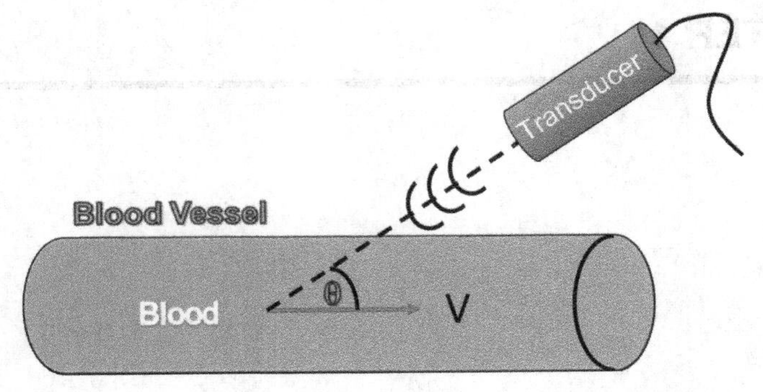

Figure 11.1. An ultrasonic transducer transmits waves toward a blood vessel through which blood flows with an instantaneous speed V. The angle between the acoustic beam and the blood vessel's axis is θ.

phenomenon is known as the Doppler effect and was first used by Doppler to explain the color shifts of stellar systems. If the relation between the velocity and the frequency shift is known and quantified and the frequency shift is measured, then one can use this effect to estimate the velocity of moving objects or measure blood flow.

In order to understand how the Doppler effect is commonly utilized in medical ultrasound, let us consider the schematic drawing shown in Fig. 11.1. Consider an ultrasonic transducer that transmits waves with a frequency f_0 into the body. The waves propagate with a speed C and encounter a blood vessel through which blood flows with an instantaneous speed V. The angle between the acoustic beam and the blood vessel's axis is θ. The "clouds" of blood cells carried by the flowing plasma have relatively high acoustic impedance and therefore reflect the waves. The reflected waves (i.e., the echoes) move back toward the transducer, but their frequency is changed to f' due to the Doppler effect.

In order to calculate the corresponding frequency shift, let us now refer to Fig. 11.2. Consider two consecutive wave fronts that propagate toward the moving target (top drawing). The distance between them (before encountering the target) is λ_0 and its value is given by

$$\lambda_0 = \frac{C}{f_0} \tag{11.1}$$

Because the target is moving toward the transducer with a relative velocity $v = V \cdot \cos(\theta)$, the time elapse between the target's encounter with these two wave fronts is given by

$$\Delta t = \frac{\lambda_0}{c + v} \tag{11.2}$$

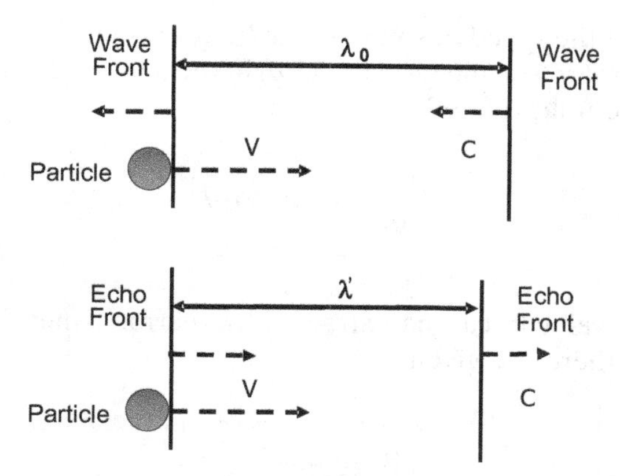

Figure 11.2. (Top) Two consecutive wave fronts move with a speed C toward a reflecting particle that is moving with a velocity V toward them. The pre-encounter wavelength is λ_0. **(Bottom)** The corresponding reflected echoes move back with a new wavelength λ'.

When the waves encounter the target, echoes are reflected. The echoes also travel with a velocity C in the medium but this time toward the transducer. If the target travels toward the transducer, then the distance between the two echo fronts will be shorter; and if the target travels away from the transducer, the distance will be longer. The distance between the two echo fronts λ' is given by

$$\lambda' = \Delta t \cdot (C - v) \tag{11.3}$$

The frequency corresponding to the new wavelength is therefore

$$f' = \frac{C}{\lambda'} \tag{11.4}$$

Hence, the Doppler frequency shift Δf is given by

$$\begin{aligned}
\Delta f &= f' - f_0 = \frac{C}{\lambda'} - f_0 = \frac{C}{\Delta t (C - v)} - f_0 = \frac{C}{\lambda_0} \cdot \frac{(C + v)}{(C - v)} - f_0 \\
\Rightarrow \Delta f &= f_0 \left[\frac{(C + v)}{(C - v)} - 1 \right] = f_0 \left[\frac{(C + v - C + v)}{(C - v)} \right] = \frac{f_0 \cdot 2v}{C - v} \\
\Rightarrow \Delta f &= \frac{f_0 \cdot 2V \cos\theta}{C - V \cos\theta}
\end{aligned} \tag{11.5}$$

The typical blood flow velocities are in the range of tens of centimeters per second (this value may be slightly higher under pathological conditions in the heart valves).

Recalling that the speed of sound in the body is about 1500 m/sec, it follows that $V << C$ and therefore the term $(V \cos \theta)$ in the denominator can be ignored. This leads to the following relation:

$$\boxed{\Delta f = \frac{f_0 \cdot 2V \cos\theta}{C}} \tag{11.6}$$

The blood flow velocity can be extracted by rearrangement from Eq. (11.6) and its value is therefore given by

$$\boxed{V = \frac{\Delta f \cdot C}{2 f_0 \cos\theta}} \tag{11.7}$$

Equation (11.7) is very useful. However, the value of the angle θ between the acoustic beam and the blood vessel is usually not known. Hence, it is required to obtain first an anatomical image using, for example, the B-scan method, from which the angle θ can be extracted. Another important point to note is that when developing this equation, we have ignored the fact that the speed of sound is not constant in the body. Thus, if the average speed of sound assumed for the tissue (e.g., 1540 m/sec) is substantially different from that of the tissue present along the acoustic beam (e.g., speed of sound for fat is abut 1450 m/sec), then an error that is proportional to the speed of sound ratio is induced.

11.2 VELOCITY ESTIMATION

In order to describe the flow velocity within an artery or a vein as a function of time, the Doppler shift must be calculated with a sufficiently high rate (at least a few dozens of times per heart beat). From the obtained results we need to calculate the corresponding instantaneous flow velocity and display it for the physician as a continuously updating graph. Let us define the central frequency as f_0 and define the signal obtained from the transducer after the Doppler shift as $s(t)$. The ultrasonic pulses are transmitted with a pulse repetition frequency of PRF. As explained in the following, the PRF is actually our sampling rate. Thus, recalling the Shannon–Whittaker sampling theorem [1], it is important to set the PRF to be at least twice as high as the maximal expected Doppler shift, that is, $\text{PRF} \geq 2 \cdot \Delta f_{\max}$.

The process through which the Doppler frequency shift is estimated is similar to a process commonly used in radio communication which is called *quadrature demodulation*. This process can be implemented by hardware using a relatively simple electric circuit. For clarity the process is schematically outlined in the block diagram depicted in Fig. 11.3.

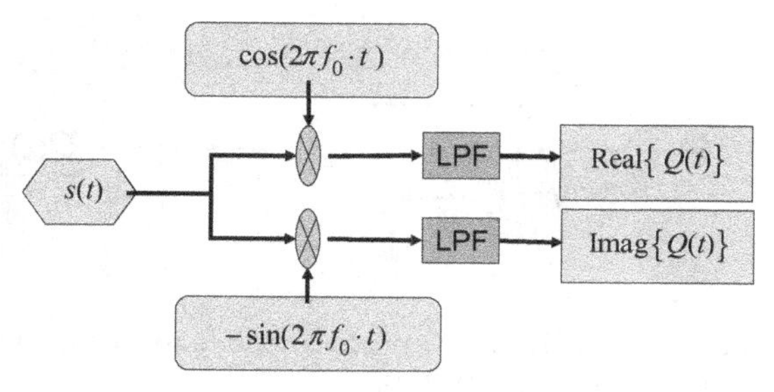

Figure 11.3. Block diagram depicting the quadrature demodulation process utilized for extracting the Doppler signal.

At the first step of the process the detected signal $s(t)$ is mixed with a reference signal $s_0(t)$. The mixing is carried out simultaneously in two parallel channels. In the first channel the signal is mixed with $\cos(2\pi t \cdot f_0)$. The output of this channel is considered the real part of our Doppler signal. The second channel is mixed with $-\sin(2\pi t \cdot f_0)$ and is considered the imaginary part of the Doppler signal. This mixing process can be carried out by hardware. The mathematical description of this mixing stage is given by

$$s'(t) = s(t) \cdot [\cos(2\pi f_0 \cdot t) - j \cdot \sin(2\pi f_0 \cdot t)] = s(t) \cdot e^{-j \cdot 2\pi f_0 \cdot t} \tag{11.8}$$

where $s'(t)$ is the post-mixing signal. In the case of a transmitted Gaussian pulse where we consider $t = 0$ at the signal's peak, the following signal is obtained:

$$\begin{aligned} s'(t) &= A \cdot e^{-\beta t^2} \cdot \cos(2\pi t \cdot f) \cdot e^{-j \cdot 2\pi t \cdot f_0} \\ &= \frac{A}{2} \cdot e^{-\beta t^2} \cdot \left[e^{j \cdot 2\pi t \cdot f} + e^{-j \cdot 2\pi t \cdot f} \right] \cdot e^{-j \cdot 2\pi t \cdot f_0} \\ &= \frac{A}{2} \cdot e^{-\beta t^2} \cdot \left[e^{j \cdot 2\pi t \cdot (f - f_0)} + e^{-j \cdot 2\pi t \cdot (f_0 + f)} \right] \end{aligned} \tag{11.9}$$

Studying the expression within the square brackets, we can note that it is comprised of two frequency-dependent terms. One has a very high frequency component $(f_0 + f) \approx 2f_0$, and the other has a very low frequency $(f - f_0) \triangleq \Delta f$ component that is actually the required Doppler shift. Since we are interested only in the Doppler shift, we can remove the high-frequency component from the mixed signal by applying a low-pass filter (LPF). The filtered signal may be considered as a complex number and we shall designate it here as $Q(t)$. Alternatively, the Doppler signal can be obtained by applying the Hilbert transform [see Eq. (9.25)] to the mixed signal. This procedure is schematically presented in the block diagram shown in Fig. 11.4. In the case of a Gaussian pulse the value of $Q(t)$ is given by

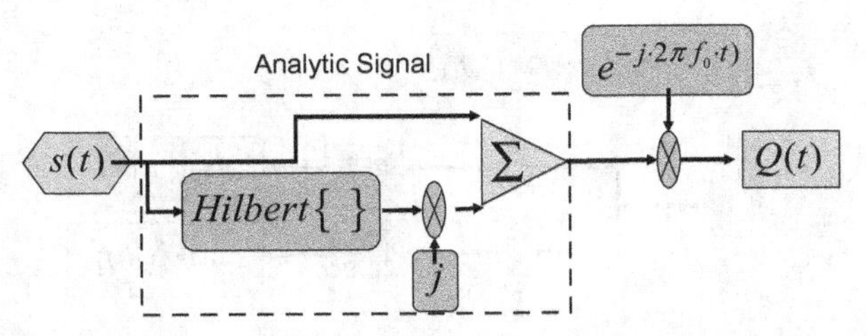

Figure 11.4. Block diagram depicting the process applied for obtaining the $Q(t)$ signal using the Hilbert transform [see Eq. (9.25)].

$$Q(t) = A' \cdot e^{-\beta t^2} \cdot e^{j\cdot 2\pi t\cdot(\Delta f)} = A' \cdot e^{-\beta t^2} \cdot e^{j\cdot 2\pi t\cdot\left(f_0 \frac{2v}{C}\right)} \tag{11.10}$$

where A' represents the amplitude of the signal after applying the filter.

As can be noted, the velocity of the target v whose value we seek is contained within the second exponential term. In order to estimate the value of v at a certain range X_0 from the transducer, we need to sample the signal around time point $t_0 = 2X_0/C$, where C is the average speed of sound in the medium and the time is measured relative to the pulse transmission time. Using the PRF as our sampling rate, the obtained discrete Doppler signals $Q_i(t_0)$ are registered and stored. After acquiring N samples, a complex vector of Doppler signals is obtained, that is, $\bar{Q}(t_0) = [Q_1, Q_2, \ldots, Q_N]$. By applying a discrete Fourier transform to this vector, the spectral distribution corresponding to $\bar{Q}(t_0)$ is obtained, from which one can estimate the Doppler frequency shift Δf.

In order to display the flow velocity as a function of time, the instantaneous Doppler shifts $\Delta f(t)$ need to be transformed into velocities via Eq. (11.7). Then, as explained in Chapter 9, these values are transformed into gray levels. Using a display mode similar to the M-mode (see Chapter 9), the flow velocity as a function of time can be obtained. A typical image is shown in Fig. 11.5. This image is continuously updated in a manner similar to that used in the M-mode. The "freshest" value is added to the right side (as a column in the image), and the "oldest" one (the most left column) is removed. (It should be pointed out that there are some scanners where the refresh direction is reversed.)

11.3 FREQUENCY SHIFT ESTIMATION

The main difficulty (in our context) with the Doppler effect measurement is the frequency shift estimation. The problem stems from two clinical requirements: The first is the natural medical need to know the exact spatial location

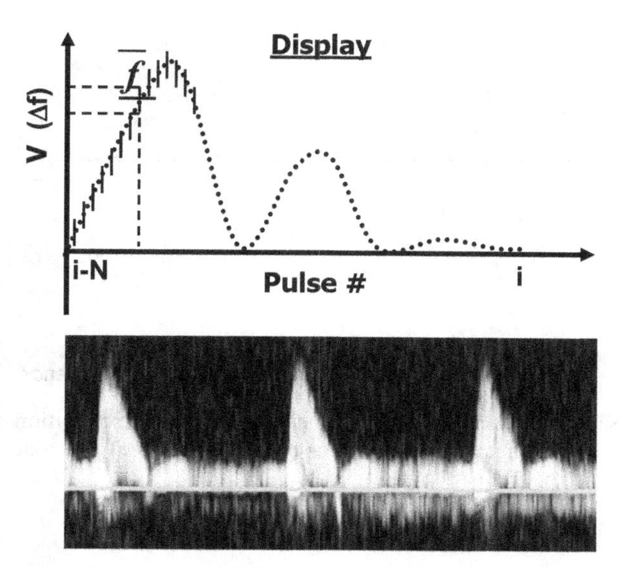

Figure 11.5. **(Top)** Schematic depiction of the ultrasonic Doppler flow display. Each column represents a spectrogram of the Doppler shift which is related to the velocity. The intensity of each frequency is represented by gray levels (strongest intensity is white). The horizontal axis is the continuously updating time axis. **(Bottom)** Typical blood flow profile in the aorta. The triangular shape corresponds to the systolic phase where the flow reaches its maximal value.

where the flow measurement takes place. The second is the need for "real-time" information. As explained in Chapter 9, spatial mapping with the pulse echo method is based on measurement of the time elapse between the transmission of the short ultrasonic pulse and the detection of the relevant echo. In order to achieve good spatial (axial) resolution, the transmitted pulse has to be as short as possible. However, in order to measure accurately the frequency shift, the spectral band has to be as narrow as possible. The problem is that these two requirements contradict each other. As recalled, a short pulse in time domain has a broad band in the frequency domain and a narrow spectral band corresponds to a broad pulse in the time domain. This is illustrated schematically in Fig. 11.6 for clarity.

We are therefore obliged to choose between an accurate temporal, and hence spatial, localization and accurate frequency shift estimation. If, for example, we choose to compromise and use a medium-sized Gaussian pulse for transmission, the result will be an unbalanced frequency shift, biased toward the higher frequencies. This can be understood by studying Eq. (11.6), which linearly relates the Doppler shift and the transmitted frequency f_0. It follows from this relation that the higher frequencies will be more shifted than the low frequencies. Consequently, the spectral band profile will be unevenly

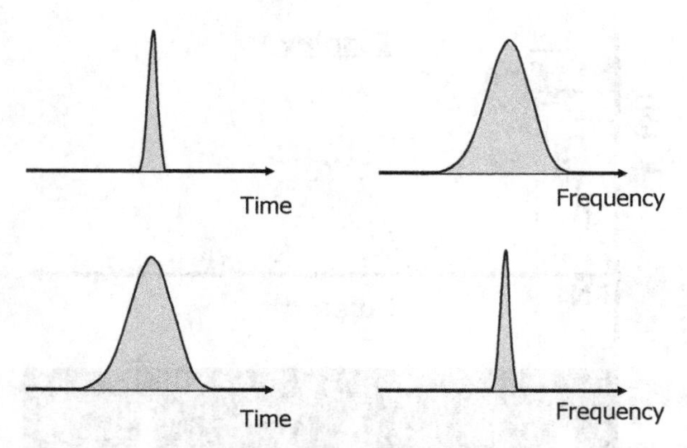

Figure 11.6. Schematic depiction of the time–frequency inverse relation. **(Top)** A short pulse in time domain has a broad frequency band. **(Bottom)** A broad pulse in time domain has a narrow frequency band.

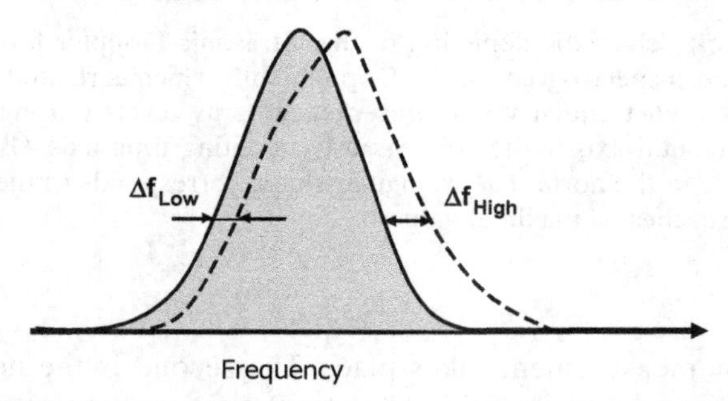

Figure 11.7. Schematic depiction of the unbalanced Doppler frequency shift for a Gaussian pulse. The Doppler shift is greater for the higher frequencies. Hence, the shifted spectral envelope will be biased toward the higher frequencies.

distorted. The basic Gaussian symmetry will be lost and the profile will be skewed and the central frequency will be shifted toward the upper side of the frequency band. This is demonstrated graphically in Fig. 11.7. Hence, an estimation based on the central frequency may be biased in this case.

Another issue that needs to be addressed is the computation time required for achieving a satisfactory spectral estimation. Naturally, the clinical requirement is for a real-time estimation of the flow velocity. The most straightforward approach is of course to use the Fourier transform. However, in order to obtain an accurate spectral analysis, the Fourier transform requires a substantial amount of data points. Seemingly, this should not be a significant obstacle, since when using the low megahertz range for transmission frequencies, the typical Doppler shift is in the range of a few kilohertz. Thus, using a

PRF that exceeds the required sampling frequency according to the Shannon–Whittaker (Nyquist) condition is possible. However, when the PRF is increased, the time left for calculations of the spectral changes is decreased. Using current equipment, the calculation of the instantaneous frequency shift for a single spatial point can be done in real time using the FT without much difficulty. However, if a 2D or 3D flow mapping of a few thousand pixels in real time is required, an alternative and much faster approach is needed.

Finally, it is important to note that a high-pass filter is commonly applied to the Doppler signal in order to avoid wall motion effects. As recalled, the cardiovascular system is flexible and the blood vessels may move or change their diameter during the cardiac cycle. This type of motion is slow and hence can be avoided by filtering the lowest Doppler shifts. This type of filter is known as the *wall motion filter*. Nevertheless, there are some scanning systems that provide a *tissue Doppler imaging* (TDI) mode [2]. With this type of imaging, the blood flow effects are removed by low-pass filtering the Doppler data and only wall motion is depicted. This is done mainly in cardiac imaging to demonstrate normal or akinetic myocardial wall motion.

11.4 DUPLEX IMAGING (COMBINED B-SCAN AND COLOR FLOW MAPPING)

For clinical purposes, it is commonly required to obtain simultaneously an anatomical image using a 2D or 3D B-scan atop which a map depicting the blood flow is shown. This combined mode is called *duplex imaging*. For clarity, the anatomy is shown in black and white using gray levels, and the flow is shown in a colored graphic overlay. This colored flow mapping (CFM) utilizes two basic colors: commonly blue and red. One of the colors is used for designating blood that flows toward the ultrasonic transducer (a positive Doppler shift), while the other designates blood that flows away from the transducer (a negative Doppler shift). The operator can choose which color corresponds to which direction.

In addition to the two basic flow mapping colors, commonly shades of green are added according to the variance in the velocity. The velocity variance is indicative of the nature of the flow within the sample volume. In a laminar plug flow where the flow is uniform, its value will be small. Contrary to that, in a turbulent flow (which may occur, for example, around pathologic heart valves or around arterial stenoses and occlusions) the variance will be large. The addition of the green color yields intermediate colors (e.g., green, cyan, and yellow). Hence, at regions where the variance is low, the blue or red colors will be strong. On the other hand, when the variance is high the intermediate colors will be dominant.

As the duplex imaging mode needs to be displayed in "real time," the amount of information to be processed is very large and the available time slot for computation is quite small. This calls for a very rapid method for the frequency shift estimation. The most popular method used today is based on the relation between the autocorrelation function and the spectrum. There are of course alternative methods, but in this book we shall limit our discussion to the autocorrelation method. For a very thorough explanation the interested reader is referred to the book by Jørgen Arendt Jensen [3], which is dedicated to this topic.

Let us start by rewriting the expression for $Q(t)$:

$$Q(t) = A' \cdot e^{-\beta t^2} \cdot e^{j \cdot (2\pi \cdot \Delta f \cdot t + \phi_0)} \tag{11.11}$$

As can be noted, the expression in the brackets of the exponential term is actually the phase of the signal $\varphi(t)$. It includes a constant phase ϕ_0 that is set by the transmitter and an additive term that is time-dependent. The temporal derivative of the phase is therefore given by

$$\frac{d\varphi(t)}{dt} = 2\pi \cdot \Delta f = 2\pi \cdot f_0 \cdot \frac{2v}{C} \tag{11.12}$$

As can be noted, this term includes the required velocity v. Thus, if we could measure this temporal derivative, the velocity of the target could be calculated. The simplest approach is to approximate this phase derivative by a finite difference. Consider two consecutive sampled signals $Q(t)$ and $Q(t + T_{PRF})$, where T_{PRF} is the time interval between two consecutive transmissions. (T_{PRF} is given by the reciprocal value of the PRF). Using a simple approximation, the corresponding temporal derivative is given by

$$\frac{d\varphi(t)}{dt} \approx \frac{\Delta\varphi}{\Delta t} = \frac{[\varphi(t + T_{PRF}) - \varphi(t)]}{T_{PRF}} \tag{11.13}$$

The required target velocity is therefore given by

$$\boxed{v \approx C \cdot \frac{\Delta\varphi}{4\pi \cdot f_0 \cdot T_{PRF}}} \tag{11.14}$$

In practice, however, the signals are first digitized and converted into discrete numbers. Thus, after acquiring N samples of the demodulated signal, we have a vector of size N which contains the complex values of these signals, that is, $\bar{Q}(t_0) = [Q_1, Q_2, \ldots, Q_N]$. Here, Q_i and $Q_{(i+1)}$ designate two sampled signals acquired at time t_0 after transmission, hence, corresponding to the same spatial location along the beam. The Q_i signals are consecutively acquired at every T_{PRF}. Since these numbers are considered complex numbers each signal may be written as

$$Q_i = a(i) + j \cdot b(i) \tag{11.15}$$

where $a(i)$ and $b(i)$ designate the real and imaginary parts of Q_i, respectively. The phase of this signal is therefore given by

$$\varphi(i) = \arctan\left[\frac{b(i)}{a(i)}\right] \tag{11.16}$$

Hence, the phase difference between two consecutive signals is given by

$$\Delta\varphi = \varphi(i+1) - \varphi(i) = \arctan\left[\frac{b(i+1)}{a(i+1)}\right] - \arctan\left[\frac{b(i)}{a(i)}\right] \tag{11.17}$$

Using some trigonometric identities, Eq. (11.17) can be rewritten as

$$\Delta\varphi = \arctan\left[\frac{b(i+1)a(i) - b(i)a(i+1)}{a(i+1)a(i) + b(i)b(i+1)}\right] \tag{11.18}$$

Substituting Eq. (11.18) into Eq. (11.14) yields

$$\boxed{v \approx C \cdot \frac{1}{4\pi \cdot f_0 \cdot T_{PRF}} \cdot \arctan\left[\frac{b(i+1)a(i) - b(i)a(i+1)}{a(i+1)a(i) + b(i)b(i+1)}\right]} \tag{11.19}$$

As was shown by Kasai et al. [4], this approximation can be further improved by using the autocorrelation function. Consider the discrete autocorrelation function for the acquired signals at a shift of one temporal unit, that is, $R(1) \equiv R(t = T_{PRF})$. The approximated value of this autocorrelation (unnormalized) function is given by

$$R(1) \approx \sum Q_{(i+1)} \cdot Q_i^* \tag{11.20}$$

where * designates a complex conjugate value. Writing Eq. (11.20) explicitly yields

$$Q_{(i+1)} \cdot Q_i^* = [a(i+1)a(i) + b(i)b(i+1)] + j[b(i+1)a(i) - b(i)a(i+1)] \tag{11.21}$$

As can be noted, the real part of this expression is identical to the expression in the denominator of Eq. (11.19), while the imaginary part is identical to the numerator of Eq. (11.19). Hence, the velocity of the target can be estimated by

$$\boxed{v \approx C \cdot \frac{1}{4\pi \cdot f_0 \cdot T_{\mathrm{PRF}}} \cdot \arctan\left[\frac{\mathrm{Imag}\{R(1)\}}{\mathrm{Real}\{R(1)\}}\right]} \tag{11.22}$$

where the discrete autocorrelation function at a unit temporal step shift $R(1)$ can be approximated by

$$R(1) \approx [Q_1^*, Q_2^*, \ldots, Q_{N-1}^*] \cdot [Q_2, Q_3, \ldots, Q_N]^T \tag{11.23}$$

This approximation has two advantages: (i) The required calculations are simple and robust, and (ii) this approximation yields good results even when the number of sampled data points N is small. In certain applications, even $N = 4$ may suffice. These two advantages enable very rapid velocity mapping, even for a large number of pixels within a very short time (real time rate).

In order to obtain a full color mapping of the flow, the variance of the flow needs to be estimated as well. This variance is naturally equal, up to a constant, to the spectral variance $\sigma(f)$. By definition, this variance is given by

$$\sigma(f)^2 = \frac{\int_{-\infty}^{\infty} (f - \bar{f})^2 P(f) df}{\int_{-\infty}^{\infty} P(f) df} = \frac{\int_{-\infty}^{\infty} f^2 P(f) df}{\int_{-\infty}^{\infty} P(f) df} - \bar{f}^2 \tag{11.24}$$

where f and $\bar{f}$ are the frequency and the average frequency, respectively, and $P(f)$ is the power spectral density (PSD). The average frequency $\bar{f}$ is given by

$$\bar{f} = \frac{\int_{-\infty}^{\infty} f \cdot P(f)\, df}{\int_{-\infty}^{\infty} P(f)\, df} \tag{11.25}$$

The spectral variance can be estimated using the autocorrelation function. This function $R(\tau)$ is related to the inverse Fourier transform of the PSD by [5]

$$R(\tau) = \int_{-\infty}^{\infty} P(f) \cdot e^{(j2\pi f \cdot \tau)} df \tag{11.26}$$

The first derivative of the autocorrelation function by τ is given by

$$\dot{R}(\tau) \triangleq \frac{dR(\tau)}{d\tau} = \int_{-\infty}^{\infty} j2\pi f \cdot P(f) \cdot e^{(j2\pi f \tau)} df \tag{11.27}$$

while the second derivative is given by

$$\ddot{R}(\tau) \triangleq \frac{d^2 R(\tau)}{d\tau^2} = -\int_{-\infty}^{\infty} (2\pi f)^2 \cdot P(f) \cdot e^{(j2\pi f \tau)}\, df \tag{11.28}$$

In the case where $\tau = 0$, the exponential term in the three equations above will be equal to 1. Comparing these terms to the terms in Eq. (11.25), it can be noted that the following relation can be written:

$$\bar{f} = \frac{\dot{R}(0)}{j2\pi \cdot R(0)} \tag{11.29}$$

Comparing these terms to the terms in Eq. (11.24), it can also be noted that the following relation can be written:

$$\sigma(f)^2 = -\frac{\ddot{R}(0)}{4\pi^2 R(0)} - \left[\frac{\dot{R}(0)}{j2\pi \cdot R(0)}\right]^2 = \frac{1}{4\pi^2}\left[\left(\frac{\dot{R}(0)}{R(0)}\right)^2 - \frac{\ddot{R}(0)}{R(0)}\right] \tag{11.30}$$

As was shown in Jørgen Arendt Jensen [3], by applying a Taylor series expansion around $\tau = 0$ and neglecting high-order terms, the angular frequency variance $\sigma(\omega)^2$ can be estimated by

$$\boxed{\sigma(\omega)^2 = \frac{2}{T_{\mathrm{PRF}}^2} \cdot \left[1 - \frac{|R(1)|}{R(0)}\right]} \tag{11.31}$$

where the autocorrelation function at zero shift $R(0)$ can be estimated from the sampled signal vector $\bar{Q}(t_0)$ by applying the following approximation:

$$\begin{aligned} R(0) \approx {} & \frac{1}{2}[Q_1, Q_2, \ldots, Q_{N-1}] \cdot [Q_1^*, Q_2^*, \ldots, Q_{N-1}^*]^T \\ & + \frac{1}{2}[Q_2, Q_3, \ldots, Q_N] \cdot [Q_2^*, Q_3^*, \ldots, Q_N^*]^T \end{aligned} \tag{11.32}$$

Utilizing Eqs. (11.22) and (11.31), color flow mapping (CFM) in "real time" can be obtained. However, in some clinical applications the physician is not interested in the direction of the flow but merely wants to know whether flow exists or not in the imaged region of interest. In such case the imaging method used is called *power Doppler imaging*. The mapped physical property with this method is the spectral power of the corresponding Doppler shift. The commonly used color mapping in this case consists of shades of yellow/gold. Higher power is depicted by brighter colors.

To conclude this section, two exemplary images are depicted. In Fig. 11.8, a CFM image depicting the blood flow in the liver portal vein. The anatomy that was obtained by the B-scan method is depicted in shades of gray. The direction of the flow is indicated by the red and blue colors, indicating flow toward and away from the transducer.

The scond image shown in Fig. 11.9 depicts a power Doppler image of flow within the kidney. As can be noted in this case, only one color is used because the imaging is nondirectional.

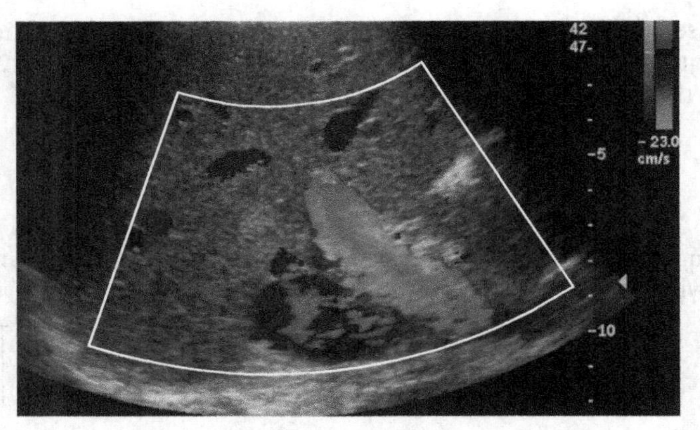

Figure 11.8. Duplex image (CFM) of the liver portal vein. The anatomy is depicted in shades of gray. The flow direction toward and away from the ultrasonic transducer is indicated by the blue and red colors. (Courtesy of Dr. Diana Gaitini.) See insert for color representation of this figure.

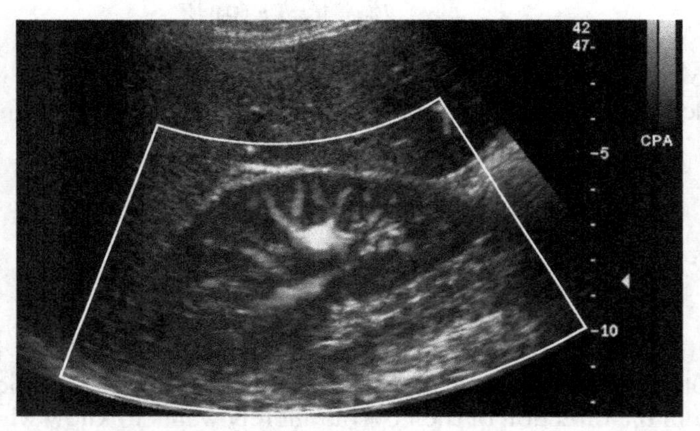

Figure 11.9. Another type of duplex imaging using the power Doppler technique. As can be noted, shades of only one color are used in this case because the flow mapping is nondirectional. (Courtesy of Dr. Diana Gaitini.) See insert for color representation of figure.

REFERENCES

1. Bellanger M, *Digital Processing of Signals*, 2nd edition, John Wiley & Sons, Tiptree, Essex, Great Britain, 1989.
2. Yu CM, Sanderson JE, Marwick TH, and Oh JK, Tissue Doppler imaging a new prognosticator for cardiovascular diseases, *J Am College Cardiol* **49**(19):1903–1914, 2007.

3. Jensen JA, *Estimation of Blood Velocities Using Ultrasound: A Signal Processing Approach*, Cambridge University Press, New York, 1996.
4. Kasai C, Namekawa K, Koyano A, and Omoto R, Real time two dimensional blood flow imaging using an autocorrelation technique, *IEEE Trans Son Ultrason* **32**:458–463, 1985.
5. Papoulis A, *Probability, Random Variables and Stochastic Processes*, 2nd edition, McGraw-Hill, New York, 1984.

CHAPTER 12

SAFETY AND THERAPEUTIC APPLICATIONS

Synopsis: In this chapter the main effects induced by the ultrasonic radiation within the tissue are discussed. These effects include temperature elevation, cavitation bubbles, and acoustic streaming. The currently accepted safety levels associated with these effects are introduced. Then, several common applications of ultrasound in medicine are briefly presented. These applications include physiotherapy, lithotripsy, and thermal ablation by high-intensity focused ultrasound (HIFU). In addition, drug delivery, gene therapy, and cosmetic applications are concisely noted.

12.1 EFFECTS INDUCED BY ULTRASOUND AND SAFETY

12.1.1 Thermal Effects

Using acoustic waves, heat can be selectively induced within the tissue. The heat generation stems mainly from the absorption mechanisms within the tissue which transform the acoustic energy into local temperature elevation. Hence, this phenomenon depends on tissue properties. As recalled from Chapter 5, the acoustic attenuation coefficient varies from one tissue type to another. Moreover, its value is frequency-related with the tendency to increase when the frequency is increased. This dependency on frequency is commonly nonlinear, and its parameters also change according to the tissue type (see Appendix A). In hard (calcified) tissues—that is, bones and teeth—the attenu-

Basics of Biomedical Ultrasound for Engineers, by Haim Azhari

ation coefficient is very high. As a quantitative comparison, it is noted that the typical attenuation coefficient for soft tissues is on the order of 0.5 [db / cm / MHz] as compared to values in the range of 10–20 [db / cm / MHz] in cortical bone tissues. Therefore, hard tissues have higher ultrasonic-induced warming potential.

The biological effects caused by the ultrasonically induced heating depend mainly on the following three factors:

(a) The magnitude of the temperature elevation, which is determined by the acoustic beam intensity and the irradiated tissue properties.
(b) The duration of the acoustic radiation.
(c) The sensitivity of the tissue to heating.

Because the first two parameters are related, they are commonly used together to define the safety margins. As a general rule, it can be stated based on reports in the literature [1] that:

> A temperature elevation of 1.5 °C above the normal body temperature (37 °C) does not cause any damage. On the other hand, a temperature elevation of 4 °C for a time duration exceeding 5 minutes may induce damage in fetus.

The American Institute for Ultrasound in Medicine (AIUM) has defined a formula that relates the maximal allowable temperature elevation and the duration of the ultrasonic radiation (published March 26th, 1997). This formula is given by

$$\Delta T^{\circ}_{\max} \leq 6 - \frac{\log_{10}(t)}{0.6} \tag{12.1}$$

where t is the radiation time in minutes and $\Delta T^{\circ}_{\max}$ is the maximal allowable temperature elevation given in Celsius thermal degrees. For example, based on this formula, an ultrasonic radiation which causes a temperature elevation of 5.5 °C is limited to a radiation period of 2 min.

In principle, any temperature elevation in excess of 6 °C above the normal body temperature (i.e., to a temperature $T \geq 43$ °C) will almost certainly cause damage to the tissue. This damage will be accumulated during the radiation time. As reported, based on in-vitro experiments, exposure to a temperature of 43 °C for 240 min is lethal. Thus, it was suggested [2] to assess the thermal dose using a cumulative equivalent minutes at 43 °C. The commonly used thermal dose, which accounts for a temperature which changes in time, is thus given by

$$D = \int_0^{\Delta t} R^{[43-T(t)]}\, dt \tag{12.2}$$

where D is the equivalent exposure time in minutes at 43 °C, Δt is the actual exposure time, $T(t)$ is the time-dependent temperature, $R = 0.25$ for $T(t) < 43$ °C, and $R = 0.5$ for $T(t) \geq 43$ °C. It should be pointed out, however, that this thermal dose was developed based on findings in the temperature range of 42 °C to 50 °C. Thus, one should be aware of this fact when applying this equation to higher temperatures.

In most soft tissues, necrosis is obtained for thermal dose values in the range of $D = 50$–240 min [3, 4]. A temperature elevation of 20 °C or higher—that is, to temperatures in excess of 57 °C—will cause an immediate death to the tissue cells.

A simple method for assessing the expected temperature elevation ΔT is to estimate first the amount of heat Q_{in} generated by the acoustic energy absorbed per given volume of tissue during the radiation time Δt. Then, estimate the amount of heat Q_{out} removed from the same volume of tissue during this time. The difference between these two terms is approximately equal to the following term:

$$[Q_{in} - Q_{out}] = \mathbb{C} \cdot \rho \cdot V \cdot \Delta T \tag{12.3}$$

where $\mathbb{C}$ is the specific heat constant for that tissue (the typical value for soft tissue is 4.186 J / gr/°C), ρ is the average tissue density in the region of interest (assumed homogeneous), and V is the tissue volume exposed to the acoustic radiation. For simplicity, it will be assumed that the cross-sectional area of the ultrasonic beam is A and the region of interest length is L. Therefore, we can define the volume by $V = A \cdot L$.

The amount of heat generated as a result of the acoustic radiation can be estimated by assuming that the wave is planar. If the transmitted intensity is I_0 and the region of interest is located at a distance x from the transmitter, then the difference between the intensity entering the tissue and the intensity leaving the tissue after a distance L is given by

$$\Delta I(x) = I(x) - I(x+L) = I_0 e^{-2\alpha x}\left(1 - e^{-2\alpha L}\right) \tag{12.4}$$

As recalled from Chapter 3, the intensity is defined as the acoustic power per unit area. Therefore, the amount of energy absorbed per unit volume E' after time Δt is given by

$$E' = \frac{\Delta I(x) \cdot A \cdot \Delta t}{A \cdot L} \tag{12.5}$$

Substituting Eq. (12.4) into this term yields

$$E' = \frac{I_0 e^{-2\alpha x}\left(1-e^{-2\alpha L}\right)\cdot \Delta t}{L} \tag{12.6}$$

The amount of heat generated during time Δt in a given volume is therefore given by

$$Q_{\text{in}} = E' \cdot V \tag{12.7}$$

Defining the heat removed by the body per unit time per unit volume as $\dot{Q}'_{\text{out}}$ and substituting into Eq. (12.3) yields

$$\begin{aligned}[Q_{\text{in}} - Q_{\text{out}}] = E' \cdot V - \dot{Q}'_{\text{out}} \cdot \Delta t \cdot V &= \left[\frac{I_0 e^{-2\alpha x}\left(1-e^{-2\alpha L}\right)}{L} - \dot{Q}'_{\text{out}}\right] \cdot \Delta t \cdot V \\ &= \mathbb{C} \cdot \rho \cdot V \cdot \Delta T\end{aligned} \tag{12.8}$$

Or after rearrangement,

$$\boxed{\Delta T = \left[\frac{I_0 e^{-2\alpha x}\left(1-e^{-2\alpha L}\right)}{L} - \dot{Q}'_{\text{out}}\right] \cdot \frac{\Delta t}{\mathbb{C} \cdot \rho}} \tag{12.9}$$

Next, let us assume that the region of interest is infinitely small so that $L \to 0$. Hence, the exponential term can be approximated by

$$e^{-2\alpha L} \approx 1 - 2\alpha L \tag{12.10}$$

Substituting Eq. (12.10) into Eq. (12.9) and recalling that $I(x) = I_0 e^{-2\alpha x}$, the general expression for the local temperature elevation resulting from the ultrasonic radiation is given by

$$\boxed{\Delta T(x) = \left[2\alpha \cdot I(x) - \dot{Q}'_{\text{out}}(x)\right] \cdot \frac{\Delta t}{\mathbb{C} \cdot \rho}} \tag{12.11}$$

If the heat conductivity of the tissue and the arterial blood perfusion to the region of interest are known, then $\dot{Q}'_{\text{out}}$ can be replaced by more explicit terms using the well-known "bioheat equation" suggested by Pennes [5]. Thus, by taking $\Delta t \to 0$ and rearranging Eq. (12.11), we obtain

$$\boxed{\mathbb{C} \cdot \rho \cdot \frac{\partial T}{\partial t} = 2\alpha \cdot I - \mathbb{C}_{\mathbb{B}} \cdot \rho_{\mathbf{B}} \cdot W_B \cdot (T - T_A) + \nabla \cdot \mathbb{K} \nabla T} \tag{12.12}$$

where $\mathbb{C}_{\mathbb{B}}$ is the specific heat constant for the blood, $\rho_{\mathbf{B}}$ is the blood density, W_B is a coefficient representing the blood perfusion to the region of interest, T_A is the arterial blood temperature, and $\mathbb{K}$ is the thermal conductivity of the tissue. (Note that the spatial coordinate was omitted for clarity.)

In most modern ultrasonic scanners used for imaging, a temperature elevation safety index is provided by the manufacturer or continuously displayed for the user. This index, which is called the *thermal index*, was defined by the American National Council for Radiation Protection and Measurement [6]. The *thermal index* (Ti) is defined as the ratio of the transmitted acoustic power (commonly set by the user during the scan) to the power required to raise the irradiated tissue temperature by 1 °C. Naturally, the actual reference power is not known since the scanned tissue differs from one scan to another. Hence, a model of the exposure conditions is used to estimate it. In these calculations, a constant attenuation coefficient α for soft tissues is assumed. Using this model and Eq. (12.12), the intensity required to maintain a temperature elevation of 1 °C can be estimated.

Because the Ti merely represents an estimated value, it is advisable to keep its magnitude as low as possible. The American Food and Drug Administration (FDA) has limited its maximal value (in the 510 K track) for ophthalmology applications to $Ti \leq 1$. For other applications the recommended maximal value is $Ti \leq 6$.

Another safety index for assessing the potential hazard from the transmitted ultrasonic device is the time-averaged intensity designated as I_{TA}. This index is proper for unfocused beams. However, in many ultrasonic imaging systems the acoustic field changes both temporally and spatially. It is therefore essential to map the acoustic field in space and time and look for the worst-case scenario. The spatial position where the I_{TA} is maximal (usually at the field's focal point) is used to define another safety index, called the *spatial peak temporal average intensity*, which is designated as I_{SPTA}. The American FDA has set its maximal allowable value (in the 510 K track) for ophthalmology devices to $I_{SPTA} \leq 50\,\text{mW/cm}^2$. For other applications the maximal value is limited to $I_{SPTA} \leq 720\,\text{mW/cm}^2$. Another safety index is determined by the maximal intensity transmitted in a pulse mode. Its maximal value is limited to $I_{\text{spatial peak pulse average}} \leq 150\,\text{W/cm}^2$.

Because modern ultrasonic imaging systems offer the user a variety of alternative imaging modes, one should be aware of the fact that the safety indices do change according to the selected imaging mode. Referring to the expected I_{SPTA} values, the imaging modes can be generally sorted in the following ascending order—that is, from lowest to highest risk [1]:

(i) A- or M-mode
(ii) B-scan
(iii) Color Doppler
(iv) Power Doppler

As for the susceptibility of the different tissue types to damage induced by the temperature elevation, it can be naturally appreciated that embryonic cells are the most vulnerable cells. Therefore, when using ultrasonography for pregnancy monitoring or gynecological treatment during pregnancy, all safety indices should be set to the minimal values possible. Other susceptible regions are eyes, neural and brain tissues, lungs, and bone–tissue interfaces.

12.1.2 Cavitation Bubbles

Cavitation bubbles is a phenomenon which occurs when sudden negative pressure is applied within fluidic mediums. Such sudden negative pressure is generated when a strong negative pulse or an intensive sinusoidal wave travel through the medium. The gases which are dissolved within the medium leave the solution mainly around small solid elements floating within the medium. These elements are called *cavitation seeds*. When this happens, tiny gas bubbles are formed. The typical radius of these bubbles is in the order of a few microns.

In a quasi-static state the radius of these bubbles is described by the Rayleigh–Plesset Equation. This differential equation is given in its implicit form by [7]

$$R\ddot{R}+\frac{3}{2}\dot{R}^2+\frac{1}{\rho_L}\left(P_\infty-P_L+\frac{4\mu\dot{R}}{R}\right)=0 \tag{12.13}$$

where $\dot{R}$ and $\ddot{R}$ are the first and second temporal derivatives, respectively, P_L is the liquid pressure at the bubble surface, and P_∞ is the pressure at a location distal to the bubbles. This pressure includes the instantaneous added pressure resulting from the acoustic field. ρ_L is the liquid density, and μ is the dynamic viscous coefficient for the fluid.

Similar to gas bubbles in soda water, these bubbles may grow, shrink or collapse. The collapse of these bubbles is a very aggressive process. At the collapse climax, the generated peak pressure and temperature may reach extremely high values. Pressures exceeding several hundred atmospheres and temperatures on the order of several thousand Kelvin degrees have been reported (see, for example, Leighton [8]).

Furthermore, an asymmetric collapse of the bubbles (see Fig. 12.1) generates a very powerful fluid jet. As reported in the literature, the velocity of

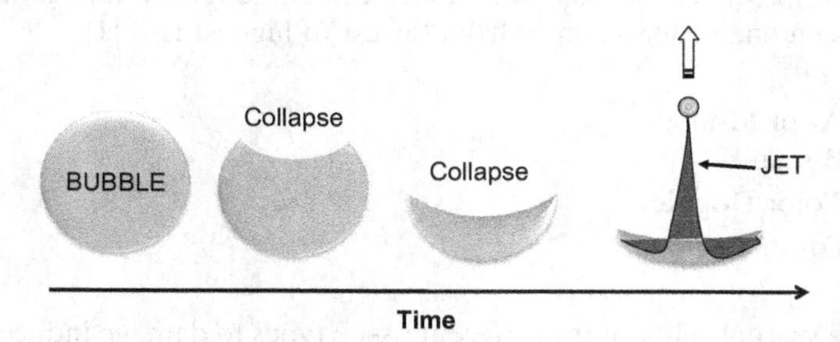

Figure 12.1. Schematic depiction of an asymmetric collapse of a cavitation bubble. The formed jet may reach a velocity of several hundred meters per second.

these jets may reach values close to 1000 [m / sec] (see, for example, Brujan et al. [9]).

Although the spatial scale of the collapse process is very small, the resulting effects on tissue cells can be deadly. The strong mechanical stresses applied on the cell membranes and the associated high temperature may easily destroy it. Moreover, as a result of the aggressive collapse, free radicals may be generated. These free radicals can harm nearby cells as well, further increasing the damage to the tissue.

The risk of cavitation bubbles generation during the ultrasonic radiation depends on the magnitude of the negative pressure associated with the propagating wave and the wave's frequency [10]. A mathematical relation named the *mechanical index* (Mi) was suggested. The mechanical index is define by

$$\boxed{\mathrm{Mi} = \frac{\mathrm{Max}\{P_{\mathrm{negative}}\}}{\sqrt{f}}} \tag{12.14}$$

where $\mathrm{Max}\{P_{\mathrm{negative}}\}$ is the maximal negative pressure measured in megapascals, and frequency f is measure in megahertz. It should be pointed out that this relation was developed for short (a few cycles) pulses and a low duty cycle (<1%).

This index has been adopted as a safety index by all regulatory institutes dealing with medical ultrasound. As reported, ultrasonic transmission with $\mathrm{Mi} < 0.7$ even into a medium containing bubbles, did not induce a growth of these bubbles (a process leading eventually to their collapse). Thus, it could be stated that the cavitation risk for ultrasonic radiation below this value is negligible.

The American FDA institute has limited (for a 510K route, track 3) the use of ultrasonic radiation in ophthalmology to a maximal value of $\mathrm{Mi} \leq 0.23$. For all other applications the maximal allowable value is set to $\mathrm{Mi} \leq 1.9$.

12.1.3 Additional Effects

In addition to the two major effects noted above, there are experimental evidences for other mechanisms that may affect biological tissues. These are referred to as the "nonthermal noncavitational" effects. Perhaps the most noted one among these effects is "acoustic streaming." As the name implies, fluids within biological tissues may move as a result of the ultrasonic radiation. The source for this fluidic motion is the pressure gradients that appear within the acoustic field. In order to understand the origin of these gradients, let us refer to Fig. 12.2.

Consider initially (Fig. 12.2, left) a planar wave propagating in a medium that has an attenuation coefficient of α. As explained in Chapter 5, the intensity of this wave will decay exponentially with the traveling distance. Thus, if at location x the corresponding intensity of the wave is $I(x)$, then after traveling a short distance Δx its value will be reduced to

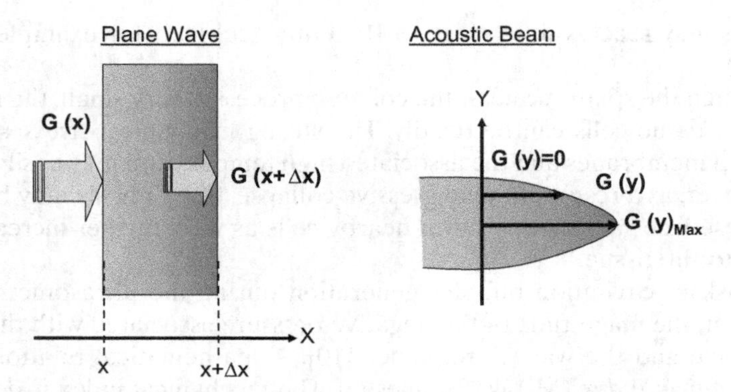

Figure 12.2. Schematic depiction of pressure gradients associated with acoustic beams.

$$I(x+\Delta x) = I(x)e^{-2\alpha\Delta x} \tag{12.15}$$

As recalled from Chapter 3, the passage of the wave through a fluidic medium is accompanied by a radiation pressure whose value is given by

$$\Gamma(x) = \frac{I(x)}{C} \tag{12.16}$$

where C is the corresponding speed of sound. The radiation pressure gradient along distance Δx will be therefore given by

$$\frac{\Gamma(x+\Delta x)-\Gamma(x)}{\Delta x} = \frac{I(x+\Delta x)-I(x)}{C\cdot\Delta x} = \frac{I(x)[e^{-2\alpha\Delta x}-1]}{C\cdot\Delta x} \tag{12.17}$$

Setting this distance to be infinitely small (i.e., $\Delta x \to 0$), the pressure gradient will be given by the following derivative:

$$\boxed{\frac{\partial(\Gamma(x))}{\partial x} = -\frac{2\alpha\cdot I(x)}{C}} \tag{12.18}$$

As can be noted, the radiation pressure gradient depends on the regional attenuation coefficient and the local intensity. However, practically used acoustic fields (see Chapter 8) have nonuniform beam profiles as schematically depicted in Fig. 12.2, right. They may change in space and in time. Furthermore, the attenuation coefficients and the speed of sound may vary from one tissue type to another. Thus, the gradients in the most general case is given by

$$\boxed{\bar{G}(\bar{r},t) = \bar{\nabla}\Gamma[I(\bar{r},t),\alpha(\bar{r}),C(\bar{r})]} \tag{12.19}$$

where $\overline{G}(\overline{r}, t)$ corresponds to the pressure gradients and $\overline{r}$ to the spatial coordinates.

By studying the physical values for $\overline{G}$ it can be noted that it has the units of a "body-force", i.e. force per unit volume. Hence the pressure gradients force the fluids within the tissue to move. And since the gradients are negative—that is, the pressure is reduced with the range from the transducer—the motion will be generally *away* from the transducer. This motion can be visualized in the laboratory where velocities in the order of a few centimeters per second were measured. However, there are also evidences for acoustic streaming under in vivo conditions. (The velocities are on the order of a few millimeters per second.) For example, Nightingale et al. [11] have observed acoustic streaming within breast cysts.

The acoustic radiation can also induce transient stresses within solid tissues (see Chapter 4). This may affect the tissue but not necessarily in a negative manner. For example, Heckman et al. [12] have reported that ultrasonic radiation may accelerate fractures healing in the tibia.

12.2 ULTRASONIC PHYSIOTHERAPY

As stated above, not all effects induced by ultrasonic radiation can jeopardize the tissue. Ultrasound has been used for many years as an aid for healing and acceleration of recovering processes. There are many reports on positive effects of ultrasonic radiation at low intensities. This includes acceleration of bone fracture healing and sports medicine. As explained above, the major effect at lower intensities is the thermal effect. The heat generated within the tissue leads to vasodilatation (widening of blood vessels). As a result, the blood perfusion to the irradiated region is augmented. As a result of the temperature elevation and the increased perfusion (and maybe also due to other effects), the metabolic rate is increased and the healing process is accelerated.

An ultrasonic physiotherapy system is comprised of three major elements:

(i) *A Power Supply Unit.* This includes a signal generator that commonly produces a single-frequency sinusoidal signal and power amplifier.

(ii) *A Control Unit.* This unit sets the transmission intensity levels. (Importantly, current regulations limit the maximum temporal average effective ultrasonic intensity to $I \leq 3$ [W/cm^2].) It also sets the mode of transmission (CW or pulses) and radiation duration. Such units must include a safety control circuit that ensures adequate tissue–transducer contact. In some machines this is done simply by using a pulse-echo cycle embedded within the therapeutic transmission.

(iii) *Piezoelectric Transducers.* The ultrasonic therapy transducers are usually quite large, with a surface area of several squared centimeters. Of course there are also smaller ones for special applications.

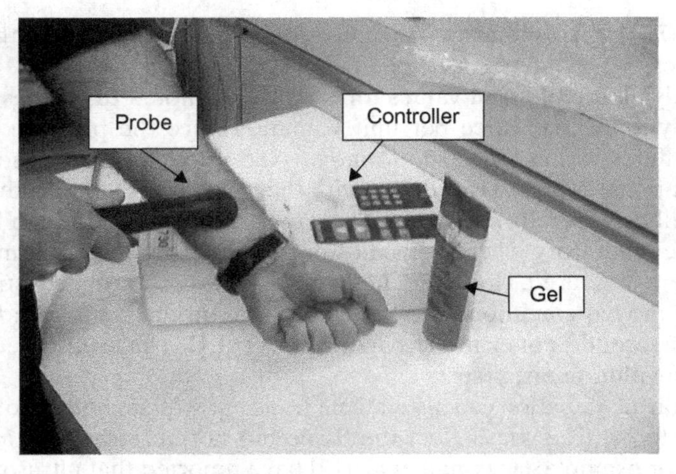

Figure 12.3. Ultrasonic physiotherapy system. In order to use it, gel is first spread atop the transducer surface and the skin. Then the transducer is gently pressed against the skin and is moved slowly in circular or straight strokes. The radiation duration and intensity are set using the controller unit.

A typical ultrasonic physiotherapy system is depicted in Fig. 12.3. Acoustic coupling with the body is achieved by spreading a special gel on the skin, or by immersing the treated organ (hand or legs) in water. Radiation is done in either (a) a constant wave (CW) where a continuous sinusoidal signal is transmitted or (b) a burst mode where short sinusoidal trains, a few milliseconds in temporal length, are transmitted. Each short burst is followed by short pause. Commonly, the transmission-to-pause ratio is 1:4, that is, a duty cycle of 20%. Treatment sessions are commonly on the order of several minutes. In order to prevent skin heating and to spread the radiation, the transducers are slowly moved manually by the operator in small circles around the region of interest.

12.3 LITHOTRIPSY

Kidney stones (renal calculi) are hard minerals and crystalline materials formed within the kidney or urinary tract. They form when there is an excess of dissolved minerals in the urine. Dehydration through reduced fluid intake without adequate fluid replacement increases the risk of their formation. They are associated with severe pains and may cause internal bleeding which appears in the urine. Commonly, with ample fluid intake and some medication, the kidney stones eventually dissolve and/or pass through the urinary tract. However, if their size exceeds about 1 cm, some intervention is required. Although there are surgical options for removing these stones, ultrasound

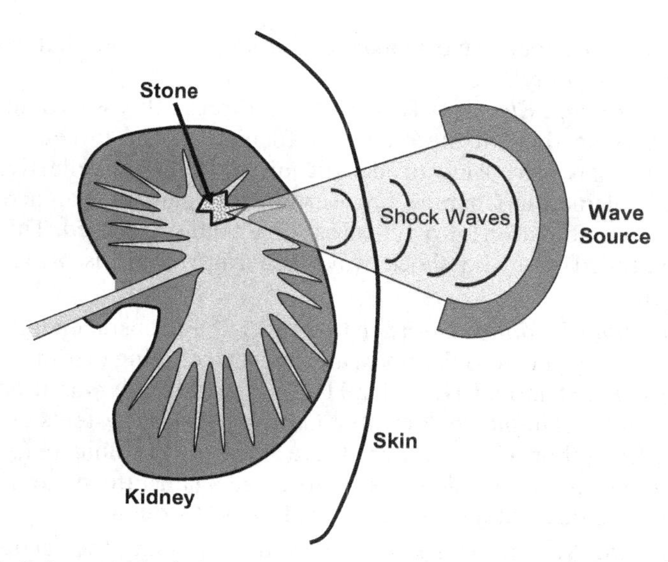

Figure 12.4. Schematic depiction of the lithotripsy procedure. A wave generator transmits shock waves through a coupling medium (water). The high-energy waves that are focused on the kidney stone cause it to break into small pieces.

offers a noninvasive alternative called *extracorporeal shock wave lithotripsy* (ESWL), or lithotripsy for short. [The name is composed of the terms Litho (stone) and Trip (to crush).]

The machine used for lithotripsy is called the *lithotripter*. The lithotripter is actually an "acoustic gun" that fires high-intensity focused shock waves at the kidney (see Fig. 12.4). The "acoustic gun" is located outside the body (hence the term extracorporeal) and its focal point is directed at the kidney stone. The spatial aiming and monitoring is commonly done using an X-ray fluoroscopy system. The intensive shock waves shatter the kidney stone (or gallstone). The fragments are washed out through the urinary tract. This procedure avoids the complications that may be associated with surgery. Hospitalization time is short and the recovery is rapid. The first reported successful application of this technique to crush kidney stones in a human was done on February 9, 1980 using a Dornier HM1 model [13].

12.3.1 Principles of Operation

The lithotripter is comprised of four major elements:

(i) *A Shock Wave Generator.* This element releases high energy within a very short time at a small point in the coupling medium (commonly degassed water). The released energy causes a small explosion (or rapid push of a surface in some devices) that gives rise to a shock wave. The shock wave propagates initially at a speed that exceeds the speed of sound.

(ii) *A Focusing Element.* This element directs the waves toward the kidney so that all the energy is focused at the target. Commonly, focusing is done using an acoustic mirror (a shaped reflective surface) as explained in Chapter 7. In certain configurations an acoustic lens or a combination of a reflector with a lens are used. This element combined with the shock wave generator comprises the "acoustic gun."

(iii) *Coupling Medium.* In order to deliver the acoustic energy from the "acoustic gun" into the body, an acoustic coupling medium is needed. In the first model (Dornier HM1) a water bath was used and the patient was immersed in water. Current modern systems utilize localized coupling via a rubber sleeve filled with a fluid (e.g., degassed water). The rubber sleeve is compressed against the patient skin close to the kidney, allowing the waves to pass through.

(iv) *Imaging System.* This system is used for guiding the "acoustic gun." Commonly, an X-ray fluoroscopy system is used. It provides projection images of the kidney stone from different viewing angles, thus allowing accurate spatial localization of the stone. The stone coordinates are fed into the control unit, which aims the "acoustic gun" so that its focal point is set on the target. Since the stone is clearly visible in the X-ray pictures (due to it high attenuation), the system can also monitor the stone fragmentation procedure.

There are basically four types of "acoustic guns." These basic types are described schematically in Fig. 12.5 [14]. The first type is an electrohydraulic generator combined with a semi-ellipsoidal acoustic mirror. Electrical energy is accumulated in a storage device and then is rapidly released at a spark-gap located at the first (closer to the reflector) focal point of the ellipsoidal reflector. The rapid release of the energy produces gas bubbles that expand faster than the speed of sound. This rapid expansion generates a spherical shock wave. The rear part of the wave reaches the ellipsoidal surface and is reflected back toward the front side. The geometry of the reflective surfaces causes the wave to refocus at the second focal point (farther from the reflector) of the ellipsoidal shape. Thus, if the second focal point is properly positioned on the kidney stone, it will cause it to crack and crash. The problem with the system is that the spark gap is worn out after a few thousand shock; as a result, the focal point may migrate [15].

The second type is a large piezoelectric transducer. The transducer is shaped to have a focal spot at adequate distance. Rapid electric activation of the piezoelectric transducers induces brisk displacement of the surface (as explained in Chapter 8). This displacement generates a highly intensive wave, which converges at the transducer's focal point. At the focal point the intensity is sufficiently high to fragment the stone.

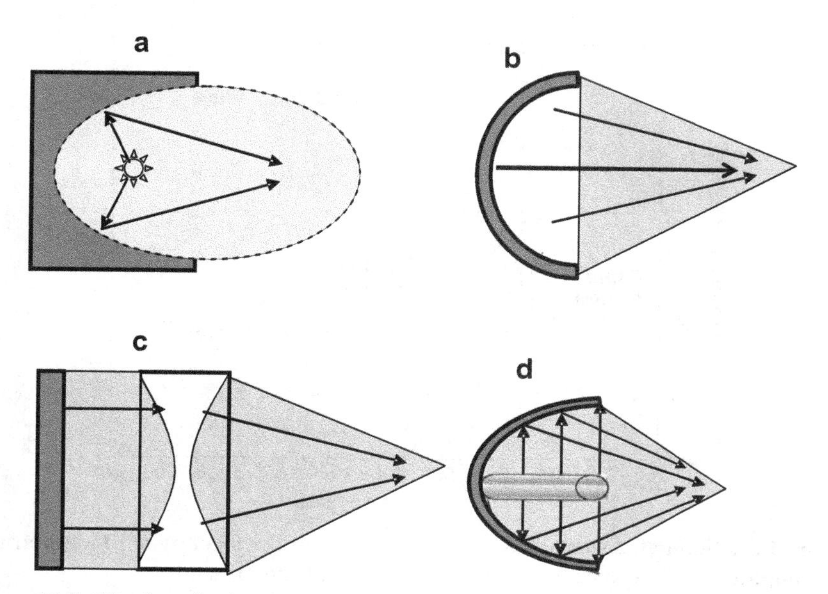

Figure 12.5. The four basic types of "acoustic guns" utilized for lithotripsy: **(a)** Spark generator with an ellipsoidal reflector. **(b)** Focused piezoelectric transducer. **(c)** Planar wave generator with an attached acoustic lens. **(d)** Cylindrical wave generator with a parabolic reflector.

The third type utilizes an electromagnetic device (somewhat resembling a Gaussian gun). A metal plate is positioned atop a spiral coil. When the switch is turned on, high voltage is applied to the coil. As a result, the metal plate is abruptly repelled, giving rise to a planar shock wave [16]. Focusing is then obtained by using an acoustic lens (see Chapter 7).

The fourth type also utilizes a similar electromagnetic concept, but in this case the membrane is shaped as a hollow cylinder. Focusing is obtained by a parabolic reflector. The advantage of this seemingly cumbersome configuration is that an ultrasonic imaging probe can be safely inserted through the central cylindrical hole. This allows hazardless ultrasonic guidance and monitoring of the procedure (avoiding X-ray radiation).

In order to provide some quantitative values, consider the second system type (Fig. 12.5b), where a spherically focused transmitter is used. For this configuration the pressure gain G_p at the focal zone is given analytically by [17]

$$G_p = \frac{\pi}{2} \cdot \frac{F}{C_0 \cdot \Delta t} \cdot [1 - \cos(\alpha_0)] \tag{12.20}$$

where C_0 is the reference (rest) speed of sound, F is the focal distance, α_0 is the cone head angle of the focused acoustic field, and Δt is the pulse duration. It should be pointed out that this equation ignores nonlinear effects.

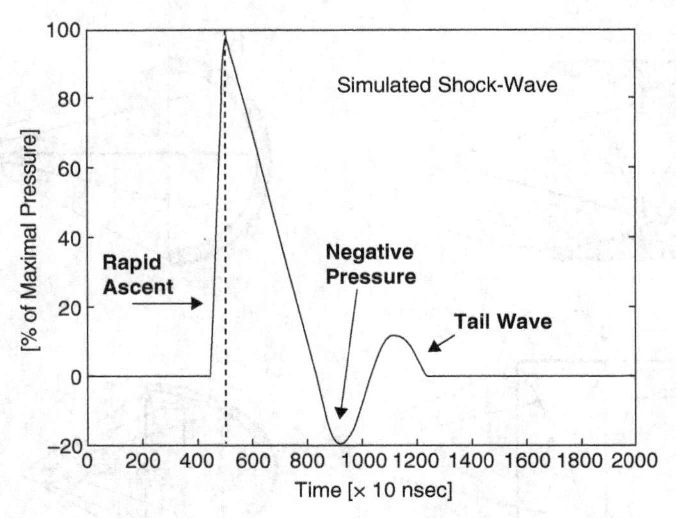

Figure 12.6. Simulated shock wave pressure profile at the focal point of a lithotripter. The amplitude refers to changes around the equilibrium pressure.

For the case where $\alpha_0 \leq 40°$, the ratio of the length to width of the focal zone (i.e., the –6-db contour) can be estimated by [17]

$$\left(\frac{\text{Length}}{\text{Width}}\right)_{-6db} = \frac{\alpha_0}{0.61 \cdot [1 - \cos(\alpha_0)]} \tag{12.21}$$

In order to design a lithotripter, one should know the typical performance characteristic of such a system. Some general reference numbers are the typical "yielding" pressure for the kidney stone, which is about 8 MPa, and the attenuation of the wave from the source to the stone, which is on the order of 30–50% [18].

The shock wave at the focal zone has a characteristic profile which is shown schematically in Fig. 12.6. It starts with a very rapid ascent, reaching the peak within a few hundred nanoseconds. The maximal obtained pressure is commonly in the range of 50–150 MPa. After the rapid ascent begins, there is a slower (within about 1 msec) pressure descent that eventually turns into a negative pressure (relative to the equilibrium). This negative pressure part may sustain for a few milliseconds. Finally, a positive pressure "tail" wave with a relatively small amplitude completes the complex.

During the lithotripsy procedure, from a few hundred to a few thousand shock waves are fired toward the stone (depends on its size) before it is adequately fragmented. It is important to note that during the process healthy tissue may be damaged as well. Local hemorrhage is common but recovery is

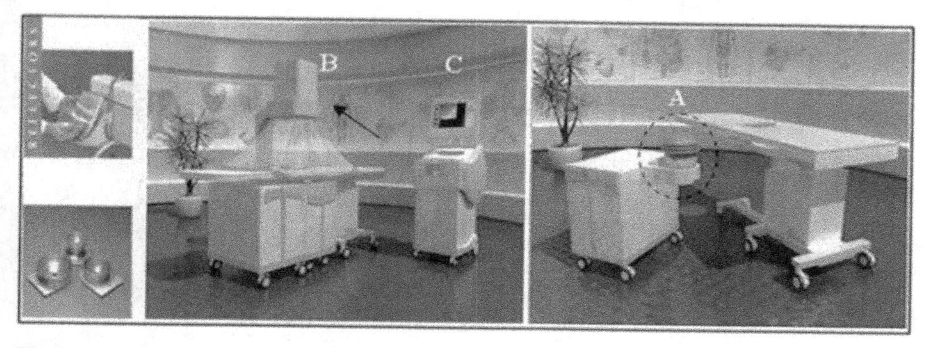

Figure 12.7. Lithotripter system manufactured by Initia and distributed by Direx. **(Right)** "Acoustic gun" marked by the broken circle **A**. Note the hole in the special treatment bed shown on the right. **(Middle)** Entire system assembled for operation. The imaging system **B** is positioned atop the treatment bed. The X-ray camera is marked by the arrow. The controlling computer console is shown on the right. **(Left)** Close-up look at the reflector. Courtesy of the Direx company, copyright © Direx.

usually complete and rapid. An exemplary system manufactured by Initia and distributed by Direx is presented in the pictures shown in Fig. 12.7.

12.4 HYPERTHERMIA HIFU AND ABLATION

As can be understood from the above, high-intensity ultrasonic radiation bears risks to the tissue; and if the high intensity is focused on a small region, it may thermally ablate it. In most applications, one would try to avoid such conditions. However, there are cases when we are interested in destroying specific tissue regions. For example, when a defined pathologic tumor is located within a healthy tissue, causing pains or danger to the patient, it is desired to selectively remove the tumor without damaging the surrounding healthy tissue. Commonly, this is done in a surgical procedure (lumpectomy). But naturally, the surgeon's knife must reach the tumor by cutting through healthy tissue along the way. This is associated with wound formation, bleeding, pain, and scarring. There is also a risk of infection. And finally the cost of such a procedure may be high. In recent years, minimal invasive surgery has gained popularity. In such procedures a special device, such as an RF needle, or a laser diffuser, is inserted into the tumor and sufficient heat is generated at the tip to thermally ablate ("cook") it. Alternatively, in a cryogenic procedure, liquid nitrogen is circulated through the needle, forming an ice-ball at the tip which freezes the tumor. Both procedures are still invasive; and if several needle insertions are needed to provide adequate results, the wounds may be clinically and/or esthetically significant.

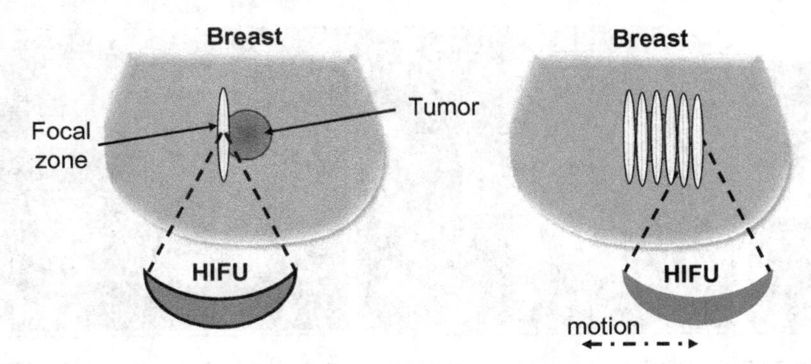

Figure 12.8. Schematics of HIFU application for breast tumor destruction. **(Left)** The HIFU transducer produces very high acoustic intensity and the focal zone (which is shaped like a cigar). This induces sufficient temperature elevation to ablate the tissue. **(Right)** In order to destroy the entire tumor, the HIFU beam is moved (mechanically or electronically) and the process is repeated as many times as required.

The idea of using high-intensity focused ultrasound (HIFU) to destroy tissues noninvasively had already been suggested in the 1940s (see historical review in reference 19). The principle of HIFU may seem quite simple. A special ultrasound transducer or an array of transducers located near the treated organ (extracorporeal or internally) is used as a virtual "knife" to destroy noninvasively a small volume of tissue within the body. The focused high acoustic energy induces a significant temperature rise at the focal point and causes local thermal ablation [see Eqs. (12.12) and (12.2)]. Because the focal zone has a typical cigar shape (see Chapter 7), the HIFU system has to be moved in a controlled manner in order to cover the entire tumor in 3D with the cigar-shaped ablated regions, as shown schematically in Fig. 12.8 for a breast tumor.

Although it was suggested a long while ago, HIFU has become very popular only in recent years. This stems from several technological complexities associated with the procedure. The first obstacle is the need for an accurate spatial localization of the tumor. The second is the need for a continuous co-registration of the HIFU beam and the tumor to be destroyed. The third is the need for an accurate motion control, mechanical and/or electronic, for steering the HIFU beam. Finaly, some kind of thermal monitoring is needed to ensure that sufficient energy has been delivered to the treated area and in order to avoid boiling and charring of the tissue (which may lead to complications). This calls for a combination of robotic devices with a sophisticated imaging modality. Image-guided HIFU surgery has been recently offered as a noninvasive alternative to conventional lumpectomy [20] in the breast [21–24] and in other organs [25–27]. In fact it is currently the most mature technology for noninvasive ablation of tumors. Thus, it may be considered the surgical knife of the future [28].

Almost all the major equations needed for designing the acoustic part of the HIFU were given in the previous chapters and above. For clarity the major steps are listed here:

(i) Calculate the acoustic field transmitted from the HIFU using the equations provided in Chapter 8.

(ii) If a reflecting or focusing device is needed utilize the equations outlined in Chapter 7.

(iii) From the acoustic field calculated above determine the intensity $I(x,y,z,t)$. Then, use Eq. (12.12) to determine the temperature elevation in the region of interest.

(iv) Optimize the transmission protocol to reach the required thermal dose [Eq. (12.2)] needed for tissue ablation in the desired time frame.

An example of a HIFU transducer before and during operation is shown in Fig. 12.9. The transducer has a diameter of 100 mm. It is a spherically focused with a focal distance of 80 mm. It was placed inside a water tank with its focal point close to the surface. As can be noted, the high-intensity acoustic beam creates a water jet and "fog."

The effect of HIFU on a tissue (in vitro) is depicted in Fig. 12.10. The target was a turkey breast. The specimen was cut after applying the HIFU at the focal plane. As can be noted, the ablated tissue has turned white.

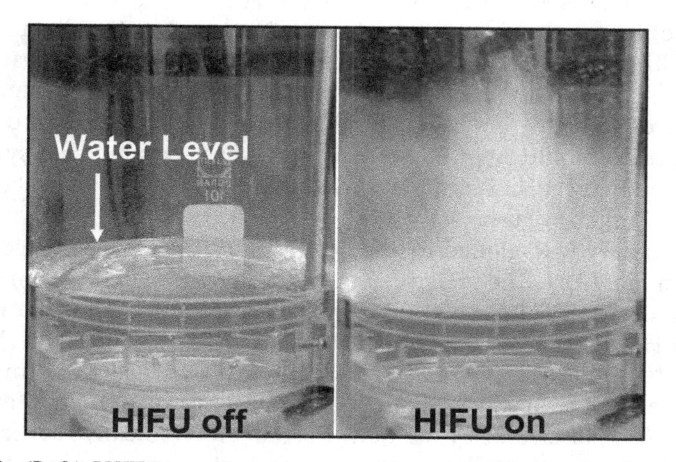

Figure 12.9. (Left) HIFU transducer immersed in water. The focal point is set at the water level. **(Right)** The HIFU transducer is turned on. Note the geyser-like effect comprised of a water jet and "fog."

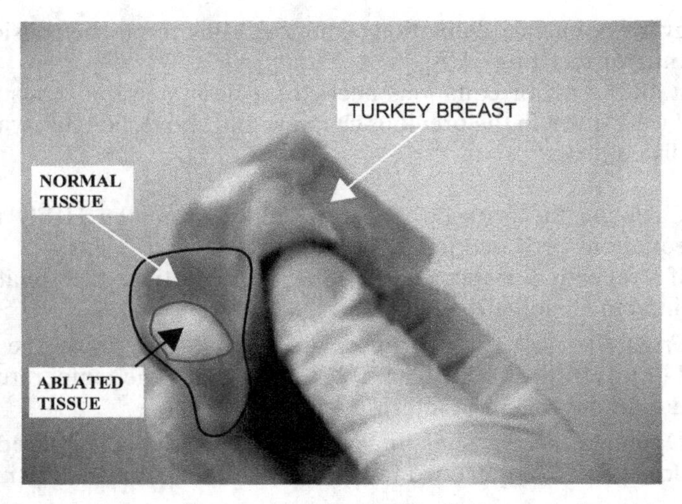

Figure 12.10. A demonstrative image of the HIFU effect on a turkey breast specimen (in vitro). The specimen was cut at the plane containing the focal zone. As can be noted, the tissue at the focal zone was ablated and its color changed to white.

Currently, the therapeutic ultrasonic HIFU beam is guided using ultrasound or an MRI scanner (e.g., references 21–24). MRI-guided HIFU surgery has been and still is under extensive research. The major advantages of MRI are its ability to provide high-quality anatomical images and noninvasively thermal maps of the treated area during the procedure. The disadvantages, however, are significant: The cost and operating expenses of the MRI scanning system are very high, and access to the patient within the scanner is quite limited. Ultrasound guidance, on the other hand, is much more cost-effective. However, image quality is inferior to that of MRI, and the applied acoustic windows are limited by bones and air. Furthermore, it currently provides only crude thermal monitoring. Nevertheless, several commercial systems for ultrasonic-guided HIFU surgery are available today for treating the prostate, thyroid, and other organs.

An exemplary MR-guided focused ultrasound (MRgFUS) surgery system is shown in Fig. 12.11. The system manufactured by InSightec is clinically used today mainly for noninvasive ablation of uterine fibroids as shown in Fig. 12.12. The HIFU system is embedded within the MRI scanner table. It includes a small water tank covered by a thin membrane. Piezoelectric motors are used to move the HIFU transducer in the X–Y plane while focal zone steering along the Z axis (perpendicular to the bed) is done electronically (as explained in Chapter 8). The patient lies prone on the bed as shown. The anatomical images acquired by the MRI scanner are used for planning the ablation procedure and thermal images for monitoring it.

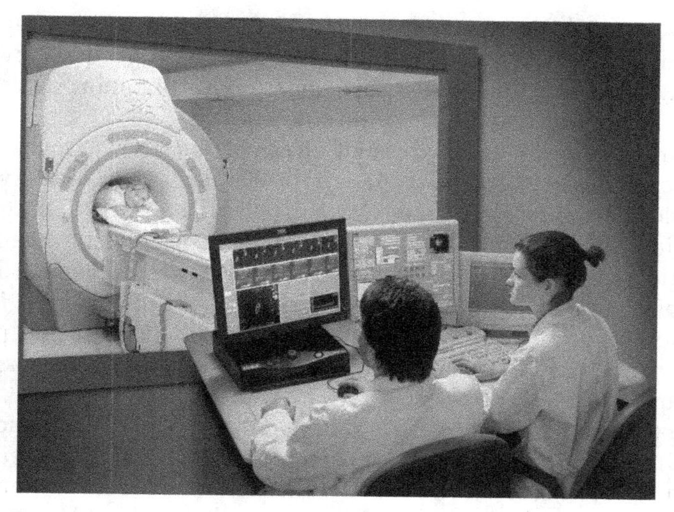

Figure 12.11. A general view of the ExAblate 2000 system for MR-guided focused ultrasound (MRgFUS) surgery manufactured by InSightec. As can be observed (in the back), the patient lies inside the MRI magnet. The console with two operators is shown in the front. Courtesy of the InSightec company, copyright © InSightec.

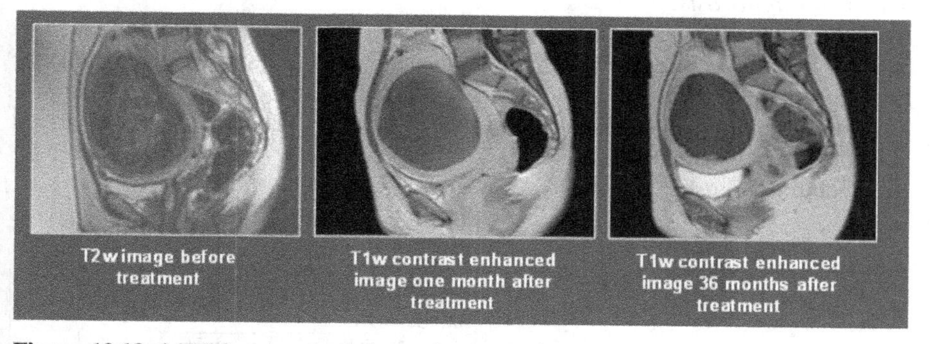

Figure 12.12. MRI images depicting a large uterine fibroid before and after HIFU treatment. Courtesy of the InSightec company, copyright © InSightec.

12.5 DRUG DELIVERY

Medication is commonly administered to the patients either orally (i.e., by swallowing pills) or intravenously (i.e., by injection). As soon as the drug reaches the blood, its concentration peaks rapidly and then starts to decay exponentially resulting from degradation or physiological process. Furthermore, without targeting, the drug will be carried by the cardiovascular system to almost every organ in the body mostly to irrelevant regions. Hence, drug delivery efficiency in low and high concentrations are needed. Several techniques utilizing ultrasound have been suggested for obtaining more localized and efficient ways to deliver drugs into the body.

Contrary to the dominant thermal effect of the HIFU, devices that are based on cavitation effects can induce controlled damage to tissues which may be invisible yet significant. By studying Eq. (12.14), it can be deduced that the transmission of ultrasonic waves at *low* frequencies can lead to the formation of cavitation bubbles even at *low* intensities. As was explained above, the collapse of the cavitation bubbles is very significant to living cells. It can tear membranes, separate cells, and alter the characteristic diffusion properties. This fact can be used to create controlled "cold" damage to the tissue, which in turn can be utilized for better transfer of chemicals.

One application is for transdermal drug delivery [29]. This application was suggested by the Sontra® company (merged with "Echo Therapeutic Inc."). By attaching an ultrasonic transducer to the skin and transmitting waves at the very low ultrasonic frequency of 55 kHz, cavitation bubbles can be generated in the dermal layers. When the bubbles collapse in the intercellular regions, they separate the cells from each other. As a result, "channels" of about 100 μm are formed. These channels open a pathway from the skin to the blood (see Fig. 12.13). These micro pathways allow the transmission of fluids and chemical substances into and out of the body. The procedure is called *sonophorosis*.

The micro pathways can be used to deliver drugs into the body and also to sample the blood without using a needle. This can be very useful for diabetic patients, for example. The ultrasonic procedure is short (about 30 sec), is painless, and the damage induced is reversible. The micro channels are closed within 24 hr. The instrument itself is compact and user friendly as can be appreciated from Fig. 12.14.

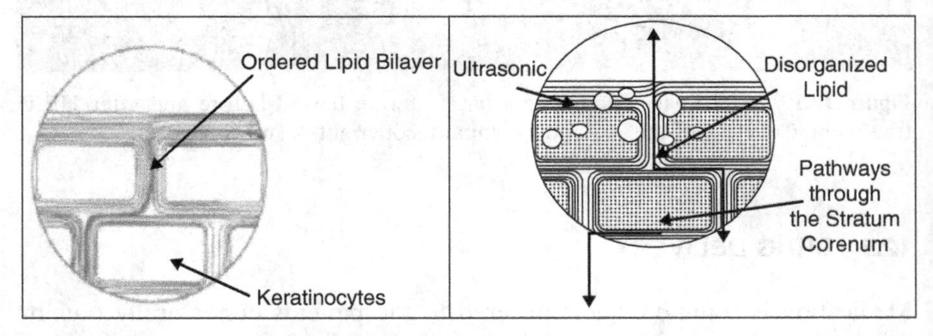

Figure 12.13. (Left) The skin is comprised of tightly arranged layers of cells. These layers isolate the body fluids from the external surroundings. **(Right)** Application of ultrasonic radiation generates cavitation bubbles. The collapse of these bubbles leads to intercellular pathways formation. Copyright © Sontra Medical Corporation.

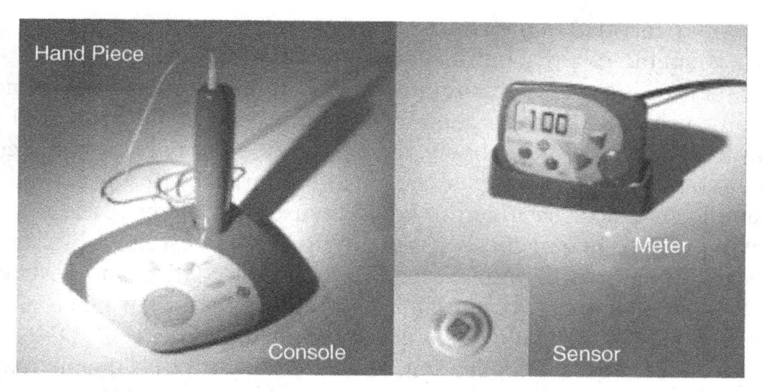

Figure 12.14. (Left) An ultrasonic system for creating micro transdermal channels. **(Right)** A glucose level display with its dedicated sensor. The sensor is attached to the skin after the sonophoresis procedure to measure the glucose in a noninvasive manner. Courtesy of the Sontra Medical Corporation, copyright © Sontra Medical Corporation.

Another challenge in drug delivery is the blood–brain barrier (BBB). This physiological barrier complicates the delivery of drugs to the brain. Currently available techniques are either invasive or nontargeted. Ultrasound, on the other hand, can be focused and used for breaking the BBB. It was shown in several studies that the BBB can be reversibly disrupted [30]. Thus, MRI guided focused ultrasound has been suggested as a tool for locally disrupting the BBB while administering drugs to the brain.

Another approach suggested in ultrasonic aided drug delivery is to use thermosensitive liposomes for encapsulating drugs [31]. The encapsulated drug is injected into the blood and reaches most of the regions in the body. However, due to the encapsulation, the drug remains inactive. When exposed to high-intensity ultrasonic pulses, these liposomes break and release their cargo at the site of exposure. Thus, using a suitable imaging modality for guidance, such as ultrasound or MRI, the high-intensity ultrasonic beam can be focused at desired regions such as tumors or other specific targets. The high-intensity ultrasonic radiation induces a drug release in the irradiated site. The drug concentration is therefore high only at the treated region. This improves the efficiency of the drug delivery and reduces the commonly associated danger to normal sites.

12.6 GENE THERAPY

The heredity information of every living cell is encoded in its genes. Genes are specific sequences of bases that contain instructions for producing specific proteins. In certain pathologic conditions, (i.e., genetic disorders), a particular gene may be altered or absent. Thus, the needed protein whose production information is encoded in the altered or missing gene is altered or missing and the body is unable to carry out its normal functions.

Gene therapy is a treatment method that attempts to replace a malfunctioning gene or provide a missing gene by delivering externally generated genes into the treated cells. Alternatively, the delivered gene may be connected to some targeting molecules and may be used to interfere with the cell function. For example, in cancer treatment it could be used to deliver genes to tumors, which will produce angiogenesis (formation of blood vessels) inhibitors. This will obstruct their growth and will lead to their degradation and destruction.

The delivered gene may be distributed in a plasmid form, which is an extra-chromosomal DNA molecule—that is, not connected to the cell's DNA, which is capable of autonomous replication. The externally produced gene may also be attached to some other molecules that may carry and target it or augment its function. The whole gene delivery complex is called a *vector*.

The major technical problem in gene therapy is to provide a method for inserting the gene into the nucleus of the treated cells without damaging the cells. One thoroughly investigated method is to use a virus as a vector. The plasmid is inserted into the virus, which carries it to the target cells. The virus penetrates the cell's membrane and inserts the gene or plasmid into the cell as it would normally do with its own genes. This method has been used in numerous animal studies and even in some human studies (in a research mode) and was shown to be successful in some cases. However, there were also several reported cases of severe adverse effects with this technique. This and the fear of viral mutations hinder its authorization for daily clinical use. Currently, there is no FDA approval for this method.

As a safer alternative to viral transfection, ultrasonic radiation has been suggested. As was noted above, exposure of living cells to intensive ultrasonic radiation for a sufficiently long duration may induce damage to the cells. One of the damaging effects is the alteration of the cells membrane properties. The membrane can become highly permeable. This allows the genes to enter the cytoplasm and even the nucleus as shown schematically in Fig. 12.15. If

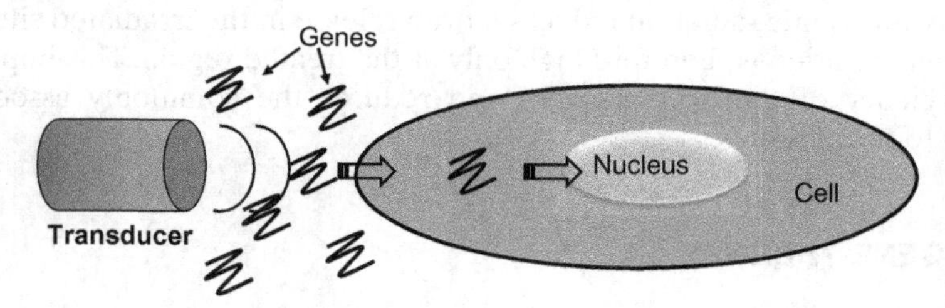

Figure 12.15. Schematic depiction of ultrasonic gene delivery to living cells. The acoustic waves alter the cell membrane properties, allowing the genes to penetrate the cell. (The transducer was drawn just for clarity. Its size is naturally much bigger than the cell.)

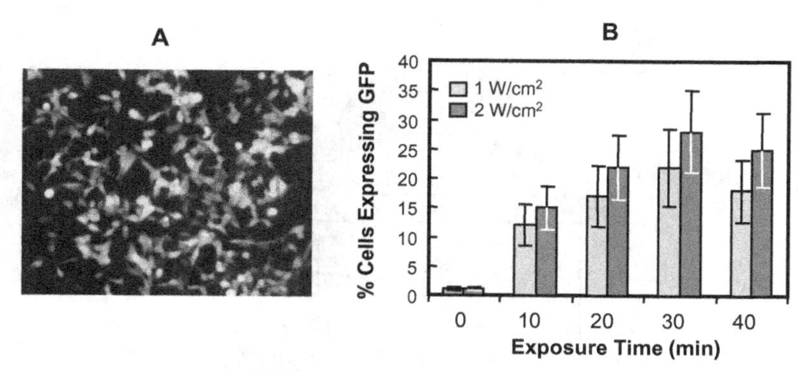

Figure 12.16. **(Left)** Fluorescence microscopy image of cells (BHK cells transfected with pGFP) into which genes were delivered by ultrasonic radiation. **(Right)** Bar graph depicting the percentage of successful gene penetration as a function of exposure duration for $1\,W/cm^2$ and $2\,W/cm^2$, intensities. Courtesy of Marcelle Machluf.

the intensity of the acoustic waves is not too high, the damage is reversible. The cells may commonly resume normal properties within 24 hours post exposure to the ultrasonic radiation.

In order to verify the success of this process, fluorescence molecules are attached to the genes. When the procedure of gene insertion is completed, the cells can be scanned by fluorescence microscopy. With a special laser illumination the fluorescence molecules produce light at a different wavelength than the illuminating light. By using proper filtering, this fluorescence light can be imaged. If the genes have indeed penetrated the cells, then the cells will become fluorescence. An example of such a scan is shown in Fig. 12.16 (left). The bright spots are clusters of cells into which the genes have penetrated. The intensity and duration of the ultrasonic radiation affect the percentage of the successful penetration. This is demonstrated in Fig. 12.16 (right), where a bar graph depicting the percentage of successful gene penetration as a function of exposure duration for $1\text{-}W/cm^2$ and $2\text{-}W/cm^2$ intensities is shown.

The exact mechanism by which the genes penetrate the cells when exposed to ultrasonic radiation is not fully known [32]. However, experimental findings do indicate that the exposure parameters has some optimal conditions [33]. For example, in Fig. 12.16 (right) it can be noted that for ultrasonic intensities of $1\,W/cm^2$ and $2\,W/cm^2$ for transmission at 1 MHz and with a duty cycle of 30%, the optimal exposure duration is 30 min. Furthermore, it was also shown [34] that the use of contrast materials augments the transfection efficiency.

12.7 COSMETIC APPLICATIONS

High-intensity ultrasonic radiation has found applications also in the field of cosmetics. Several patents and companies have suggested the use of

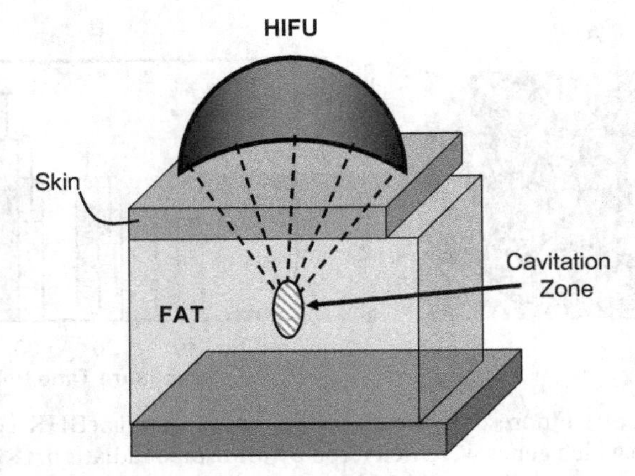

Figure 12.17. Schematic depiction of a commercially available system for ultrasonic "body-contouring." The high-intensity ultrasonic beam is focused at the subcutaneous fatty layer. Cavitation bubbles are formed at the focal zone. Their collapse leads to the destruction of the fat cells.

ultrasonic-based devices to achieve cosmetic effects. The first application is for hair removal. By using different techniques, the acoustic energy is directed to the roots of the hair. The heat generated by the acoustic energy loosens the hair and to some extent prevents it from re-growing.

The second application of ultrasound in cosmetics which has gained popularity in recent years is "body contouring." Using intensive ultrasonic waves, subcutaneous fat cells at specific regions are noninvasively destroyed. This is done as shown schematically if Fig. 12.17, by focusing the beam at the fatty tissue. The cells are destroyed by cavitation or thermal effects as was explained at the beginning of this chapter. The fat cells are ruptured and their fat contents are released into the blood and intercellular regions. This fat is slowly absorbed and removed by the body. Since this procedure is applied selectively, the body contour is altered to make the person look nicer (hopefully).

REFERENCES

1. ter Haar G and Duck FA, editors, *The Safe Use of Ultrasound in Medical Diagnosis*, British Institute of Radiology, London, 2000.
2. Sapareto SA and Dewey WC, Thermal dose determination in cancer therapy, *Int J Radiat Oncol Biol Phys* **10**(6):787–800, 1984.
3. Damianou C and Hynynen K, The effect of various physical parameters on the size and shape of necrosed tissue volume during ultrasound surgery, *J Acoust Soc Am* **95**(3):1641–1649, 1994.

4. Damianou CA, Hynynen K, and Fan X, Evaluation of accuracy of a theoretical model for predicting the necrosed tissue volume during focused ultrasound surgery, *IEEE Trans Ultrason Ferroelectr Freq Control* **42**(2):182–187, 1995.
5. Pennes HH, Analysis of tissue and arterial blood temperatures in the resting human forearm, *J Appl Physiol* **1**:93–122, 1948.
6. The American National Council for Radiation Protection and Measurement, Report 113, Bethesda, MD, 1992.
7. Brennen CE, *Cavitation and Bubble Dynamics*, Oxford University Press, New York, 1995.
8. Leighton TG, *The Acoustic Bubble*, Academic Press, London, 1994.
9. Brujan EA, Kester N, Shmidt P, and Vogel A, Dynamics of laser induced cavitation bubble near an elastic boundary, *J Fluid Mech* **433**:251–281, 2001.
10. Apfel RE and Holland CK, Gauging the likelihood of cavitation from short pulse duty cycle diagnostic ultrasound, *Ultrasound Med Biol* **17**:179–188, 1991.
11. Nightingale KR, Kornguth PJ, Walker WF, McDermott BA, Trahey GE, A novel technique for differentiating cysts from solid lesions: Preliminary results in breast, *Ultrasound Med Biol* **21**:745–751, 1995.
12. Heckman JD, Rayby JP, McCabe J, Frey JJ, and Kilcoyne RF, Acceleration of tibial fracture healing by non-invasive low intensity pulsed ultrasound, *J Bone Joint Surg Am* **76**:26–34, 1994.
13. Chaussy Ch, Brendel W, and Schmiedt E, Extracorporeally induced destruction of kidney stones by shock waves, *Lancet* **2**:1265, 1980.
14. Wilbert DM, A comparative review of extracorporeal shock wave generation, *BJU Int* **90**:507–511, 2002.
15. Chow GK and Streem SB, Extracorporeal lithotripsy—Update on technology, *Urol Clin N Am* **27**:315–322, 2000.
16. Wilbert DM, Reichenberger H, Noske E, and Hohenfellner R, New generation shock wave lithotripsy, *J Urol* **138**:563–565, 1987.
17. Reichenberger H, *Proc IEEE* **76**(9):1236–1246, 1988.
18. Srivastava RC, Leutloff D, Takayama K, and Gronig H, editors, *Shock Focusing Effect in Medical Sciences and Sonoluminescence*, Springer-Verlag, Berlin, 2003.
19. Kremakau FW, Cancer therapy with ultrasound: A historical review, *J Clin Ultrasound* **7**:287–300, 1979.
20. Randal J, High intensity focused ultrasound makes its debut, *Nat Cancer Inst* **94**(13):962–864, 2002.
21. Hynynen K, Pomeroy O, Smith D, Huber P, McDannold N, Kettenbach J, Baum J, Singer S, and Jolesz F, MR imaging-guided focused ultrasound surgery of fibroadenomas in the breast: A feasibility tudy, *Radiology* **219**:176–185, 2001.
22. Gianfelice D, Khiat A, Amara M, Belblidia A, and Boulanger Y, MR imaging-guided focused ultrasound ablation of breast cancer: Histopathologic assessment of efficacy—Initial experience, *Radiology* **227**(3):849–855, 2003.
23. Huber PE, Jenne JW, Rastert R, Simiantonakis I, Sinn HP, et al. A new noninvasive approach in breast cancer therapy using magnetic resonance imaging-guided focused ultrasound surgery, *Cancer Res* **61**(23):8441–8447, 2001.

24. Gianfelice D, Abdesslem K, Boulanger Y, Amara M, and Belblidia A, MR imaging-guided focused ultrasound surgery (MRIGFUS) of breast cancer: Correlation between dynamic contrast-enhanced MRI and histopathologic findings, *RSNA* 2002.
25. Tempany CMC, Stewart EA, McDannold N, Quade B, Jolesz F, and Hynynen K, MRI guided focused ultrasound surgery (FUS) of uterine leiomyomas: A feasibility study, *Radiology* **227**:897–905, 2003.
26. Wu F, Chen WZ, Bai J, Zou JZ, Wang ZL, Zhu H, and Wang ZB, Tumor vessel destruction resulting from high-intensity focused ultrasound in patients with solid malignancies, *Ultrasound Med Biol* **28**(4):535–542, 2002.
27. Madersbacher S, Schatzl G, Djavan B, Stulnig T, and Marberger M, Long-term outcome of transrectal high-intensity focused ultrasound therapy for benign prostatic hyperplasia, *Eur Urol* **37**:687–694, 2000.
28. Kennedy JE, Ter Haar GR, and Cranston D, High intensity focused ultrasound: Surgery of the future? *Br J Radiol* **76**:590–599, 2003.
29. Lavon I and Kost J, Ultrasound and transdermal drug delivery, *Drug Discovery Today* **9**(15):670–676, 2004.
30. Kinoshita M, Targeted drug delivery to the brain using focused ultrasound, *Topics Magn Reson Imag* **17**(3):209–215, 2006.
31. Dromi S, Frenkel V, Luk A, Traughber B, Angstadt M, Bur M, Poff J, Xie J, Libutti SK, Li KC, and Wood BJ, Pulsed-high intensity focused ultrasound and low temperature-sensitive liposomes for enhanced targeted drug delivery and antitumor effect. *Clin Cancer Res* **13**(9):2722–2727, 2007.
32. Duvshani-Eshet M, Baruch L, Kesselman E, Shimoni E, and Machluf M, Therapeutic ultrasound mediated DNA to cell and nucleus: Bioeffects revealed by confocal and atomic force microscopy, *Gene Therapy* **13**:163–172, 2006.
33. Duvshani-Eshet M and Machluf M, Therapeutic ultrasound optimization for gene delivery: A key factor achieving nuclear DNA localization, *J Cont Release* **108**:513–528, 2005.
34. Duvshani-Eshet M and Machluf M, Efficient transfection of prostate tumor facilitated by therapeutic ultrasound in combination with contrast agent: From *in vitro* to *in vivo* setting, *Cancer Gene Therapy* **14**:306–315, 2007.

APPENDIX A

TYPICAL ACOUSTIC PROPERTIES OF TISSUES

Table A.1. Typical Density, Speed of Sound, and Acoustic Impedance Values Obtained From [1–6]

Tissue or Material	Density (g/cm^3)	Speed of Sound (m/sec)	Acoustic Impedance $[kg/(sec \cdot m^2)] \times 10^6$
Water	1	1480	1.48
Blood	1.055	1575	1.66
Fat	0.95	1450	1.38
Liver	1.06	1590	1.69
Kidney	1.05	1570	1.65
Brain	1.03	1550	1.60
Heart	1.045	1570	1.64
Muscle (along the fibers)	1.065	1575	1.68
Muscle (across the fibers)	1.065	1590	1.69
Skin	1.15	1730	1.99
Eye (lens)	1.04	1650	1.72
Eye (vitreous humor)	1.01	1525	1.54
Bone axial (longitudinal waves)	1.9	4080	7.75
Bone axial (shear waves)	1.9	2800	5.32
Teeth (dentine)	2.2	3600	7.92
Teeth (enamel)	2.9	5500	15.95

Notes: These are just representative values. For more accurate numbers see literature. Speed of sound will vary with temperature in most cases.

Basics of Biomedical Ultrasound for Engineers, by Haim Azhari

Table A.2. Typical Attenuation and *B*/*A* Values Obtained From [1–6]

Tissue or Material	α (db/cm) @ f MHz	$\alpha = a \cdot f^b$		Nonlinear Parameter
		a [db/(cm MHz)]	b	B/A
Water	—	0.002	2	5.2
Blood	—	0.15	1.21	6
Fat	—	0.6	1	10
Liver	—	0.9	1.1	6.8
Kidney	—	1	1	7.4
Brain	—	0.8	1.35	6.9
Heart	2 @1 MHz	—	—	6.8
Muscle (along the fibers)	1.3 @1 MHz	—	—	(average for all directions) 7.4
Muscle (across the fibers)	3.3 @1 MHz	—	—	
Skin	9.2 @5 MHz	—	—	—
Eye (lens)	7.8 @10 MHz	2	—	—
Eye (vitreous humor)	0.6 @6 MHz	—	—	—
Bone (skull)	—	20	—	—
Bone (trabecular)	2–15 @ 0.2–1 MHz	—	—	—
Teeth (dentine)	80 @18 MHz	—	—	—
Teeth (enamel)	120 @18 MHz	—	—	—

Notes: These are just representative values. For more accurate numbers see literature.

REFERENCES

1. Duck FA, *Physical Properties of Tissue*, Academic Press, London, 1990.
2. Duck FA, Propagation of sound through tissue, in *The Safe Use of Ultrasound in Medical Diagnosis*, ter Haar G and Duck FA, editors, British Institute of Radiology, London, 2000, pp. 4–15.
3. Bacon DR, *IEEE Trans Ultrason Ferroelectr Freq Control* **35**:153–161, 1988.
4. Wells PNT, *Biomedical Ultrasonics*, Academic Press, New York, 1977.
5. Zeqiri B, *Ultrasonics* **27**:314–315, 1989.
6. Kaye & Laby, *Tables of Physical & Chemical Constants*, National Physics Laboratory, UK, website, www.kayelaby.npl.co.uk.

APPENDIX B

EXEMPLARY PROBLEMS

CHAPTER 1

1.1 A two-dimensional planar acoustic wave is defined by the function $U = A \cdot e^{j(\omega t - k_x \cdot x - k_y \cdot y)}$. This wave impinges upon a perfect reflecting line that is parallel to the X axis. The corresponding reflected wave equation is given by

A. $U = A \cdot e^{j(\omega t + k_x \cdot x - k_y \cdot y)}$

B. $U = A \cdot e^{j(\omega t - k_x \cdot x - k_y \cdot y)}$

C. $U = A \cdot e^{j(\omega t + k_x \cdot x + k_y \cdot y)}$

D. $U = A \cdot e^{j(\omega t - k_x \cdot x + k_y \cdot y)}$

1.2 If the wave propagation direction is designated by vector $\overline{K}$ and the corresponding direction of the matter's particles motion is designated by vector $\overline{V}$, then which of the following statements are incorrect:

A. For a planar longitudinal wave, $\overline{K} \cdot \overline{V} \neq 0$.

B. For a planar shear wave, $\overline{K} \cdot \overline{V} \neq 0$.

C. For a planar longitudinal wave, $\overline{K} \cdot \overline{V} = 0$.

D. For a planar shear wave, $\overline{K} \cdot \overline{V} = 0$.

Basics of Biomedical Ultrasound for Engineers, by Haim Azhari

1.3 What is the wavelength for a two-dimensional wave whose function is given by $U = A\cos(2\pi f \cdot t - x - y)$?

A. 1

B. 2

C. $\sqrt{2\pi}$

D. $\sqrt{2}$

E. None of the above

1.4 In a nondispersive medium, which of the following waves has the fastest propagation speed: $U_1 = A\cos(5\pi \cdot t - 4x - 3y)$ or $U_2 = A\cos(5\pi \cdot t + 3x - 4y)$ or $U_3 = A\cos(5\pi \cdot t + 5y)$?

A. U_1

B. U_2

C. U_3

D. All three waves have the same propagation speed.

E. It is impossible to know.

1.5 How much time would it take for the front of the wave defined by $U_1 = A\cos\left(\sqrt{2} \cdot t - x - y\right)$ to reach the point P whose coordinates are $\{X_0 = 8.66; Y_0 = 5\}$, assuming that the wave front passes through the axes origin ($x = 0, y = 0$) at time $t = 0$. (*Note*: All distances are measured in meters.)

A. 0.9659

B. 10

C. $\sqrt{2}$

D. $10\sqrt{2}$

E. None of the above.

1.6 Two one-dimensional harmonic waves are transmitted at the same time but from sources which are located a distance of Δx apart. The first wave is defined by the function

$$U_1 = A\cos(2\pi f_1 t - k_1 x)$$

and the second wave is defined by

$$U_2 = A\cos(2\pi f_2 t - k_2 x).$$

Given that the second frequency is twice as high as the first one, at what time elapse Δt will the phases of the two waves be equal for a point located at x_0?

1.7 Two continuous one-dimensional harmonic waves are transmitted at the same time but along opposite directions. The equation for the first wave is given by

$$U_1 = A\cos(12\pi \cdot t - 4\pi x)$$

and the equation for the second is given by

$$U_1 = A\cos(12\pi \cdot t + 4\pi x).$$

What is the spatial distance between two stationary points in the medium?

1.8 What is the velocity vector for a wave defined by $U = A \cdot e^{j(10t-2\cdot x+3\cdot y-4\cdot z)}$ if dispersion is negligible?

1.9 What is the amplitude of a spherical wave at location S, given that its maximal amplitude is u_0 at a distance of r_0 from the source?

1.10 Consider a hypothetical one dimensional nonuniform medium for which the speed of sound is distributed as follows:

$$C(r) = C_0 \qquad \text{for} \quad r \le 0,$$

$$C(r) = C_0 + \alpha \cdot r \qquad \text{for} \quad 0 \le r0 \le 1,$$

$$C(r) = C_0 \qquad \text{for} \quad r > 1$$

Given that the irradiating field is $U_0(r) = A \cdot \sin(2\pi \cdot r)$, write the implicit expression for the scattered acoustic field $U_S(r)$ by using the first Born approximation.

CHAPTER 2

2.1 Consider the set of pendulums described in the following figure:

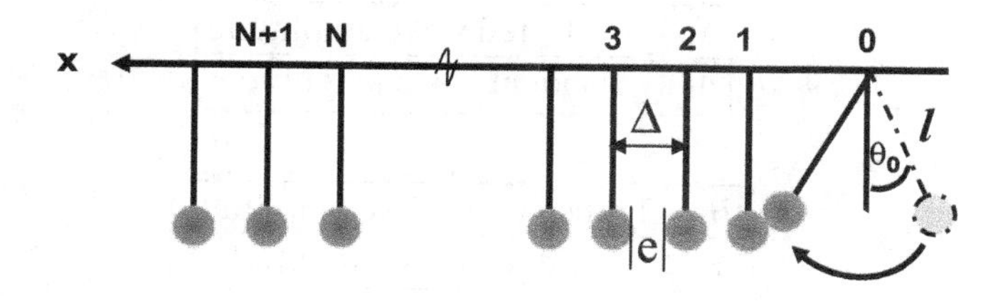

Each pendulum is comprised of a small steel ball of mass m which is hanged on a thin wire of length l. The gap between two consecutive balls is ε, and the distance between the wires is Δ.

The first pendulum (#0) is pulled by an angle θ_0 and then released. At time $t = 0$ the ball of the first pendulum hits the second pendulum (#1). As a result, a one-dimensional wave is initiated.

Assume that all the kinetic energy of each impinging ball is transferred immediately at the moment of contact, and that $\varepsilon << l$. Please answer the following questions:

A. What is the velocity of the nth ball when the wave front reaches it (i.e., what is its velocity immediately after the impact of the $(n - 1)$th ball?

B. What is the average propagation speed of this wave, between the first and the nth ball?

C. After how many balls will the wave cease to propagate?

2.2 An acoustic guitar has strings of length $l = 65\,\text{cm}$. Its A (la) string has a diameter of $d = 5\,\text{mm}$, and it is made of nylon with a density of $\rho = 0.9\,\text{g/cm}^3$. What is the required tension T in Newton in order to yield a tone of 440 Hz. (*Hint*: The sound is generated by standing waves.)

2.3 Let us define the ratios for reflection and transmission of one-dimensional waves at a point connecting two springs:

$$\frac{A_2}{A_1} = \frac{\text{Reflected wave amplitude}}{\text{Incident wave amplitude}}$$

$$\frac{B}{A_1} = \frac{\text{Transmitted wave amplitude}}{\text{Incident wave amplitude}}$$

Which of the following expressions designates energy conservation:

A. $B^2 = A_1^2 + A_2^2$

B. $\frac{B}{A_1} - \frac{A_2}{A_1} = 1$

C. $A_1^2 = B^2 + A_2^2$

D. None of the above

2.4 A string of density ρ per unit length is connected to a wave generator on its left side. A medical device, connected to its right side, pulls the string with a time-related force given by the function $F(t) = a \cdot t^2$. At time $t = 1$ sec, a transverse wave is transmitted from the left side. Where would the wave front be after 1 sec?

A. $\sqrt{\frac{a}{\rho}}$

B. $\frac{3}{2}\sqrt{\frac{a}{\rho}}$

$F(t) = a \cdot t^2$

C. $2\sqrt{\frac{a}{\rho}}$

D. None of the above

2.5 A device for disintegrating blood clots is comprised of a longitudinal wave generator and a catheter (a wire) of length L which is inserted into the body. A special tip of length T is attached to its end as shown in the following figure. The volumetric density of the catheter is ρ_1 and its modulus of elasticity is E_1. The volumetric density of the tip is ρ_2 and its modulus of elasticity is E_2. Given that the ratio for the two modulus of elasticity is $E_1 = 4E_2$, how long would it take for a wave to reach the edge of the device?

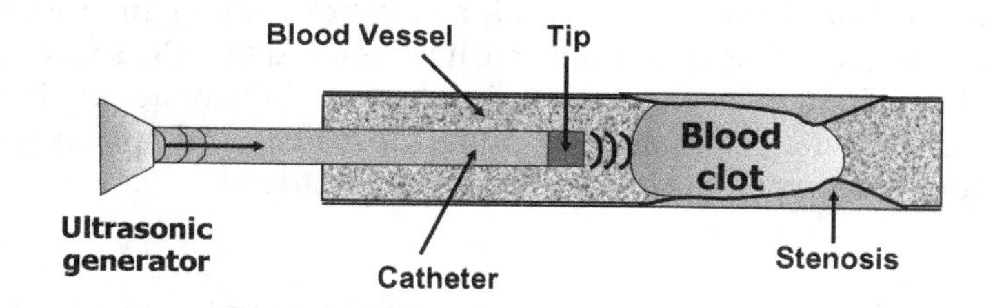

2.6 A transverse wave propagates with amplitude A1 in a string with an acoustic impedance Z1. The wave encounters a second string which has an impedance Z2 and is welded to the first string. Given that Z2 > Z1, what would be the phase θ_1 of the reflected wave (echo) and the through transmitted wave θ_2:

A. $\theta_1 = \theta_2 = 0$

B. $\theta_1 = \theta_2 = \pi$

C. $\theta_1 = 0, \theta_2 = \pi$

D. $\theta_1 = \pi, \theta_2 = 0$

E. None of the above

2.7 A device for hair removal is comprised of a transverse wave generator attached to a special short metal cable of length L_1 and density ρ_1. The cable clamps the hair at its tip. The device pulls the hair with tension T and then transmits a pulse of a transverse wave with amplitude A_0. If the hair length is L_2 and its density is ρ_2, what is the amplitude of the wave reaching the hair root?

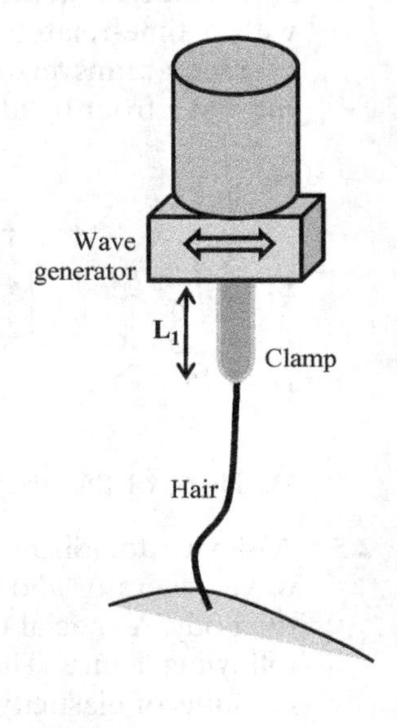

2.8 For the same device described in the previous problem, what is the maximal pulse repetition frequency (PRF) that can be used if it is required that the next pulse be transmitted only after the echo from the root has reached the wave generator?

2.9 An implanted thin metallic nail is used for bonding two sections of a broken bone. When the bone is subjected to a sudden impact, elastic waves are generated. The wave propagation speed in the bone along the direction parallel to the nail is about 4000 (m/sec). In order to minimize the risk of loosening the nail, it is required that the speed of sound in the nail will be as close as possible to that of the bone. Kindly suggest a suitable metal for manufacturing the nail (for simplicity you may ignore biocompatibility and other considerations).

2.10 A dental treatment device has an aluminum applicator for which the density is 2.7 (g/cm^3) and the speed of sound is 6300 (m/sec). The applicator touches the treated tooth and transmits acoustic energy into the tooth enamel. Assuming a perfect contact between the applicator and the tooth, what is the energy transmission efficiency?

CHAPTER 3

3.1 A piezoelectric crystal of dimensions $(x, y, z) = (L_0, L_0, L_0)$ is placed in a water tank.

The formula for the voltage across the crystal edges in direction i (for $i = x, y, z$) is $V_i = \Delta L_i \cdot \gamma$, where ΔL_i is the change in the length of the crystal in dimension i and γ is the crystal's piezoelectric constant.

An ultrasound transducer placed under the crystal transmits a continuous sinusoidal wave. As a result, the crystal is compressed along the y direction. (Assume that the change in its area A in the x, z plane is negligible.) The crystal is connected to a voltmeter that measures the voltage across its edges as shown.

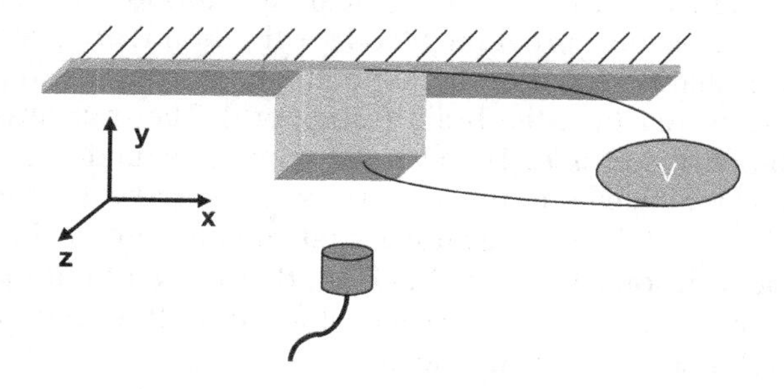

(a) Explain what causes the crystal to compress?

(b) Given that the crystal's Young modulus is E, and that $Z_{water} << Z_{crystal}$, what is the intensity of the ultrasound transducer if the voltmeter measures the voltage V_0?

3.2 A water tank has a water level meter that triggers an alarm if the water passes the red line (marked as a dashed line). Initially the water level is d [mm] below the red line. An ultrasound transducer is also placed in the water tank.

Given that ρ_{water} is the water density, C_{water} is the sound velocity in water, and f is the ultrasound transducer frequency,

(a) What should be the intensity of the ultrasonic transducer to trigger the alarm?

(b) What would be the amplitude of vibration of the water molecules?

(c) Please answer **(a)** and **(b)** again given that a small amount of some synthetic liquid is spilled into the water tank. The liquid cannot mix with water and thus floats above the water. The water level remains as before (d [mm] below the red line).

It is given that $\rho_{liquid} = 0.5\rho_{water}$ and $C_{liquid} = C_{water}$.

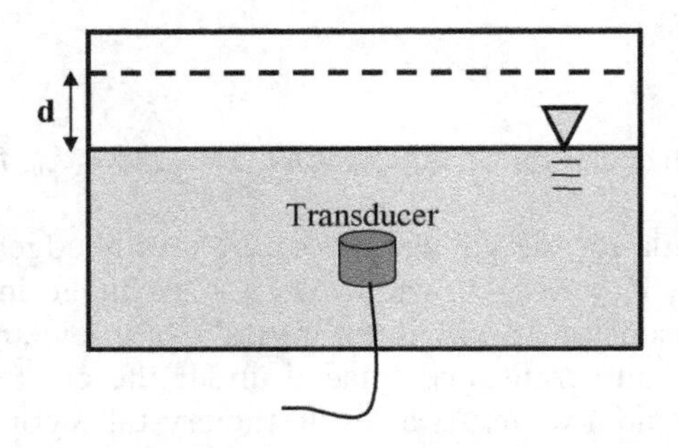

3.3 A company called Neurosonix® has developed an ultrasonic device that is attached during an open heart surgery to the aorta (the main blood vessel leaving the heart). The device deflects in a noninvasive manner small blood clots (called emboli) near the carotid bifurcation (a blood vessel junction that provides blood to the brain) by producing strong acoustic radiation force on the emboli. The purpose is to prevent these emboli from reaching the brain (see figure). The transmission intensity in the blood vessel is I_0. The average speed of sound is C_0. The average emboli area subjected to the acoustic beam is A, the average mass of the emboli is m, and the average horizontal emboli velocity (i.e., perpendicular to the ultrasonic beam) is V. Given that the width of the device is D, what is the emboli average speed along the ultrasonic beam (vertical direction) after it passes the device?

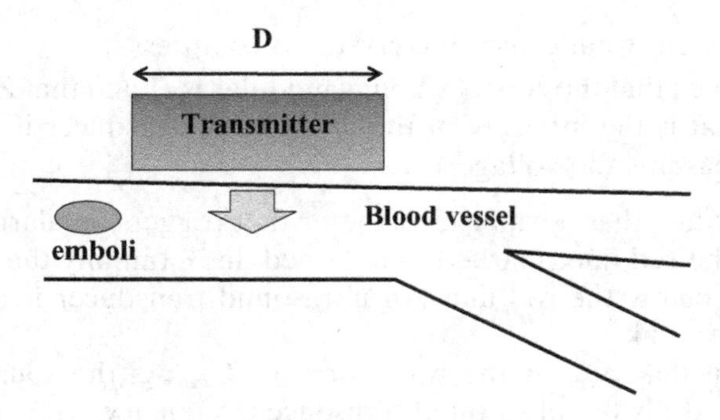

3.4 A device for measuring the radiation intensity of ultrasonic transducers is based on a pendulum that is comprised of a highly reflective plate attached to a hinge that allows it to freely swing back and forth. The pendulum is immersed in water and its initial position is naturally vertical. When an ultrasonic transducer is placed in the water tank and transmits waves toward the reflective plate, the radiation force causes the pendulum to swing away from its equilibrium condition by an angle θ. Assuming that the angle is very small, what is the relation between the angular position and the beam intensity?

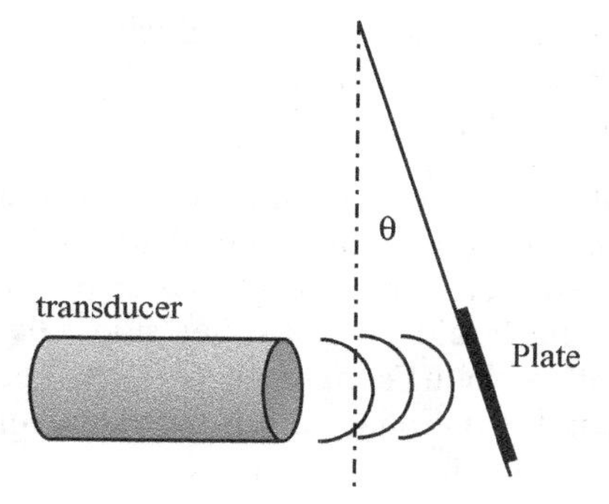

3.5 A planar harmonic wave travels in water with a displacement given by $A_0 \cdot \sin(2\pi f \cdot t - k \cdot x)$. The wave then passes into a layer of oil. The transmission coefficient for the wave amplitude on its passage from the water into the oil is T. Given that the density of water is ρ_0 and that of the oil is ρ_1, it is requested to determine the ratio between the wave energy per unit volume in the oil relative to that in the water.

CHAPTER 4

4.1 From an isotropic metal for which the two Lame constants are related as $\lambda' = 2\mu$, a very thin and long rod was built. Given that the metal density is ρ and that the velocity for shear waves in that metal is C_{shear}, what is the velocity for longitudinal waves in that rod?

A. $\sqrt{\dfrac{3\mu}{\rho}}$

B. $C_{\text{shear}}\sqrt{\dfrac{8}{3}}$

C. $C_{\text{shear}}\sqrt{3}$

D. $C_{\text{shear}}\sqrt{2}$

E. None of the above

4.2 How many types of ultrasonic waves can be propagated in a solid material along the Z axis, so that $K_3 = K$ and $K_1 = K_2 = 0$, given that the material has the elastic constants $C_{1313} = C_{2323} = \mu$, $C_{2313} = C_{1323} = \mu/2$, $C_{3333} = 4\mu$, and $C_{2333} = C_{3323} = C_{1333} = C_{3313} = 0$? Please calculate their velocities.

4.3 An isotropic material has shear wave velocity of C_{shear}. Given that the material has a Poisson's ratio of $\upsilon = 0.5$ (an incompressible material), please express the value of Young's modulus E in terms of C_{shear} and its density.

4.4 An isotropic material has shear wave velocity of C_{shear} and longitudinal wave velocity of C_{long}. Given that the density is ρ_0, what is the corresponding bulk modulus for that material?

4.5 A metal plate of thickness T is immersed in a liquid. Planar harmonic waves of frequency f are transmitted in the liquid toward the plate and impinge upon its surface at right angle. After passing through the plate, the waves enter the liquid on the other side. Given that the metal is isotropic and that its Young's modulus is E, its shear modulus is G, and its density is ρ, what is the phase difference between the waves entering the plate and the waves leaving the plate?

CHAPTER 5

5.1 An ultrasonic harmonic wave is transmitted with a relative intensity of 100 db into a tissue. What would be the relative intensity in db units if the frequency is doubled but the amplitude is kept the same? (*Hint*: $\log_{10}(2) \approx 0.3$.)

A. It will remain 100 db -for there is no connection between intensity and frequency!

B. 97 db

C. 103 db

D. 94 db

E. 106 db

5.2 A sono-micrometer device includes several point source transmitters and receivers implanted within the tissue. A spherical wave is transmitted from one of the sources through fatty tissue. The corresponding relevant tissue parameters are: density, ρ; speed of sound, C; and the attenuation coefficient, α. The amplitude of the wave (equivalent to pressure) at a distance $a/2$ from the source is measured by a receiving transducer and produces 1 V as an output signal. What is the amplitude (measured in volts) at a distance $a/2$ from the transmitter?

A. 1/2

B. 1/4

C. $0.5e^{-\alpha \cdot a/2}$

D. $0.25e^{-\alpha \cdot a/2}$

E. None of the above

5.3 For a Gaussian pulse with a central frequency f_0 traveling in a tissue for which the attenuation coefficient is $\alpha(f) = \alpha_0 + \alpha_1 \cdot f$, which of the following statements is *not* true:

A. The central frequency is the most attenuated frequency.
B. The deeper the wave penetrates the tissue, the lower is the value of the peak frequency.
C. The corresponding wave number may be considered a complex number.
D. If the frequency is doubled, then at distance L the amplitude ratio for the new amplitude and the original one will be $e^{-\alpha_1 f \cdot L}$.
E. Only one of the above answers is not correct.

5.4 An ultrasonic pulse defined by the function $S(t) = Vo \cdot \text{sinc}(\pi\Delta f \cdot t) \cdot \sin(2\pi f_0 \cdot t)$ is transmitted into a tissue for which the attenuation coefficient is $\alpha(f) = a \cdot f^b = 1 \cdot f^2$. What would be the amplitude ratio of the highest-frequency component to the lowest-frequency component after traveling a distance L within the tissue?

5.5 A system for ultrasonic disintegration of kidney stones (called lithotripter) is comprised of a high-intensity transmitter of radius R attached to an acoustic lens (see Chapter 7). The focal distance is F and the gain factor (i.e., the intensity at the focal point relative to the transmitted intensity) is given by (see Chapter 7): $4\pi R^4/(\lambda^2 \cdot F^2)$. Assume that the tissue is homogeneous with a typical speed of sound of C, density of ρ, and attenuation coefficient given by $\alpha_0 + \alpha \cdot f$. What is the optimal frequency for transmission?

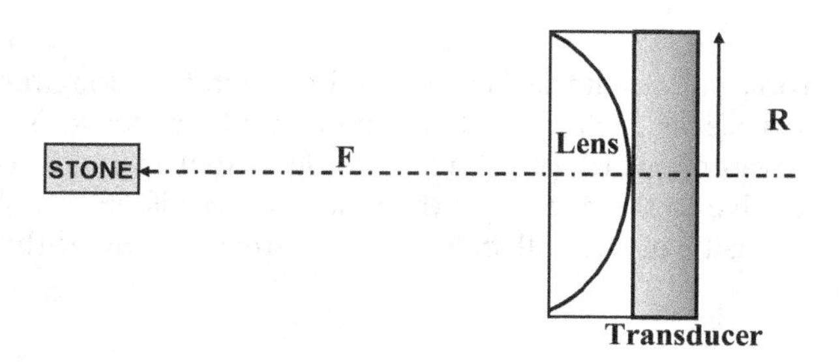

CHAPTER 6

6.1 A wave is transmitted through three layers as shown in the accompanying figure. The speed of sound values are C_1, C_2, and C_3, correspondingly. The transmission frequency is given by $(2n - 1)C_2/4d$. In order to transfer most of the energy through the

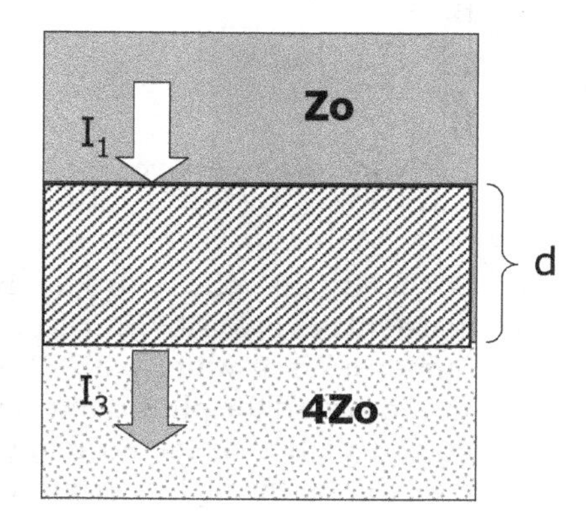

middle layer (neglecting attenuation), the density of the material in the middle layer should be

A. $\rho_2 = 2Z_0/C_2$

B. $\rho_2 = Z_0/C_2$

C. $\rho_2 = 4Z_0/C_2$

D. $\rho_2 = \sqrt{2} \cdot Z_0/C_2$

E. None of the above

6.2 A sono-micrometer device includes several point source transmitters and receivers implanted within the tissue. A spherical wave is transmitted from one of the sources through fatty tissue. The corresponding relevant tissue parameters are: density, ρ; and speed of sound, C. What is the acoustic impedance of the fat at a distance λ from the transmitting source?

A. $C \cdot \rho$

B. $j \cdot 2\pi \cdot C \cdot \rho$

C. $j \cdot 2\pi \cdot C \cdot \rho \left[\frac{1 - j2\pi}{1 + 4\pi^2}\right]$

D. $j \cdot 2\pi \cdot f \cdot \rho \left[\frac{1 - j2\pi}{1 + 4\pi^2}\right]$

E. None of the above

6.3 An isotropic solid material has a speed of sound for longitudinal waves which is twice as high as that in water. A planar wave is transmitted from the water toward a thick plate made of that material. The incident angle (relative to the normal to the plate surface) is $\theta = \pi/4$. What would be the intensity of the reflected wave relative to that of the impinging wave?

A. 0

B. 1

C. $\sqrt{2}$

D. $\sqrt{2}/2$

E. None of the above

6.4 For a certain solid isotropic material the ratio between the two Lamè constants is given by $\lambda' = 2\mu$. A thick plate made from this material is immersed in a water bath. If a longitudinal wave transmitted from the water toward the plate impinges the surface of the plate with an incidence angle (relative to the normal to the surface) of θ, the longitudinal wave that will be through-transmitted into the plate will:

A. Have an inclination angle θ_L which is equal to $2\theta_T$, where θ_T is the inclination angle for the through transmitted shear wave in the plate.

B. Have an inclination angle θ_L which relates to θ_T the inclination angle for the through transmitted shear wave in the plate as follows: $\sin(\theta_L) = \sin(\theta_T)/2$.

C. Have an inclination angle θ_L which relates to θ_T the inclination angle for the through transmitted shear wave in the plate as follows: $\sin(\theta_L) = \sin(2\theta_T)$.

D. Be the only wave, since a shear wave cannot be generated from a longitudinal wave.

E. None of the above

6.5 A group wave comprising of two planar waves is transmitted into a liquid medium. The frequency of the first wave is f_1, and the frequency of the second wave is $f_2 = 2f_1$. The acoustic impedance of the liquid is Z_1 and its thickness is S_1. Behind the liquid there is a block made of a solid material for which the acoustic impedance is Z_2 and its thickness is S_2. Behind the solid block there is vacuum as shown schematically in the accompanying figure. The attenuation coefficient for the *amplitude* within the liquid medium is $\alpha_0 + \alpha_1 \cdot f$, and within the solid it is $\gamma_0 + \gamma_1 \cdot f$. Kindly answer the following:

A. What would be the *intensity* reaching the liquid–solid interface and what would be the intensity reaching the solid–vacuum interface, given that the transmission intensity for both waves was equal and its magnitude is I_0?

B. Where would the echoes come from and what would be their intensities?

C. When would the first echo reach the transducer if the compressibility coefficient for the fluid is β?

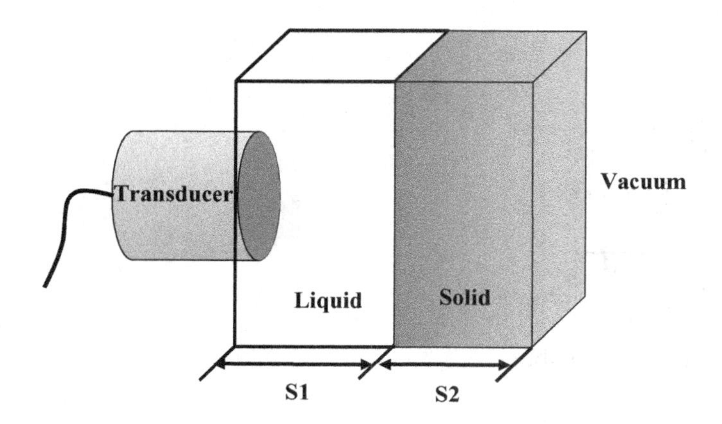

CHAPTER 7

7.1 Given that the speed of sound in the acoustic lenses is always higher than that of the surrounding medium, which pair of the acoustic lenses shown below will definitely focus sound waves:

A. E and A

B. D and B

C. E and C

D. B and C

E. None of the above

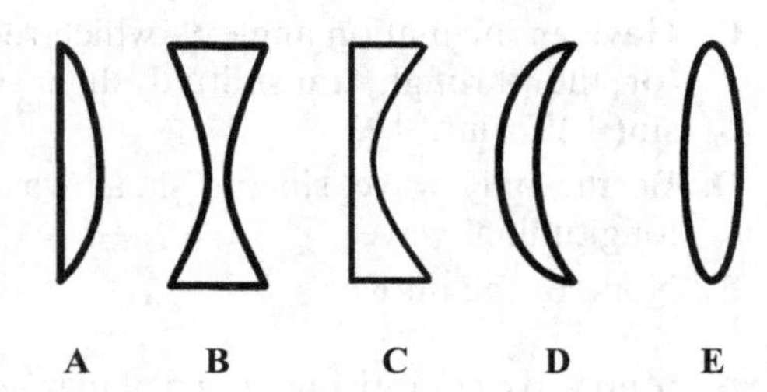

7.2 You have been asked to build an acoustic spherical lens that will focus planar waves at a distance F in water. For that purpose you were given a special plastic material for which the speed of sound is twice as high as the speed of sound in water. What is the required radius of curvature for that lens?

A. F

B. $2 \cdot F$

C. $F/2$

D. $3 \cdot F/2$

E. None of the above

7.3 What should be the aperture radius—that is, $r(\theta = \pi/2)$—for an ellipsoidal lens for which the speed of sound is C_1, given that the speed of sound in the medium is C_0 ($<C_1$) and that the distance from the directrix line is D:

A. $\frac{C_1}{C_0}D$

B. $\frac{C_0}{C_1}D$

C. $D\left(\frac{C_1}{C_0}\right)^2$

D. $D\sqrt{\left(\frac{C_1}{C_0}\right)^2 - 1}$

E. None of the above

7.4 What should the aperture radius [i.e., $r(\theta = \pi/2)$] be for a parabolic acoustic mirror for which the focal distance is F (see accompanying figure)?

A. F

B. 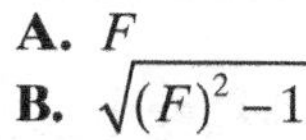 $\sqrt{(F)^2 - 1}$

C. $F/2$

D. $2F$

E. None of the above

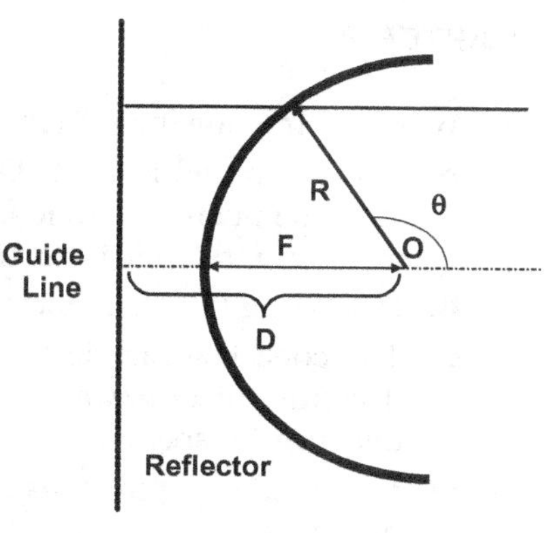

7.5 An ultrasonic transducer transmits waves with a frequency of f_0 and with an intensity of I_0 toward a spherical lens. The transducer is immersed in water and is located at a distance of S_1 from the lens. The radius of the lens is a (see accompanying figure). The lens is attached to the body (for simplicity, one can assume that the tissue is homogeneous. At a distance S_2 a very small metallic sphere with a diameter d is implanted in the tissue. Other given parameters are as follows: The speed of sound in the water is C_0, its density is ρ_0, and its attenuation in negligible. The speed of sound in the lens is C_1, its density is ρ_1, and its attenuation coefficient (for amplitude attenuation) is α_1. The density of the tissue is ρ_2 and its compressibility index is β. The average travel distance for an acoustic ray within the lens is H. The acoustic impedance of the metal is very high compared to that of the tissue.

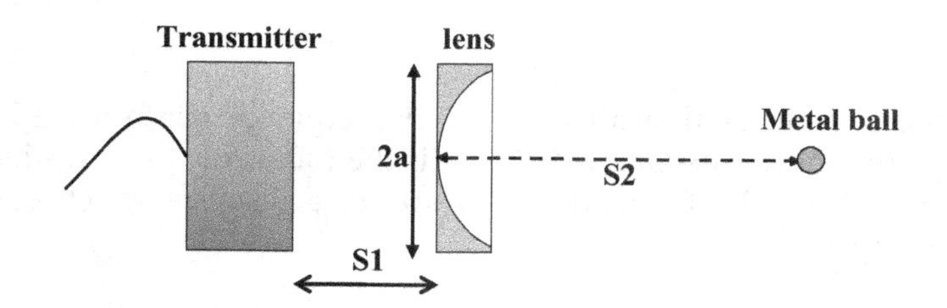

A. What is the needed radius of curvature for the lens in order to focus the beam on the metallic sphere?

B. What is the intensity reaching the metallic sphere? (You may ignore the effects stemming from the inclination angles at the lens surface and you may also neglect that differences in the acoustic ray paths to the focal point.)

C. What is the force applied on the metallic sphere? (*Note*: The sphere is much smaller than the beam width at the focal zone.)

CHAPTER 8

8.1 Which of the flowing sentences is *correct*?

A. The beam width for a disc transducer at a distance which is larger than the near-field zone is independent of the speed of sound of the medium into which it is transmitted.

B. Increasing the frequency will decrease the near-field zone.

C. The cone head angle for the main lobe of the acoustic beam for a disc transducer in water is smaller than that angle for the same transducer in Perspex.

D. For imaging purposes a transducer with a high Q-Factor is preferred.

E. None of the above

8.2 Which of the flowing statements is *incorrect* for a linear phased array transducer of length L, given that there are N elements, the transmission frequency is f, and the speed of sound in medium is C.

A. In order to steer the beam by an angle β, the phase for the first element should be $-2\pi f \cdot L \cdot \sin(\beta)/2C$

B. In order to avoid side lobes, it is required that $L/(N-1) \le C/f$.

C. The maximal steering angle is $\sin^{-1}\left[\frac{C \cdot (N-1)}{f \cdot L} - 1\right]$.

D. Only two of the above answers are correct.

8.3 A disc transducer transmits a wave of frequency f into a medium for which the speed of sound is C. What is the radius of that transducer if it is known that the far field zone starts at a distance which equals 25 wavelengths.

A. $\frac{\sqrt{99}}{2} \cdot \frac{C}{f}$

B. $25 \cdot \frac{C}{f}$

C. $5 \cdot \sqrt{\frac{C}{f}}$

D. $\frac{\sqrt{101}}{2} \cdot \frac{C}{f}$

E. None of the above

8.4 An array that is comprised of three point sources arranged as shown in the following figure transmits harmonic waves. The central source transmits at a frequency of f_1 while the other two sources transmit at f_2. The medium is a fluid with a density of ρ and compressibility index of β. All three sources produce an acoustic filed with a pressure amplitude of P_0 at distance a.

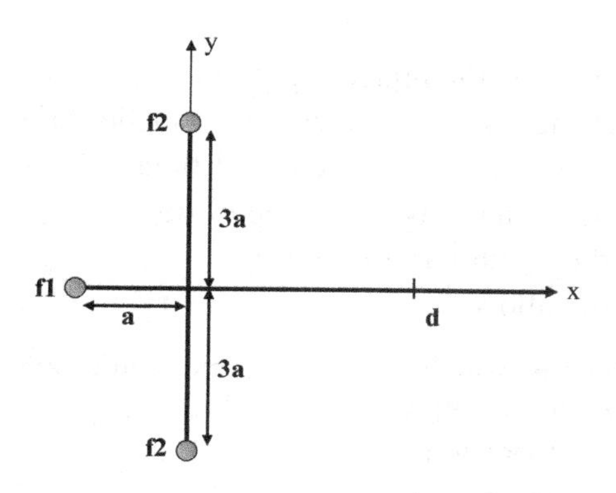

A. Please write the explicit term for the acoustic pressure at point d located on the X axis. (You may neglect attenuation.)
B. What is the pressure at point $d = 4a$ given that $f_1 = f_2$? Include also the effect of attenuation, given that $\alpha(f) = \alpha_0 + \alpha_1 \cdot f$.
C. What is the particles velocity at this point?
D. What is the intensity at this point?

8.5 A disc transducer of radius a transmits waves toward a target that is at distance R from the center of the transducer and at offset distance d from the beam axis (see accompanying figure). Given that the speed of sound is C and that the target is at the far-field zone, what is the required maximal frequency in order to obtain an echo from that target (side lobes can be ignored)?

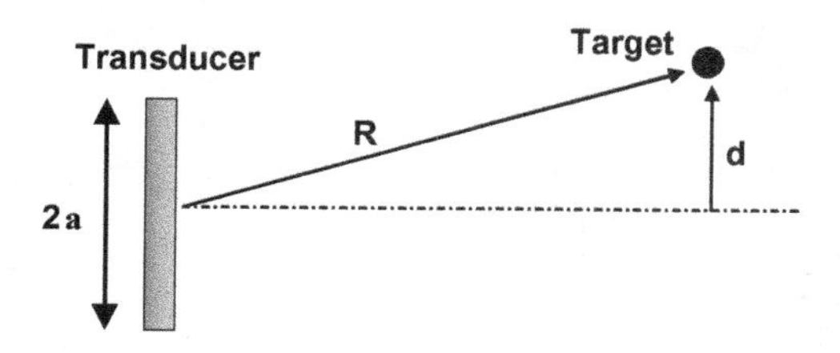

8.6 What is the directivity function at an angular azimuth of $\theta = 30°$ for a linear transducer (not an array!), given that its length is $L = 20\lambda$?

8.7 Two linear transducers of the same length and thickness are placed at a right angle to each other so that the left point of the horizontal transducer touches the bottom point of the second. Write the expression for the acoustic field.

CHAPTER 9

9.1 In M-mode imaging, the image represents:
A. The targets acoustic impedance as a function of time
B. The targets velocity as a function of time
C. The targets distance as a function of time
D. The speed of sound up to the target
E. None of the above

9.2 In B-scan imaging, which of the following could be done with a phased array and not with a single transducer:
A. Control the focal zone distance
B. Transmit through bones
C. Measure flow velocity with the Doppler effect
D. Control the transmission frequency
E. None of the above

9.3 In order to image distal objects with the B-scan mode, we should:
A. Choose high PRF and low frequency
B. Choose low PRF and high frequency
C. Choose high PRF and high frequency
D. None of the above

9.4 It is required to measure the diameter of the heart's left ventricular cavity (see accompanying figure) using A-mode. After transmitting a pulse four echoes where obtained at times t_1, t_2, t_3, and t_4. Given that the average speed of sound is C, the diameter is

A. $\frac{C}{4} \cdot \sum t_i$

B. $C \cdot \sum \frac{t_i}{2}$

C. $(t_4 - t_1)\frac{C}{2}$

D. $(t_3 - t_2)\frac{C}{2}$

E. None of the above

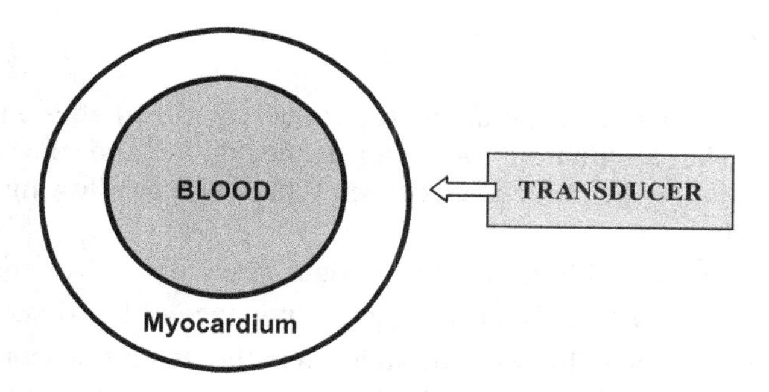

9.5 What is the expected best lateral resolution in the far field zone for a disc transducer of radius a which transmits waves at a frequency f_0 into a homogenous medium that has a speed of sound C (you may ignore side lobes)?

9.6 Using the simple Raleigh scattering model, what is the average scatterer radius a, if a scattering coefficient of 0.0111 was measured at 25-mm distance and at angle of 120°, for a frequency of 1.5 MHz and speed of sound of 1500 m/sec.

9.7 Defining the axial resolution as the minimal distance between two echoes at their 50% amplitude (full width at half-maximum) after applying the Hilbert transform to the echo train, use a programming language (C or Matlab) to estimate the the expected axial resolution for a Gaussian pulse with central frequency of 3 MHz and pulse width parameter $\beta = 40$ (1/sec^2).

9.8 Use a programming language (C or Matlab) and the plate model to simulate the A-line from a target comprised of: water–medium 1–medium 2–water. Every layer is 10 mm thick. The speed of sound in water is 1500 m/sec and its attenuation is negligible. The speed of sound for medium 1 is 1550 m/sec and its attenuation is 0.6 db/cm. The speed of sound for medium 2 is 1450 m/sec and its attenuation is 0.5 db/cm.

9.9 For the same model above, suggest an optimal TGC function.

9.10 Use a programming language (C or Matlab) to simulate the point spread function (PSF) 50 mm away from a line transducer (not an array) 20 mm in length and 2 mm in thickness transmitting into water ($C = 1500$ m/sec).

CHAPTER 10

10.1 An impediogram is obtained for tissue sample that is immersed in water. The attenuation of water is negligible and the attenuation coefficient of the tissue is unknown. Which of the following statements is correct?

A. It is not possible to obtain the right impedance of the tissue.

B. It is possible to obtain the right impedance of the tissue.

C. Regardless of the wave attenuation in the tissue, the acoustic impedance of the water after the tissue will have the same value as the water before the tissue.

D. Even if the tissue sample thickness is known, it is impossible to determine the tissue attenuation coefficient from the measure signal.

10.2 A tissue phantom is comprised of two horizontal layers of equal thickness placed on a table. A pressure of P pascals is applied vertically on the phantom. The longitudinal wave velocity is the same for the two layers, but shear wave velocity for the top layer is slightly higher than that of the bottom one. Pulse-echo measurements of longitudinal waves are done along the vertical direction (top to bottom), before and after the load is applied. Which of the following statements is correct *after* applying the load?

A. The time elapse between the echo from the top of the phantom to the echo from the boundary between the two layers will equal the time elapse between the echo from the boundary between the two layers and the bottom.

B. The time elapse between the echo from the top of the phantom to the echo from the boundary between the two layers will be longer than the time elapse between the echo from the boundary between the two layers and the bottom.

C. The time elapse between the echo from the top of the phantom to the echo from the boundary between the two layers will be shorter than the time elapse between the echo from the boundary between the two layers and the bottom.

D. It is impossible to predict the ratio between the two above-mentioned time elapses.

10.3 Consider a 3×3 pixels checkerboard pattern (upper left corner is black) where white color represents an echo (= 1) and a black color represents no-echo (= 0). This pattern is located in the middle of a larger black region. Then consider another image where this checkerboard pattern has been moved by one pixel horizontally and one pixel vertically. Using the equations given in Chapter 10, please calculate the correlation and the SAD coefficients for each of the eight (one pixel shift) directions.

10.4 When applying a two-dimensional autocorrelation to the figure shown herein, how many peaks are expected to appear in the output image?

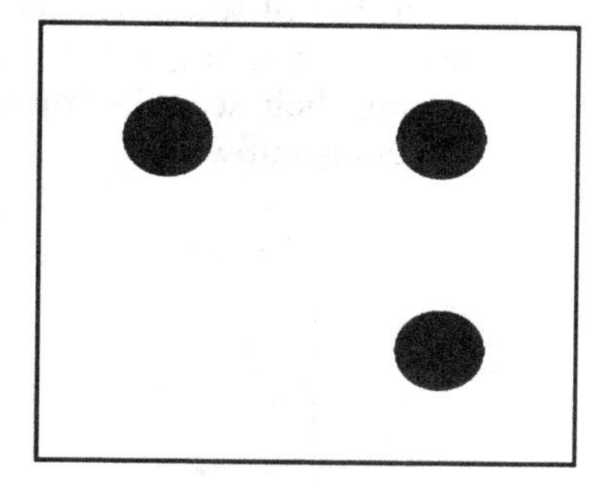

A. 3
B. 7
C. 8
D. 9
E. None of the above

10.5 An ultrasound signal is sent from Transducer "a" to Receiver "b" through a cylindrical ultrasound phantom full of water in which the speed of sound is C_2. A plastic rectangular block of thickness L is placed in the phantom as shown in the diagram. The speed of sound in plastic is C_1 and is faster than the speed of sound in water $C_1 > C_2$. How much sooner does a signal arrive when traveling from Transducer "a" to Receiver "b" than from Transducer "c" to Receiver "d"?

10.6 For the same setup, the echoes received by Transducer "a" were also recorded. The time elapse between the registered first two echoes was

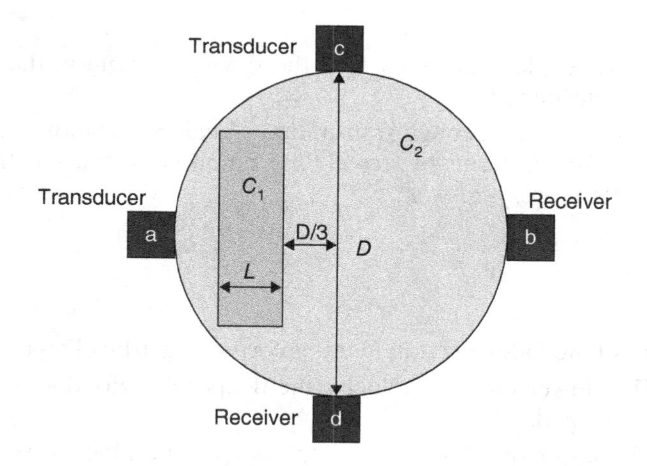

A. $(D - D/3 - L) \cdot 1/C_2 + L/C_1$
B. $(D - D/3 - L) \cdot 2/C_2 + 2L/C_1$
C. L/C_1
D. $2L/C_1$
E. None of the above

10.7 Consider the experimental setup shown herein, where an L-shaped object is immersed in water. The speed of sound in the object is higher than that of water. A pair of transducers are scanning the object from both sides along the Y axis. The transducers retain the light of sight during their scanning motion. The distance between the pair of transducers is known.

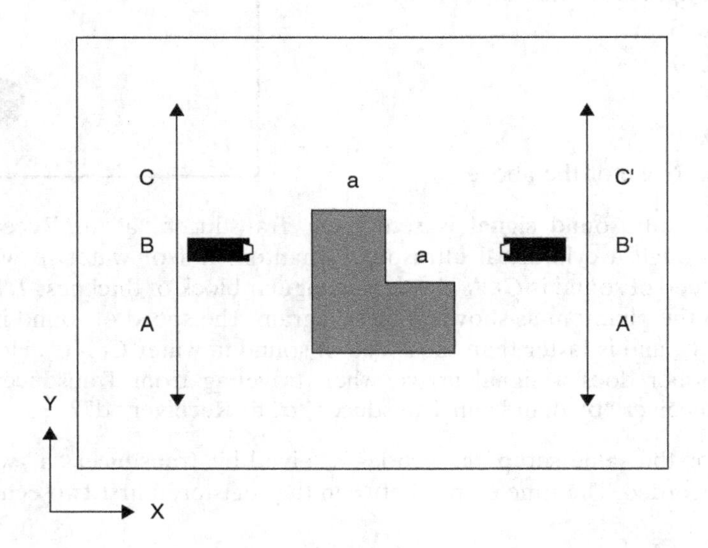

A. Please explain explicitly how the speed of sound in the object can be determined.

B. Compare the through-transmitted signals at locations A–A′, B–B′, and C–C′ in terms of arrival time, amplitudes, and spectral pattern. Explain every answer!

CHAPTER 11

11.1 Which of the following statements is correct regarding Doppler imaging?

A. The slower the flow velocity, the deeper the blood vessels that can be imaged.

B. The faster the flow velocity, the deeper the blood vessels that can be imaged.

C. The deeper the blood vessels that need to be imaged, the higher the required PRF.

D. The PRF must be at least twice as high as the transmitted frequency.

E. There are two correct answers above.

11.2 What is the maximal velocity that could be measured using echo-Doppler for a target located at a distance D inside the tissue, given that the transmission frequency is f, the speed of sound is C, and there are other reflecting objects along the way?

A. $\frac{C^2}{2Df}$

B. $\frac{C^2}{8Df}$

C. $\frac{2D}{1/f}$

D. $D\sqrt{\frac{C}{2D}f}$

E. None of the above

11.3 In a special Doppler measurement, a group wave which is comprised of two frequencies is transmitted. The first frequency is f_1 and the second is $f_2 = 2f_1$. The wave encounters a blood vessel for which the flow velocity is V. The average detected frequency will be

A. $\left[\frac{2V \cdot f1 \cdot \cos(\theta)}{C}\right]$

B. $\frac{3}{2} \cdot f1$

C. $\frac{3}{2} f1 \cdot \left[1 + \frac{2V\cos(\theta)}{C}\right]$

D. $f1 \cdot \left[1 + \frac{2V\cos(\theta)}{C}\right]$

E. None of the above

11.4 In color Doppler imaging after applying quadrature demodulation for the signals acquired from a certain pixel, the following values were obtained: $Q = [1 + j, 2 + j, 3 + j]$. In this setup, red color designates a negative flow and blue color designates a positive flow. Also, a threshold is set on flow velocity so that if $|V| \leq \text{const} \cdot 0.001$ (where the constant includes all relevant factors as explained in Chapter 11), this pixel is assigned a "no-flow" value. Which of the following statements is correct?

A. The pixel will be colored red.

B. The pixel will be colored blue.

C. The pixel will be assigned a "no-flow" value.

D. The flow velocity in that pixel is $V = \text{const} \cdot \arctan(5)$.

E. It is impossible to know.

11.5 In cardiac research it is desired to measure the velocity of the left ventricular endocardial wall (inner wall) during systole (ejection phase). For that purpose a special metal bead was implanted in the endocardium. The wall contracts in the short axis plane with a radial velocity that is given by $V_R = V_0 \sin(\pi t/\tau)$ for $0 \le t \le \tau$ and along the longitudinal plane with a velocity that is given by $V_L = \frac{V_0}{3}\sin(\pi t/\tau)$ for $0 \le t \le \tau$, where τ is the systolic duration period. An ultrasonic transducer is placed in the short axis plane so that the radial motion of the bead is "head-on" with reference to the acoustic beam. Given that the transmission frequency is f_0 and that speed of sound is C, what will be the maximal frequency shift detected?

CHAPTER 12

12.1 In an ultrasonic treatment of an organ, a temperature elevation of 3 °C was measured in the treated tissue. Then the blood flow to that organ was temporarily stopped and a temperature elevation of 4 °C was measured. What is the ratio between the allowed transmission duration before and after stopping the blood?

A. $10^{0.6}$

B. The ratio is undefined since under temperature elevation of 6 °C there is no danger to the tissue.

C. $\log_{10}(4/3)$

D. None of the above

12.2 The danger of cavitation bubble formation:

A. Increases as the frequency used is increased

B. Decreases as the frequency used is increased

C. Is not related to the frequency at all

D. Depends only on the negative pressure produced

E. None of the above

12.3 An ultrasonic device for "body contouring" (a cosmetic procedure for ultrasonic fat removal) is based on the formation of cavitation bubbles within the fatty tissue. The cavitation bubbles collapse and tear the fatty cells membranes. As a result, the fatty cells die and the fat is removed from the treated area. The acoustic impedance of the fatty tissue is Z. You have four available transmission options to choose from. Which is the most suitable one for the task (assume that the transmission is a CW sinusoidal wave)?

A. $I = 100\,(\text{W/cm}^2)$ $f = 1\,\text{MHz}$

B. $I = 64\,(\text{W/cm}^2)$ $f = 2\,\text{MHz}$

C. $I = 81\,(\mathrm{W/cm^2})$ $f = 0.5\,\mathrm{MHz}$
D. $I = 36\,(\mathrm{W/cm^2})$ $f = 0.25\,\mathrm{MHz}$
E. All four options are equally suitable.

12.4 Which of the following statements regarding ultrasonic safety is not true?

A. The danger of acoustic induced heating in the tissue is increased as the transmission frequency is increased.
B. The lower the frequency used, the higher the allowed intensity.
C. Higher I_{SPTA} values are likely to occur when using "power Doppler" imaging.
D. Safety indices are more limiting when applying ultrasonic imaging to the eyes.
E. There is only one correct answer above.

12.5 A high-tech company has developed a special ultrasonic device for stopping extensive tissue bleeding during surgery. The device is comprised of an ultrasonic transducer attached to a conical "waveguide." The transducer transmits a continuous sinusoidal wave into the conical component. The conical component is attached on its other side to a needle that is inserted into the treated tissue. Given that: The upper surface area of the cone is A_0 and its lower area is A_1. The cone height is H. Its acoustic impedance is Z_1 and its attenuation coefficient (for amplitude) is α_1. The diameter of the needle is $\varnothing d$ and its length is L. The acoustic impedance of the needle is Z_2 and its attenuation coefficient (for amplitude) is α_2. The typical acoustic impedance for the tissue is Z_3.

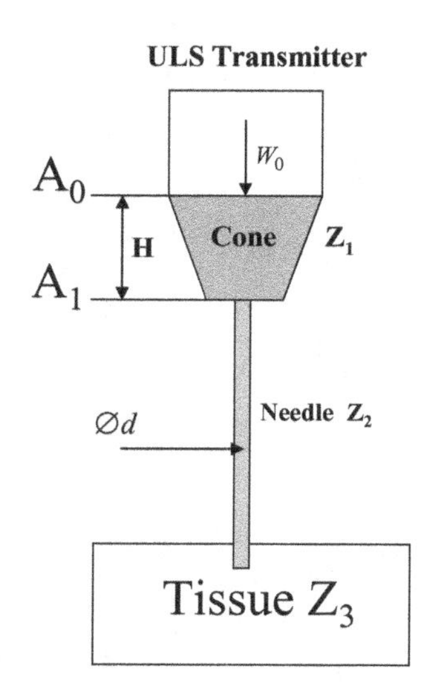

A. If the acoustic power inserted into the conical component is W_0, what is the acoustic power entering the tissue?

B. How much power will remain in the conical component?

C. What will be the temperature elevation of the needle after Δt seconds, if its density is ρ and its specific heat is $\mathbb{C}_p$?

D. What is the velocity of the needle tip (the side inserted into the tissue)?

APPENDIX C

ANSWERS TO EXEMPLARY PROBLEMS

CHAPTER 1

1.1 Answer: D
Explanation: Propagation along the positive direction of an axis is characterized by a negative sign of the wave number. Accordingly, the impinging wave is propagating along the positive X and Y direction, while the reflected wave will be propagating along the positive X direction and the *negative* Y direction.

1.2 Answer: B and C
Explanation: The direction of propagation for a shear wave is orthogonal to the direction of the particle's motion. Hence, the scalar product of the vectors equals 0. On the other hand, the direction of propagation for a longitudinal wave is parallel to the direction of the particle's motion. Hence, in this case the scalar product of the vectors will not equal 0.

1.3 Answer: C
Explanation: Recalling the general form for a harmonic wave, $u = Ae^{j[\omega \cdot t - \bar{K} \cdot \bar{R}]}$, we can write for this particular wave the relation $\bar{k} \cdot \bar{R} = K_x x + K_y y = x + y$, from which we conclude that $K_x = 1$ and $K_y = 1$. Thus the magnitude of the corresponding wave number is given by $|\bar{K}| = \sqrt{K_x^2 + K_y^2} = \sqrt{2}$. Finally, using the relation $k = 2\pi/\lambda$, it is concluded that: $\lambda = 2\pi/K = \sqrt{2}\pi$.

Basics of Biomedical Ultrasound for Engineers, by Haim Azhari

1.4 Answer: D

Explanation: Using the same derivation as in the previous question, it can be obtained that all three waves have the same wave number $K = 5$, and since the propagation speed in a nondispersive medium is given by $C = \omega/K$, it follows that all three waves have the same speed, which is given by $C = 5\pi/5 = \pi$.

1.5 Answer: A

Explanation: From the relations listed above, the corresponding wave number is given by $|\bar{K}| = \sqrt{K_x^2 + K_y^2} = \sqrt{2}$ and the speed of propagation is given by $C = \omega/K = \sqrt{2}/\sqrt{2} = 1$.

The distance that the wave has to travel in order to reach point P is the projection of its distance from the origin along the wave propagation direction. In order to calculate this projection, we first have to find the unit vector along the wave propagation direction, which value is given by $\hat{n}_K = (\sqrt{2}/2)\hat{\imath} + (\sqrt{2}/2)\hat{\jmath}$. Using vector notation, the point is described by the location vector $\bar{R} = 8.66\hat{\imath} + 5\hat{\jmath}$. Hence the distance that the wave front has to travel in order to reach this point is given by $S = \hat{n}_K \cdot \bar{R} = 9.659$. And since the speed of propagation is 1, it follows that the required time is 9.659 sec.

1.6 Answer: $\Delta t = \dfrac{(x_0 + 2\Delta x)}{c}$

Explanation: As the phase of the wave is given by $2\pi f \cdot t - k \cdot x$, the required condition for equal phases is given by

$$2\pi f_1 \cdot \Delta t - k_1 x_0 = 2\pi f_2 \cdot \Delta t - k_2(x_0 + \Delta x)$$

Recalling that the wave number is given by $k = 2\pi/\lambda$ and that $2f_1 = f_2$, it follows that

$$k_1 = \frac{2\pi}{C} \cdot f_1 \quad \text{and} \quad k_2 = \frac{4\pi}{C} \cdot f_1$$

Substituting these values into the above equation yields

$$2\pi f_1 \cdot \Delta t - \frac{2\pi}{C} \cdot f_1 \cdot x_0 = 4\pi f_1 \cdot \Delta t - \frac{4\pi}{C} \cdot f_1 \cdot (x_0 + \Delta x)$$

Dividing both sides by $2\pi f_1$ and multiplying by c yields

$$c \cdot \Delta t - x_0 = 2c \cdot \Delta t - 2(x_0 + \Delta x)$$

from which we derive that the required time elapse is given by

$$\boxed{\Delta t = \frac{(x_0 + 2\Delta x)}{c}}$$

1.7 Answer: $\Delta x = \frac{1}{4}$

Explanation: Since both waves have the same temporal and spatial frequencies, a standing wave is formed. As explained, the standing wave equation is comprised of a spatial function multiplied by a temporal function. Referring to the first function, the following condition should hold for any stationary point:

$$\begin{aligned} &\cos(4\pi x) = 0 \\ &\Rightarrow 4\pi x = \pi\left(n + \frac{1}{2}\right), \qquad n = 0, 1, 2, \ldots \end{aligned}$$

Taking any two consecutive values for n and subtracting the obtained distances yields the above answer.

1.8 Answer: $C \cdot \hat{n}_K = \frac{20}{29}\hat{i} - \frac{30}{29}\hat{j} + \frac{40}{29}\hat{k}$

Explanation: First let us calculate the norm for the corresponding wave number

$$|\bar{K}| = \sqrt{K_x^2 + K_y^2 + K_z^2} = \sqrt{2^2 + 3^2 + 4^2} = \sqrt{29}$$

and using the relation $C = \omega/K$ for a nondispersive medium the corresponding magnitude of the wave velocity is given by $C = 10/\sqrt{29}$.

The direction of propagation is given by a unit vector along the wave number, that is,

$$\hat{n}_K = \frac{+2}{\sqrt{29}}\hat{i} + \frac{-3}{\sqrt{29}}\hat{j} + \frac{+4}{\sqrt{29}}\hat{k}$$

(Note that a wave propagating along the positive direction of an axis is designated by a negative sign for the corresponding wave number component.)

The corresponding vector of propagation is therefore given by

$$C \cdot \hat{n}_K = \frac{20}{29}\hat{i} - \frac{30}{29}\hat{j} + \frac{40}{29}\hat{k}$$

1.9 Answer: $u(t,S) = \frac{A \cdot r_0}{S} \cdot e^{j(\omega \cdot t - k \cdot S)}$

Explanation: First let us write the corresponding expression for this wave at the specified reference location:

$$u(t, r_0) = \frac{A}{r_0} \cdot e^{j(\omega \cdot t - k \cdot r_0)} = u_0 \cdot e^{j(\omega \cdot t - k \cdot r_0)}$$

From this expression we can derive the maximal amplitude which is given by

$$u_0 = \frac{A}{r_0}, \qquad \Rightarrow A = u_0 \cdot r_0$$

Hence, at location S, its value is given by

$$u(t,S) = \frac{A \cdot r_0}{S} \cdot e^{j(\omega \cdot t - k \cdot S)}$$

1.10 Answer: $u_S(r) = \int_0^1 \frac{e^{jK_0|r-r'|}}{4\pi|r-r'|} \cdot k_0^2 \left[\left(\frac{C_0}{C_0 + \alpha \cdot r'} \right)^2 - 1 \right] \cdot A \sin(2\pi \cdot r') \cdot dr'$

Explanation: Substituting for the refraction index $n(r) = C_0/C(r)$, we obtain the following object function:

$$F(r) = 0 \qquad \text{for} \quad r \le 0$$

$$F(r) = k_0^2 \left[\left(\frac{C_0}{C_0 + \alpha \cdot r} \right)^2 - 1 \right] \qquad \text{for} \quad 0 \le r \le 1$$

$$F(r) = 0 \qquad \text{for} \quad r > 1$$

Substituting this term and the expression for the irradiating field, the first Born approximation is implicitly given by

$$u_S(r) = \int_0^1 \frac{e^{jK_0|r-r'|}}{4\pi|r-r'|} \cdot k_0^2 \left[\left(\frac{C_0}{C_0 + \alpha \cdot r'} \right)^2 - 1 \right] \cdot A \sin(2\pi \cdot r') \cdot dr'$$

CHAPTER 2

2.3 Answer: D

Explanation: The energy of a propagating transverse wave in a string is given by

$$E = \frac{1}{2}\rho \cdot A^2 \cdot \omega^2$$

Hence, it depends on the density of each string as well as on the amplitude of each wave. The exact term for energy conservation as shown in Chapter 2 is given by

$$Z_1 A_1^2 = Z_1 A_2^2 + Z_2 B^2$$

where Z_1 and Z_2 designate the acoustic impedance ($Z \equiv \rho \cdot C$) of the two strings, respectively.
Thus none of the given terms is correct.

2.4 Answer: B
Explanation: The instantaneous traveling speed for a transverse wave in a tensed string is given by

$$C(t) = \sqrt{\frac{\text{Tension}}{\text{Density}}}$$

Substituting the values for this specific case yields

$$C(t) = \sqrt{\frac{a \cdot t^2}{\rho}} = \sqrt{\frac{a}{\rho}} \cdot t$$

The corresponding distance which this wave will travel in the specified time elapse is therefore given by

$$S = \int_1^2 C(t) \cdot dt = \sqrt{\frac{a}{\rho}} \cdot \int_1^2 t \cdot dt = \frac{3}{2}\sqrt{\frac{a}{\rho}}$$

2.5 Answer: A
Explanation: Considering the ratio between the two modulus of elasticity it follows that

$$C_1 = \sqrt{\frac{E_1}{\rho_1}} = 2\sqrt{\frac{E_2}{\rho_1}} \quad \text{and} \quad C_2 = \sqrt{\frac{E_2}{\rho_2}}$$

The traveling time for the wave will therefore be

$$t = \frac{L}{C_1} + \frac{T}{C_2} = \frac{L}{2\sqrt{\frac{E_2}{\rho_1}}} + \frac{T}{\sqrt{\frac{E_2}{\rho_2}}} = \frac{1}{\sqrt{E_2}} \cdot \left(\frac{L}{2}\sqrt{\rho_1} + T\sqrt{\rho_2}\right)$$

2.6 Answer: D

2.10 Answer: 99.9%

CHAPTER 3

3.1 Answer: $I = \dfrac{V \cdot E \cdot c}{2L_0 \cdot \gamma}$

Explanation: From Hooke's Law it follows that

$$\tau = E \cdot \varepsilon$$

$$\varepsilon = \frac{\Delta L}{L_0}, \qquad \text{where} \quad \tau = \frac{F}{A} = \Gamma$$

(and where A = surface area, F = force, Γ = acoustic pressure)

$$\Rightarrow \Delta L = \varepsilon \cdot L_0 = \frac{\Gamma}{E} \cdot L_0$$

Thus, the corresponding voltage is related to the acoustic pressure through

$$V_0 = \frac{\Gamma}{E} \cdot L_0 \cdot \gamma$$

Since the acoustic impedance of the crystal is much higher than that of water, we can assume that it is a prefect reflector and that the acoustic pressure is given by

$$\Gamma = \frac{2I}{c}$$

Substituting this value for the intensity yields

$$I = \frac{\Gamma \cdot c}{2} = \frac{V \cdot E \cdot c}{2L_0 \cdot \gamma}$$

3.2 Solution:

(a) Since the acoustic impedance of air is negligible as compared to that of the water, the interface serves as a perfect reflector for the ultrasonic waves. Thus the corresponding acoustic pressure will cause the water level at the beam incidence location to ride. Equating the resulting hydrostatic presure to the acoustic pressure yields

$$P_{\text{water}} = \rho \cdot g \cdot d = \Gamma = \frac{2 \cdot I}{C} \Rightarrow I = \frac{\rho \cdot g \cdot C \cdot d}{2}$$

(b) The corresponding amplitude is given from the following relation:

$$I = \frac{1}{2} \cdot \rho \cdot \xi^2 \cdot (2\pi f)^2 \cdot c$$

$$\Rightarrow \xi = \sqrt{\frac{2I}{\rho \cdot (2\pi f)^2 \cdot c}} = \sqrt{\frac{2 \cdot \frac{\rho gcd}{2}}{\rho \cdot (2\pi f)^2 \cdot c}} = \sqrt{\frac{gd}{(2\pi f)^2}} = \frac{\sqrt{gd}}{2\pi f}$$

(c) Hint for solving this part: Note that now the water surface is not a perfect reflector, thus only part of the energy is through-transmitted to the liquid upper surface.

3.3 Solution: $(I_0 \cdot A \cdot D)/(C_0 \cdot m \cdot V)$
Explanation: From Chapter 3 we know that the radiation pressure is given by

$$\Gamma = \frac{I_0}{C_0}$$

Therefore, the acoustic radiation force applied on the emboli along the beam direction is given by

$$F = \frac{I_0}{C_0} \cdot A$$

This force will accelerate the emboli along the vertical direction by

$$a = \frac{F}{m} = \frac{I_0}{C_0} \cdot \frac{A}{m}$$

The resulting velocity that the emboli will accumulate is proportional to the time it will be subjected to the acoustic radiation force, and this time it is given by

$$\Delta t = \frac{D}{V}$$

Consequently, the accumulated vertical velocity of the emboli will be

$$V_y = a \cdot \Delta t = \frac{I_0}{C_0} \cdot \frac{A}{m} \cdot \frac{D}{V}$$

3.4 Solution: $I = C_0 \cdot m \cdot g \cdot sin(\theta)/2 \cdot A$

3.5 Solution: $T2 \cdot 1/\rho 0$

CHAPTER 4

4.1 Answer: B

Explanation: The propagation velocity for longitudinal waves in a thin rod is given by (see Chapter 2)

$$C_{\text{long}} = \sqrt{\frac{E}{\rho}}$$

And from the relations given in Chapter 4, we know that for an isotropic material, Young's modulus is given by

$$E = \frac{\mu(3\lambda' + 2\mu)}{(\lambda' + \mu)}$$

Substituting the given relation of

$$\lambda' = 2\mu$$

we obtain

$$E = \frac{\mu(3 \cdot 2\mu + 2\mu)}{(2\mu + \mu)} = \frac{8\mu}{3}$$

From Chapter 4 we also know that the velocity for shear waves in an isotropic material is given by

$$C_{\text{shear}} = \sqrt{\frac{\mu}{\rho}}$$

Thus we can find the shear modulus from

$$\mu = \rho \cdot C_{\text{shear}}^2$$

which leads to the answer

$$C_{\text{long}} = \sqrt{\frac{8}{3}} \cdot C_{\text{shear}}$$

4.2 Answer: Three type of waves

Explanation: Using Eq. (4.42) and substituting for the elastic constant values, the following determinant is obtained:

$$\begin{vmatrix} (\mu\cdot K^2-\omega^2\cdot\rho) & \frac{\mu}{2}\cdot K^2 & 0 \\ \frac{\mu}{2}\cdot K^2 & (\mu\cdot K^2-\omega^2\cdot\rho) & 0 \\ 0 & 0 & (4\mu\cdot K^2-\omega^2\cdot\rho) \end{vmatrix}=0$$

Writing the terms explicitly yields

$$(\mu\cdot K^2-\omega^2\cdot\rho)(\mu\cdot K^2-\omega^2\cdot\rho)(4\mu\cdot K^2-\omega^2\cdot\rho) -\left(\frac{\mu}{2}\cdot K^2\right)\left(\frac{\mu}{2}\cdot K^2\right)(4\mu\cdot K^2-\omega^2\cdot\rho)=0$$

which, after rearrangement, can be written as

$$\left[(\mu\cdot K^2-\omega^2\cdot\rho)+\frac{\mu}{2}\cdot K^2\right]\left[(\mu\cdot K^2-\omega^2\cdot\rho)-\frac{\mu}{2}\cdot K^2\right]\cdot(4\mu\cdot K^2-\omega^2\cdot\rho)=0$$

This implies that there are three different solutions for K or actually three different types of waves. Recalling that $C^2=(\omega/K)^2$, the first wave velocity can be obtained from

$$\left[(\mu\cdot K^2-\omega^2\cdot\rho)+\frac{\mu}{2}\cdot k^2\right]=0,$$
$$\Rightarrow C_1=\sqrt{\frac{3\mu}{2\rho}}$$

The second wave velocity can be obtained from

$$\left[(\mu\cdot K^2-\omega^2\cdot\rho)-\frac{\mu}{2}\cdot K^2\right]=0,$$
$$\Rightarrow C_2=\sqrt{\frac{\mu}{2\rho}}$$

The third wave velocity can be obtained from

$$(4\mu\cdot k^2-\omega^2\geq\rho)=0$$
$$\Rightarrow C_3=2\sqrt{\frac{\mu}{\rho}}$$

4.3 Answer: $E = 3\rho \cdot C^2_{Shear}$

Explanation: As recalled from Chapter 4, the speed of sound for shear waves in an isotropic material is given by

$$C_{\text{shear}} = \sqrt{\frac{\mu}{\rho_0}}$$

Also we know that the shear modulus μ is related to Young's modulus E through the relation

$$\mu = \frac{E}{2(1+\upsilon)}$$

where υ is Poisson's ratio. Substituting a Poisson's ratio of 0.5, we obtain

$$\mu = \frac{E}{3}$$

Thus, from this relation we obtain

$$E = 3\rho \cdot C^2_{\text{shear}}$$

4.4 Answer: $B = \rho \cdot C^2_{\text{long}} - \frac{4}{3} C^2_{\text{shear}}$

Explanation: As recalled, the speed of sound for shear waves in an isotropic material is given by

$$C_{\text{shear}} = \sqrt{\frac{\mu}{\rho_0}}$$

Hence, the shear modulus is given by

$$\mu = \rho_0 \cdot C^2_{\text{shear}}$$

The speed of sound for longitudinal waves in an isotropic material is given by

$$C_{\text{long}} = \sqrt{\frac{(\lambda' + 2\mu)}{\rho_0}}$$

Thus, it follows that

$$\lambda' = \rho_0 \cdot C^2_{\text{long}} - 2\rho_0 \cdot C^2_{\text{shear}}$$

Recalling that the bulk modulus is given by

$$B = \lambda' + \frac{2}{3}\mu$$

We obtain the answer,

$$B = \rho \cdot \left(C_{\text{long}}^2 - \frac{4}{3} C_{\text{shear}}^2 \right)$$

4.5 Answer: $\Delta\varphi = \dfrac{2\pi f \cdot T}{\sqrt{\dfrac{1}{\rho} \cdot \left[\left(\dfrac{2G - E}{E - 3G} \right) - 2G \right]}}$

CHAPTER 5

5.1 Answer: E
Explanation: In Chapter 3 we have the following relation for the intensity of a ultrasonic harmonic wave:

$$I = \frac{1}{2} \rho \cdot \eta_0^2 \cdot \Omega^2 \cdot c$$

where η_0 is the amplitude and Ω is the angular frequency. From this relation we can see that if the frequency is doubled while the amplitude is kept the same, then the intensity is augmented by fourfold, that is, $I_{\text{new}} = 4I_{\text{old}}$.
Thus, the new value in decibels is given by

$$I_{\text{new}}[db] = 10 \log_{10}\left(\frac{4I_{\text{old}}}{I_{\text{reference}}} \right) = 10 \log_{10}(4) + 10 \log_{10}\left(\frac{I_{\text{old}}}{I_{\text{reference}}} \right) = 106[db]$$

5.2 Answer: C

5.3 Answer: A

5.4 Answer: $e^{-2 f_0 \cdot \Delta f \cdot L}$
Explanation: In order to find the highest- and lowest-frequency component of the transmitted signal, we must first study its spectrum by applying a Fourier transform, that is, $F\{S(t)\}$. From the spectrum we realize that the signal in the frequency domain is a RECT function for which the lowest frequency is $f_{\min} = f_0 - \Delta f/2$ and the highest frequency is $f_{\max} = f_0 + \Delta f/2$ and both have the same amplitude (say A_0). After traveling a distance L within the tissue, the amplitude of the highest-frequency component will be

$$A(f_{\max}, L) = A_0 e^{-L \cdot (f_0 + \Delta f/2)^2}$$

The amplitude of the lowest-frequency component will be

$$A(f_{\min}, L) = A_0 e^{-L\cdot(f_0-\Delta f/2)^2}$$

Thus, the ratio for both components is given by

$$\frac{A(f_{\max}, L)}{A(f_{\min}, L)} = e^{-2L\cdot f_0\cdot\Delta f}$$

5.5 Answer: $f_{\text{optimal}} = \dfrac{2}{\alpha_1\cdot F}$

Explanation: Assuming that the transmission intensity is I_0 and that the traveling distance for each acoustic ray to the stone is approximately F, then the intensity at the stone (the focal point) is given by

$$I_{\text{focus}} = I_0 \frac{4\pi R^4}{\lambda^2\cdot F^2}\cdot e^{-(\alpha_0+\alpha_1 f)F}$$

Substituting for λ and rearranging, we obtain

$$I_{\text{focus}} = I_0 \frac{4\pi R^4}{C^2\cdot F^2}\cdot e^{-\alpha_0 F}\cdot f^2\cdot e^{-\alpha_1 f\cdot F} \triangleq B\cdot f^2\cdot e^{-\alpha_1 f\cdot F}$$

Taking the derivative of this relation with respect to the frequency f and equating to zero for maximal value yields

$$\frac{\partial(I_{\text{focus}})}{\partial f} = B\cdot\left[2f\cdot e^{-\alpha_1 f\cdot F} - \alpha_1\cdot F\cdot f^2\cdot e^{-\alpha_1 f\cdot F}\right] = 0,$$
$$\Rightarrow 2f = f^2\cdot\alpha_1\cdot F$$

Hence, the optimal frequency is

$$f_{\text{optimal}} = \frac{2}{\alpha_1\cdot F}$$

CHAPTER 6

6.1 Answer: A
Explanation: Considering the conditions for an optimal matching layer, we know that one of the requirements is that the thickness of such a layer should be

$$d = (2n-1)\frac{\lambda}{4}$$

Substituting this condition for the wave frequency yields

$$f = (2n-1)\frac{C_2}{4d}$$

which is exactly the case described in this question. From Chapter 6, we also know that the corresponding acoustic impedance should be

$$Z_2 = \sqrt{Z_1 \cdot Z_3} = 2 \cdot Z_0$$

Thus, the required density is

$$\rho_2 = 2Z_0/C_2$$

6.2 Answer: C

6.3 Answer: B
Explanation: From Snell's Law it can be easlity derived that the ratio between the incident and transmistted angles is given by

$$\sin(\theta_T) = 2 \cdot \sin(\theta_i)$$

Thus, the first critical angle is at

$$\theta_i = \pi/6 = (30°)$$

And since in our case

$$\theta_i = \pi/4 > \pi/6$$

a total reflection will occur.

6.4 Answer: E
Explanation: The speed of a longitudinal wave in a thick isotropic solid material is given by

$$C_{\text{long}} = \sqrt{\frac{(\lambda' + 2\mu)}{\rho_0}}$$

and the shear wave velocity is given by

$$C_{\text{shear}} = \sqrt{\frac{\mu}{\rho_0}}$$

From the given ratio $\lambda' = 2\mu$ it follows that

$$C_{\text{long}} = \sqrt{\frac{(2\mu+2\mu)}{\rho_0}} = 2\sqrt{\frac{\mu}{\rho_0}} = 2\cdot C_{\text{shear}}$$

Applying Snell's Law to this material yields

$$\frac{C_{\text{long}}}{\sin(\theta_L)} = \frac{2\cdot C_{\text{shear}}}{\sin(\theta_L)} = \frac{C_{\text{shear}}}{\sin(\theta_T)}$$

Hence, the correct answer is

$$\sin(\theta_L) = 2\cdot\sin(\theta_T)$$

Thus, the matching answer is "none of the above."

6.5 Item A:
The intensity reaching the liquid–solid interface is given by

$$I_1 = I_0\cdot e^{-2S_1(\alpha_0+\alpha_1\cdot f_1)}\left(1+e^{-2S_1\alpha_1\cdot f_1)}\right)$$

The intensity reaching the solid–vaccum interface is given by

$$I_2 = I_1\cdot T'_{12}\cdot e^{-2S_2(\gamma_0+\gamma_1\cdot f_1)}\left(1+e^{-2S_1\gamma_1\cdot f_1)}\right)$$

where the transmission coefficient is given by

$$T'_{12} = \frac{Z_1}{Z_2}T^2 = \frac{4\cdot Z_1\cdot Z_2}{(Z_1+Z_2)^2}$$

Item B:
The first echo will be reflected from the liquid/solid interface and its intensity is given by

$$I_{\text{echo\_1}} = I_0\cdot\left(\frac{Z_2-Z_1}{Z_2+Z_1}\right)^2\cdot e^{-4S_1(\alpha_0+\alpha_1\cdot f_1)}.\left(1+e^{-2S_1\alpha_1\cdot f_1)}\right)^2$$

The second echo will be reflected from the solid–vaccum interface and it intensity is given by

$$I_{\text{echo\_2}} = I_0\cdot\left(\frac{4Z_2Z_1}{(Z_2+Z_1)^2}\right)^2\cdot\left[e^{-4S_1(\alpha_0+\alpha_1\cdot f_1)}.\left(1+e^{-2S_1\alpha_1\cdot f_1)}\right)^2\right]$$
$$\cdot\left[e^{-4S_2(\gamma_0+\gamma_1\cdot f_1)}.\left(1+e^{-2S_2\gamma_1\cdot f_1)}\right)^2\right]$$

Item C:
The corresponding speed of sound in the liquid is given by

$$C = \frac{1}{\sqrt{\beta\rho}}$$

and since the acoustic impedance in this case is

$$Z = \rho \cdot C$$

it follows that the speed of sound in the liquid medium is given by

$$C = \frac{1}{Z \cdot \beta}$$

And the corresponding time to echo is given by

$$\Delta t = \frac{2S_1}{C} = 2S_1 \cdot Z \cdot \beta$$

CHAPTER 7

7.1 D

7.2 C
Explanation: For focusing plannar waves it is advantageous to use a lens that has a flat surface (normal angle of incidence) on the side facing the incoming waves. For such a spherical lens the relation between the radius of curvature and the focal distance is given by

$$F = \frac{R \cdot n_1}{n_1 - n_2}$$

where $n_1 = 1/C_1$ is the refraction index in the medium (water in our case) and $n_2 = 1/C_2$ is the refractive index in the lens. For the specific lens material use in this case $n_2 = 1/2C_1$, substituting this relation into the above equation and rearranging the terms yields

$$R = \frac{F(n_1 - n_2)}{n_1} = \frac{F\left(\frac{1}{C_1} - \frac{1}{2C_1}\right)}{\frac{1}{C_1}} = \frac{F}{2}$$

7.3 B
Explanation: For conics geometries the following relations hold:

$$\frac{r}{D} = \varepsilon, \qquad \Rightarrow r = \varepsilon \cdot D$$

As recalled from Chapter 7, in order for an ellipsoidal shape to serve as a focusing lens it is required that the eccentricity index should equal the refractive indices ratio, that is,

$$\varepsilon = \frac{n_2}{n_1} = \frac{1}{C_1} \cdot \frac{C_0}{1}$$

Thus, after substitution we obtain

$$\Rightarrow r = \frac{C_0}{C_1} \cdot D$$

7.4 D

7.5 Item A: $R = \dfrac{S_2 \cdot \left(\sqrt{\beta \cdot \rho_2} - \dfrac{1}{C_1} \right)}{\sqrt{\beta \cdot \rho_2}}$

Item B: $I_{\text{sphere}} = I_0 \cdot T_{01} \cdot T_{12} \cdot e^{-2\alpha_1 H} \cdot e^{-2\alpha_2 S_2} \cdot G$
where the two (energy) transmission coefficients are given by

$$T_{01} = \frac{4\rho_0 C_0 \cdot \rho_1 C_1}{(\rho_0 C_0 + \rho_1 C_1)^2},$$

$$T_{12} = \frac{4\sqrt{\dfrac{\rho_2}{\beta}} \cdot \rho_1 C_1}{\left(\sqrt{\dfrac{\rho_2}{\beta}} + \rho_1 C_1 \right)^2}$$

and the gain factor of the lens is given by

$$G = 4\pi \left(\frac{a^2}{\lambda \cdot S_2} \right)^2 = 4\pi \left(\frac{a^2 \cdot \sqrt{\beta \rho_2} \cdot f_0}{S_2} \right)^2$$

Item C: Given that the acoustic impedance of the sphere is much larger than that of the tissue, we may consider the sphere as a perfect reflector. Hence, the acoustic pressure applied on the sphere is given by (see Chapter 3)

$$\Gamma = \frac{2 \cdot I_{\text{sphere}}}{1/\sqrt{\beta \cdot \rho_2}}$$

where I_{sphere} is given above.

The force acting on the sphere is generated by the pressure applied on the left half surface of the metallic sphere. In this case the "hydraulic cross section" of the sphere is simply its cross section. Thus the force is given by

$$F_{\text{sphere}} = \frac{\pi}{2} \cdot d^2 \cdot I_{\text{sphere}} \cdot \sqrt{\beta \cdot \rho_2}$$

CHAPTER 8

8.1 Answer: C

8.2 Answer: D

8.3 Answer: D

Explanation: Using the equation for the distance to the near field, we have

$$Z_{NF} = \frac{4a^2 - \lambda^2}{4\lambda} = 25 \cdot \lambda$$
$$\Rightarrow 4a^2 = 101\lambda^2$$
$$\Rightarrow a = \lambda\sqrt{\frac{101}{4}} = \frac{\sqrt{101}}{2} \cdot \frac{C}{f}$$

8.4 Solution:

Item A: As recalled, the pressure field for a point source is given by

$$P_s(r) = A_0 \cdot \frac{e^{j(\omega t - kr)}}{r} ds$$

Given the pressure amplitude at distance a, we can find the constant $A_0 \cdot ds$ from

$$|P_s(a)| = \frac{(A_0 \cdot ds)}{a} = P_0,$$
$$\Rightarrow P_s(r) = (a \cdot P_0) \cdot \frac{e^{j(\omega t - kr)}}{r}$$

Recalling that the wave number is given by

$$k = \frac{2\pi}{\lambda} = \frac{2\pi f}{C}$$

And that the speed of sound in a fluid is given by

$$C = \frac{1}{\sqrt{\beta\rho}}$$

The explicit term for the pressure at distance r is given by

$$P_s(r) = (a \cdot P_0) \cdot \frac{e^{j2\pi f\left(t - r\cdot\sqrt{\beta\rho}\right)}}{r}$$

For the central source the distance is given by

$$r = a + d$$

And for the other two sources, we have

$$r = \sqrt{9a^2 + d^2}$$

Substituting these values while ignoring attenuation, we obtain

$$P(d) = (a \cdot P_0) \cdot \frac{e^{j2\pi f_1\left(t-(a+d)\cdot\sqrt{\beta\rho}\right)}}{(a+d)} + 2(a \cdot P_0) \cdot \frac{e^{j2\pi f_2\left(t-\sqrt{9a^2+d^2}\cdot\sqrt{\beta\rho}\right)}}{\sqrt{9a^2+d^2}}$$

Item B: For the case where $d = 4a$ we obtain for all sources the same distance $5a$. And if also $f_1 = f_2$, it follows that

$$P(d) = \left(\frac{3}{5} \cdot P_0\right) \cdot e^{j2\pi f_1\left(t-(5a)\cdot\sqrt{\beta\rho}\right)}$$

When including the effect of attenuation we obtain

$$P(d) = \left(\frac{3}{5} \cdot P_0\right) \cdot e^{j2\pi f_1\left(t-(5a)\cdot\sqrt{\beta\rho}\right)} \cdot e^{-5a(\alpha_0+\alpha_1)}$$

Item C: The particles velocity is given by

$$u = \frac{P(d)}{Z}$$

where the acoustic impedance in this case is given by

$$Z = \frac{j10\pi f \cdot a \cdot \rho\left(1 - j2\pi f \cdot \sqrt{\beta\rho}\right)}{1 + 100\pi^2 \cdot f^2 \cdot \beta\rho \cdot a^2}$$

Item C: The intensity is given by

$$I = \mathrm{Re}\left\{\frac{P^2(d)}{2Z}\right\}$$

8.5 Answer: *(0.61 · C · R)/(d · a)*

CHAPTER 9

9.1 Answer: **C**

9.2 Answer: **A**

9.3 Answer: **D**
Explanation: We need low PRF to give enough time for the waves to travel back and forth, and we need low frequency to minimize attenuation.

9.4 Answer: **D**
Explanation: For the cavity diameter calculation we need the echoes from the myocardium–blood interfaces.

9.5 Answer: $\dfrac{1.22 \cdot a}{\sqrt{1-\left(\dfrac{0.61 \cdot C}{a \cdot f_0}\right)^2}}$

Explanation: The best lateral resolution can be obtained in that case at the end of the near field zone Z_{near}. The lateral resolution is determined by the width (defined as $2d$) of the main lobe at that point. Using the relation given in Chapter 8, we can write

$$\sin\theta = \frac{d}{\sqrt{d^2 + Z_{\text{near}}^2}} = \frac{0.61 \cdot \lambda}{a} = \frac{0.61 \cdot C}{a \cdot f_0}$$

where θ is the main lobe's head angle.

From this relation we can extract the value for d, as follows:

$$\frac{d^2}{d^2 + Z^2} = \left(\frac{0.61 \cdot C}{a \cdot f_0}\right)^2$$

$$\Rightarrow d^2 = d^2\left(\frac{0.61 \cdot C}{a \cdot f_0}\right)^2 + Z_{\text{near}}^2\left(\frac{0.61 \cdot C}{a \cdot f_0}\right)^2$$

$$\Rightarrow d^2\left[1 - \left(\frac{0.61 \cdot C}{a \cdot f_0}\right)^2\right] = Z_{\text{near}}^2\left(\frac{0.61 \cdot C}{a \cdot f_0}\right)^2$$

$$\Rightarrow d = \frac{Z_{\text{near}}\left(\dfrac{0.61 \cdot C}{a \cdot f_0}\right)}{\sqrt{1 - \left(\dfrac{0.61 \cdot C}{a \cdot f_0}\right)^2}}$$

The distance for the end of the near field can be calculated from

$$Z_{\text{near}} = \frac{a^2}{\lambda} = \frac{a^2 \cdot f_0}{C}$$

And since the lateral resolution equals to $2d$, the answer is

$$2d = \frac{2 \cdot \frac{a^2 \cdot f_0}{C} \cdot \left(\frac{0.61 \cdot C}{a \cdot f_0} \right)}{\sqrt{1 - \left(\frac{0.61 \cdot C}{a \cdot f_0} \right)^2}} = \frac{1.22 \cdot a}{\sqrt{1 - \left(\frac{0.61 \cdot C}{a \cdot f_0} \right)^2}}$$

CHAPTER 10

10.1 Answer: B
Explanation: Since only the first echo is needed to calculate the tissue acoustic impedance in this case, its value is not affected by the tissue attenuation.

10.2 Answer: B
Explanation: Using the following relation,

$$C_{\text{shear}} \approx \sqrt{\frac{E}{3 \cdot \rho_0}}$$

we can conclude that the upper layer is slightly stiffer. Thus, it will deform less than the second layer.

10.4 Answer: B

10.5 Answer: $L\left(\frac{1}{C_2} - \frac{1}{C_1} \right)$

10.6 Answer: D

CHAPTER 11

11.1 Answer: A

11.2 Answer: B
Explanation: Since there are other reflecting objects along the way, we cannot “interleave” pulses. Thus, the minimal time we have to wait is

$$\Delta t = \frac{2D}{C}$$

The maximal PRF is therefore

$$\text{PRF} = \frac{1}{\Delta t} = \frac{C}{2D}$$

From sampling theory it follows that the maximal frequency shift that could be detected is

$$\Delta f_{\max} = \frac{\text{PRF}}{2} = \frac{C}{4D}$$

Therefore, the maximal velocity that can be detected is given by

$$V_{\max} = \frac{\Delta f_{\max} C}{2f} = \frac{C^2}{8 \cdot D \cdot f}$$

11.3 Answer: C
Explanation: The received new frequency for the first component is

$$f_1^{\text{new}} = f_1 + \Delta f_1 = f_1 + \frac{f_1 \cdot 2\cos\theta}{C}$$

The received new frequency for the second component is

$$f_2^{\text{new}} = f_2 + \Delta f_2 = 2f_1 + \frac{2f_1 \cdot 2\cos\theta}{C}$$

Therefore, the average received new frequency is

$$\frac{3}{2} f_1 \cdot \left[1 + \frac{2V\cos(\theta)}{C}\right]$$

11.4 Answer: A
Explanation: Using the autocorrelation function approximation given in Chapter 11, the value for $R(1)$ is estimated from

$$R(1) \approx [Q_1^*, Q_2^*] \begin{bmatrix} Q_2 \\ Q3 \end{bmatrix} = Q_1^* \cdot Q_2 + Q_2^* \cdot Q_3$$

Substituting for the given values, we obtain

$$R(1) \approx 10 - 2j$$

Thus, the velocity in that pixel is given by

$$V = const \cdot \arctan(-0.2)$$

This implies that the flow has a negative value and the pixel will be colored red.

11.5 Answer: $\Delta f_{max} = \frac{2\sqrt{10}}{3} \cdot \frac{V_0 \cdot f_0}{C} \cdot \cos\left(\arctan\left(\frac{1}{3}\right)\right)$

Explanation: Since the bead has two velocity components its velocity vector has a magnitude of

$$V(t) = \sqrt{V_R^2 + V_L^2} = \frac{\sqrt{10}}{3} \cdot V_0 \sin\left(\frac{\pi t}{\tau}\right)$$

And its angular orientation is given by

$$\theta = \arctan\left(\frac{V_L}{V_R}\right) = \arctan\left(\frac{1}{3}\right)$$

The maximal velocity occurs at $t = \tau/2$ when the sin term equals 1, that is,

$$V_{max} = \frac{\sqrt{10}}{3} \cdot V_0$$

And the corresponding maximal frequency shift is given by

$$\Delta f_{max} = \frac{2\sqrt{10}}{3} \cdot \frac{V_0 \cdot f_0}{C} \cdot \cos\left(\arctan\left(\frac{1}{3}\right)\right)$$

CHAPTER 12

12.1 Answer: A

Explanation: As noted in Chapter 12, the allowed radiation duration and temperature elevation are related through the following equation:

$$\Delta T_{^\circ C} \leq 6 - \frac{\log_{10}(t)}{0.6}$$

Thus, before stopping the blood the allowed radiation time (t_3) can be calculated from

$$3 \le 6 - \frac{\log_{10}(t_3)}{0.6}$$

which yields

$$t_3 \le 10^{1.8} \text{ (minutes)}$$

Similarly, the allowed radiation time (t_4) after stopping the blood can be calculated, and its value is

$$t_4 \le 10^{1.2} \text{ (minutes)}$$

Thus, the ratio between the two is

$$\frac{t_3}{t_4} = \frac{10^{1.8}}{10^{1.2}} = 10^{0.6}$$

12.2 Answer: B
Explanation: The danger of cavitation bubble formation is inversely proportional to the square root of the frequency.

12.3 Answer: C
Explanation: In this application we would like to maximize the probability of cavitation bubbles formation. Thus we seek the combination which has the highest mechanical index (MI). Recalling the relation given in Chapter 3 for the pressure and intensity, we can calculate the pressure of the acoustic waves:

$$I = \frac{p^2}{Z}$$
$$\Rightarrow P = \sqrt{I \cdot Z}$$

Recalling the definition given in Chapter 12 for the mechanical index,

$$\text{MI} = \frac{P_{\text{negative}}}{\sqrt{f}}$$

and subsituting the value for the pressure given above yields

$$\text{MI} = \frac{\sqrt{I \cdot Z}}{\sqrt{f}}$$

(note that since the transmitted wave is sinusoidal, the maximal negative and positive pressures are equal).

By subsitituting the values given in the question, we find that the maximal MI value is obtained for the transmission combination of $I = 81\,[\text{W/cm}^2]$ and $f = 0.5\,\text{MHz}$, which equals $12.728 \cdot \sqrt{Z}$.

12.4 Answer: B

12.5 Solution and Explanation:
A. The transmitted wave intensity is given by

$$I_0 = \frac{W_0}{A_0}$$

Thus, the intensity at the bottom of the conical part is

$$I_1 = I_0 \frac{A_0}{A_1} \cdot e^{-2\alpha_1 H} = W_0 \frac{e^{-2\alpha_1 H}}{A_1}$$

The needle cross-sectional area is

$$A_2 = \pi \frac{d^2}{4}$$

Thus, the acoustic power reaching the needle upper surface area is

$$W_2 = I_1 \cdot A_2 = W_0 \cdot \frac{e^{-2\alpha_1 H}}{A_1} \cdot \pi \frac{d^2}{4}$$

The corresponding power transmission coefficient for the cone–needle interface is

$$T_{12} = \frac{4 \cdot Z_1 \cdot Z_2}{(Z_1 + Z_2)^2}$$

Therefore the power entering the needle is

$$W_3 = W_2 \cdot T_{12} = W_0 \cdot \frac{e^{-2\alpha_1 H}}{A_1} \cdot \pi \frac{d^2}{4} \cdot \frac{4 \cdot Z_1 \cdot Z_2}{(Z_1 + Z_2)^2}$$

Accounting for attenuation, the power at the tip of the needle is

$$W_4 = W_3 \cdot e^{-2\alpha_2 L} = W_0 \cdot \frac{e^{-2\alpha_1 H}}{A_1} \cdot \pi \frac{d^2}{4} \cdot \frac{4 \cdot Z_1 \cdot Z_2}{(Z_1 + Z_2)^2} \cdot e^{-2\alpha_2 L}$$

The corresponding power transmission coefficient for the tissue–needle interface is

$$T_{23} = \frac{4 \cdot Z_2 \cdot Z_3}{(Z_2 + Z_3)^2}$$

Therefore, the power entering the tissue is

$$W_5 = W_4 \cdot T_{23} = W_0 \cdot \frac{e^{-2\alpha_1 H}}{A_1} \cdot \pi \frac{d^2}{4} \cdot \frac{4 \cdot Z_1 \cdot Z_2}{(Z_1 + Z_2)^2} \cdot e^{-2\alpha_2 L} \cdot \frac{4 \cdot Z_2 \cdot Z_3}{(Z_2 + Z_3)^2}$$

B. The power remaining within the conical part is

$$W_{\text{cone}} = W_0 - W_3 = W_0 \cdot \left[1 - \frac{e^{-2\alpha_1 H}}{A_1} \cdot \pi \frac{d^2}{4} \cdot \frac{4 \cdot Z_1 \cdot Z_2}{(Z_1 + Z_2)^2} \right]$$

C. Neglecting heat loss due to the air convection, the heat remaining in the needle is given by

$$Q = (W_2 - W_5) \cdot \Delta t = \mathrm{C} \cdot \rho \cdot \pi \frac{d^2}{4} \cdot L \cdot \Delta t$$

Thus, the corresponding temperature elevation is given by

$$\Delta T = \frac{W_2 \cdot \left[1 - e^{-2\alpha_2 L} \cdot \frac{4 \cdot Z_2 \cdot Z_3}{(Z_2 + Z_3)^2} \right]}{\mathrm{C} \cdot \rho \cdot \pi \frac{d^2}{4} \cdot L} \cdot \Delta t$$

D. Assuming that the waves in the needle are planar, the needle tip particles' velocity is given by

$$u = \frac{P}{Z_2}$$

And since pressure and intensity are related, we obtain

$$I_{tip} = \frac{P^2}{2Z_2},$$
$$\Rightarrow P = \sqrt{2 \cdot I_{tip} \cdot Z_2}$$

where the intensity at the needle tip is given by

$$I_{\text{tip}} = \frac{W_2 \cdot e^{-2\alpha_2 L}}{\frac{\pi d^2}{4}}$$

And after substitution, we obtain

$$u = \frac{\sqrt{2 \cdot I_{tip} \cdot Z_2}}{Z_2} = \sqrt{\frac{2 \cdot 4 \cdot W_2 \cdot e^{-2\alpha_2 L}}{\pi d^2 \cdot Z_2}} = \frac{2}{d} \cdot e^{-.\alpha_2 L} \cdot \sqrt{\frac{2 \cdot W_2}{\pi \cdot Z_2}}$$

INDEX

A

B

Basics of Biomedical Ultrasound for Engineers, by Haim Azhari